Springer Series in Chemical Physics

Volume 123

The Springer Series in Chemical Physics consists of research monographs in basic and applied chemical physics and related analytical methods. The volumes of this series are written by leading researchers of their fields and communicate in a comprehensive way both the basics and cutting-edge new developments. This series aims to serve all research scientists, engineers and graduate students who seek up-to-date reference books.

More information about this series at http://www.springer.com/series/676

Anatoly Pravilov

Gas-Phase Photoprocesses

Springer

Anatoly Pravilov
Department of Physics
Saint Petersburg State University
Saint Petersburg, Russia

ISSN 0172-6218 ISSN 2364-9003 (electronic)
Springer Series in Chemical Physics
ISBN 978-3-030-65569-3 ISBN 978-3-030-65570-9 (eBook)
https://doi.org/10.1007/978-3-030-65570-9

This Springer imprint is published by the registered company Springer Nature Switzerland AG
The registered company address is: Gewerbestrasse 11, 6330 Cham, Switzerland

Preface

The principal purpose of this book is to provide the readers who research or study in the fields of molecular spectroscopy/molecular physics, as well as chemical physics/physical chemistry and atmospheric photochemistry, with basic frameworks and characteristics of processes, that occur in electronically-excited states of small molecules, complexes, and clusters. These states can be populated due to the absorption of a photon or collisionally; a photon can radiate in the latter case. Absorption of a photon accompanied by the decay of species (molecule, complex, cluster) excited states or a process of formation of a species excited state in a collision which is accompanied by radiation are referred to as *photoprocesses*. Different intra- and intermolecular interactions, including collision-induced processes, can occur in electronically-excited states.

Many of the data and features concerning subjects in the books have not been discussed in monographs previously published. The book can be used as a textbook for students. Problems useful for students are included in Chap. 2.

The book offers a detailed treatment of nonadiabatic perturbations in electronically-excited valence states of molecules induced by intramolecular and intermolecular interactions in collisions or optically populated weakly-bound complexes. The author is an expert in these fields of science. He studies them for more than 50 years. A significant part of the data discussed in the book has been obtained in the author's laboratory.

Saint Petersburg, Russia Anatoly Pravilov

Acknowledgements

I am very grateful to Springer International Publishing AG for the suggestion to publish this book.

Many studies necessary for the writing of this book were carried out with the assistance of my students, postgraduate students and co-workers. I very much appreciate their help and collaboration.

Finally, I wish to thank my wife Natalya for her selfless patience, kind understanding and support throughout the entire period of my researches.

Contents

Abbreviations

(aa)	$I(^2P_{3/2}) + I(^2P_{3/2})$ dissociation limit of an iodine molecule
(ab)	$I(^2P_{3/2}) + I(^2P_{1/2})$ dissociation limit of an iodine molecule
amu	Atomic mass unit
br.r.	Branching ratio
CIP	Collision-induced predissociation
CINAT	Collision-induced nonadiabatic transition
EP	Electronic predissociation
FCD	Franck-Condon density
FCF	Franck-Condon factor
FWHM	Full width of half maximum
HFI	Hyperfine interaction
HFP	Hyperfine predissociation
IP	Ion-pair state
IUPAC	International Union for Pure and Applied Chemistry
lhs	Left-hand side
LCS	Local coordinate system
MD	Nuclear magnetic-dipole interaction with the electrons
MQ	Nuclear electric-quadrupole interaction with the electrons
OODR	Optical-optical double resonance
PEC	Potential energy curve
PES	Potential energy surface
PT	Perturbation theory
RKR	Rydberg-Klein-Rees
REMPI	Resonance-enhanced multiphoton ionization spectroscopy
$R\leftrightarrow T$	Rotational-translational energy transfer process
rhs	Right-hand side
$T\leftrightarrow T$	Translational-translational energy transfer process
VP	Vibrational predissociation
vdW	Van der Waals
VD	Vibrational distribution

$V \leftrightarrow T$	Vibrational-translational energy transfer process
X_2	Homonuclear halogen molecule
XY	Heteronuclear halogen molecule

Symbols

Latin

A	Pre-exponential factor
A_{nm}	Einstein spontaneous emission coefficient
$A^{el}{}_{1,2}$	Electronic matrix element of the interaction of the states 1, 2
A_{el}	Electronic matrix element of the interaction
b	Impact parameter
B_e	Rotational constants for a $v = 0$ vibrational level
B_v	Rotational constants for a $v \neq 0$ vibrational level
B_{mn}	Einstein absorption coefficient
B_{nm}	Einstein stimulated emission coefficient
c	Concentration
c	Light velocity in vacuum
C	Heat capacity
$C_{1,2}$	Mixing coefficient of the states $\Psi_1(J)$ and $\Psi_2(J)$
d	Gas-kinetic diameter
D	Debye
D_e	Potential energy depth
D_0	Binding energy of the lowest van der Waals level
D_0	Dissociation energy for the $v = 0$ vibrational level
e	Electron (elementary) charge
E	Energy
E_k^{el}	Energy of the k-th electronic state
E_a	Activation energy
E_{SO}	Spin-orbit splitting
ΔE	Change of energy
eV	Electron-volt
f	Oscillator strength
$f_A(u_A)$	Velocity distribution function of species A

F	Partition function
$\boldsymbol{F}$	Total angular momentum of a molecule
F_t	Partition function corresponding to the translational motion
F_e	Electronic partition function
F_e	Nuclear partition function
F_r	Partition function corresponding to the rotational motion
F_v	Partition function corresponding to the vibrational motion
g	Statistical weight (degeneracy)
g	Gerade
h	Planck's constant
$\hbar$	Planck's constant
H	Perturbation matrix element
$\widehat{\boldsymbol{H}}$-	Hamiltonian
$\widehat{\boldsymbol{H}}_{SO}$	Operator of spin-orbit coupling
$\widehat{\boldsymbol{H}}_{SS}$	Spin-spin operator
$\widehat{\boldsymbol{H}}_{OR}$	Orbit-rotational operator
$\widehat{\boldsymbol{H}}_{SR}$	Spin-rotational operator
$\widehat{\boldsymbol{H}}_{FS}$-	Fine-structure operator
$\widehat{\boldsymbol{H}}_{G}$	Gyroscopic operator
$\widehat{\boldsymbol{H}}_{HFS}$	Hyperfine operator
$\widehat{\boldsymbol{H}}_{MD}$	Magnetic-dipole operator
$\widehat{\boldsymbol{H}}_{EQ}$	Electric-quadrupole operator
$\widehat{\boldsymbol{H}}_{Z}$	Zeeman operator
$\widehat{\boldsymbol{H}}_{S}$	Stark operator
I	Nuclear spin
$\boldsymbol{I}$	Nuclear spin momentum
I_{A}	Ionization potential of species A
I	Principal moments of inertia
$\widehat{\boldsymbol{I}}_{i\mathrm{A}}$	Angular momentum operator
j	Total angular momentum of electron
J	Rotational quantum number
$\boldsymbol{J}$	Total angular momentum of molecule electrons
$J_0(\tilde{v})$	The incident monochromatic light intensity at wavenumber $\tilde{v}$
$J(\tilde{v}, x)$	The intensity after a light beam has traversed a distance x [cm] in the gas medium
k	Boltzmann constant
k_v	Gyroscopic predissociation rate constant
$k(\tilde{v})$ $(k(\lambda))$	Absorption coefficient
k (k')	Rate constant of a process
$k_{ij,\ lm}$	Microscopic reaction rate constant.
k_{I}	Rate constant of unimolecular process

$k_{chl}^{I}(\lambda)$	Rate constant of chemiluminescence reactions with unimolecular kinetics
$k_{chl}^{II}(\lambda)$	Rate constant of chemiluminescence reactions with bimolecular kinetics
$k_{chl}^{III}(\lambda)$	Rate constant of chemiluminescence reactions with termolecular kinetics
k_{II}	Rate constant of bimolecular process
k_{gk}	Gas-kinetic rate constant
k_{II}^{gk}	Gas-kinetic rate constant of bimolecular process
k_{III}	Rate constant of termolecular process
k_{III}^{gk}	Gas-kinetic rate constant of termolecular process
K	Equilibrium constant
K_c	Equilibrium constant at constant concentration
K_p	Equilibrium constant at constant pressure
l	Orbital moment of electron
$\boldsymbol{L}$	Total orbital moment of molecule
m	Species mass
M	Species, third body
M	Molecule
n	Van der Waals mode
N	Concentration (number of species per unit volume)
N_A	Avogadro number
p	Momentum
p	Pressure
P	Steric factor
P	Probability
$P(v)$	Vibrational distribution
$P(i, v_i; v_f)$	Vibrational level distribution as a result of the $i,v_i \rightarrow f,v_f$ collision-induced process
$P_i(j)$	Term of a $RgI_2(E,v_E,n_E \leftarrow B,v_B,n_B)$ progression
q	Differential cross-section of scattering (reaction)
Q	Heat of reaction
Q_0^0	Reaction thermal effect at $T = 0$ K
$\boldsymbol{Q}$	Coordinates of slow subsystem
Q	Quadrupole moment
$\widehat{\boldsymbol{Q}}$-	Electric-quadrupole operator
$Q_{\alpha\beta}$	Quadrupole moment tensor
r	Process (reaction) rate
$\boldsymbol{r}$	Coordinates of fast subsystem
r_I	Rate of unimolecular process
r_{II}	Rate of bimolecular process
r_{III}	Rate of termolecular process
R	Distance between center of mass of the interacting species
$\boldsymbol{R}$	Vector between center of mass of the interacting species

r	Internuclear distance in a diatomic molecule
R	The distance between center of mass in a complex
R_e	Equilibrium distances corresponding to the potential well depth of a vdW complex
R_0	Equilibrium distances corresponding to the $n = 0$ vdW modes
R	Gas constant
r_e	Internuclear equilibrium distance
R_e^{nm}	Electronic transition matrix element
$\bar{R}_{v\prime v\prime\prime}$	R-centroid
Rg	Rare gas atom
s	Electron spin
$\boldsymbol{s}$	Electron spin momentum
S	Total spin quantum number
$\boldsymbol{S}$	Total spin angular momentum
ss	Steady-state (stationary)
T	Kinetic energy
T	Temperature
T_e	Electronic term
T_v	Oscillation period
$\hat{\boldsymbol{T}}$-	Kinetic energy operator
V	Volume
V	Velocity of the relative motion of the species
$\boldsymbol{V}_{\mathrm{rel}}$	Relative velocity vector
$\bar{V}_{AB}$	Average relative velocity of the species A and B
$\hat{V}$	Potential energy operator
$\hat{\boldsymbol{V}}$	Perturbation operator
v	Vibrational quantum number
u	Ungerade
U	Potential energy
$\boldsymbol{u}$	Velocity vector
$U(R)$	Interaction potential
$U_k(\boldsymbol{Q})$	Energy of the k_{th} state
$U_{L-J}(R)$	Lennard-Jones potential
X_i^A	Distribution over quantum number of species A
z_{II}	Number of binary collisions
z_{III}	Number of triple collisions
Z	Number of collisions needed for the complete relaxation

Greek

α	Parameter of Morse potential
α_A	Polarizability of species A

Γ	Symmetry type (species, irreducible representation)
Γ_{pred}	Predissociation rate
$\delta_{0,\Lambda}$	Kronecker symbol
ε	Potential well depth
ε	Symmetry index
ε', ε''	Free level energies
$\varepsilon(\tilde{\nu})$ $(\varepsilon(\lambda))$	Molar absorption coefficient
ε_0	Vacuum permittivity
ζ	Massey parameter
λ	Light (radiation) wavelength
λ	De Broglie wavelengths
Λ	Projection of orbital quantum number on molecular axis
μ	Reduced mass
μ	Dipole moment
$\mu_{n,m;v',v''}$	Matrix element for a electronic transition between vibronic levels
$\mu(R)$	Transition dipole moment function
$\boldsymbol{\mu}$	Dipole moment vector
$\hat{\boldsymbol{\mu}}$ -	Transition dipole moment operator
ν	Frequency
ν	Overall process order
$\nu_i(\nu'_k)$	Stoichiometric coefficient
ν, $\tilde{\nu}$	Wavenumber
ρ_{nm}	Energy per unit volume per unit wavenumber
$\rho_{nm}^{h\nu}$-	Number of photons per volume unit
$\sigma(\bar{\nu})$ $(\sigma(\lambda))$	Absorption cross-section
σ	Collision cross-section
σ	Symmetry order
σ	Spin wave function
$\boldsymbol{\sigma_v}$	Operator of a reflection in the molecule-fixed coordinate system through a plane containing the internuclear axis
Σ	Quantum number of spin component (the projection of S on the internuclear axis)
τ	Lifetime
τ_{rad}	Radiative lifetime
φ	Wave function
$\varphi_{AB}(\lambda)$	Integral absolute quantum yield of AB molecule photodecay
φ_i^{AB}	Integral absolute quantum yield for i-th process of AB molecule photodecay
$\varphi_i^{AB}(\lambda_j - \lambda_k)$	Convolution quantum yield of i-th primary photoprocess within the band $\lambda_j \div \lambda_k$
$\varphi_i^{AB}(\lambda)$	Absolute quantum yield for i-th process of AB molecule photodecay
$\Phi_{B_i}^{AB}(\lambda)$	Absolute quantum yield of the product B_i of the AB photolysis

$\Phi_{\rm lum}(\lambda)$	Absolute luminescence quantum yield of AB or photodecay product AB–B_i
$\Phi_{B_i}^{AB}$	Integral absolute quantum yield for the formation of the product of AB photodecay B_i
$\Phi_i^{AB,c}(\lambda_j - \lambda_k)$	Convolution quantum yield for the formation of the B_i photoproduct within the band $\lambda_j \div \lambda_k$
Φ	Electronic wave function
$\Phi_k(\boldsymbol{r},\boldsymbol{Q})$	Wave function of fast subsystem
χ, χ_Q, $\chi(Q)$	Vibrational wave function
χ_r	Rotational wave function
$\chi(\boldsymbol{Q})$	Wave function of slow subsystem
ψ	Molecular wave function
Ψ_{es}	Electronic wave function including spin
ω	Angular frequency
ω_e	Vibrational frequency
$\omega_e x_e$	Anharmonicity
Ω	Projection of total electronic angular momentum on molecular axis

Other Symbols

#	Rovibrational excitation
*	Rovibronic excitation

Chapter 1
Introduction

The book deals with processes that occurred in electronically-excited states of free (isolated) small molecules as well as weakly-bound complexes and clusters. These states can be populated due to the absorption of a photon (photochemistry) or collisionally. In the latter case, a photon can radiate (photochemistry backward), i.e., in photoprocesses.

Description of radiative population and decay of electronically excited states of free molecules and weakly-bound complexes (clusters), as well as intra- and intermolecular perturbations, is one of the fundamental problems of modern chemical physics/physical chemistry and molecular spectroscopy. This information is necessary for describing a wide range of processes occurring in the upper atmosphere of the Earth and other planets as well as interstellar space, gas-phase lasers, plasmas, flames, and so on. To describe gas-phase photoprocesses, one has to use basic information on:

- Kinetics for chemical reactions, collisional and intramolecular processes described in Chap. 2.
- The frameworks of the theory of elementary processes (Chap. 3).

The data given in these chapters are analyzed in a volume sufficient for understanding the kinetics and mechanism of gas-phase photoprocesses in the following chapters.

The principal goal of the book is a detailed description of:

Photolysis of free molecules, absorption and emission, intramolecular perturbation, dissociation, and electronic predissociation of di- and polyatomic molecules, the basic patterns of these processes are discussed in Chap. 4. Understanding the mechanism of the electronically-excited state perturbations in free molecules (*intramolecular* perturbation) is very useful for descriptions of nonadiabatic processes in more complex systems. Besides, a study of perturbations in free molecules stimulates developments of the theory and accurate definition of molecule spectroscopic characteristics.

A. Pravilov, *Gas-Phase Photoprocesses*, Springer Series in Chemical Physics 123,
https://doi.org/10.1007/978-3-030-65570-9_1

Chapter 5 is devoted to energy transfer of translational, rotational, rovibronic, and electronic excitation in collisions. The chapter focuses on so-called collision-induced nonadiabatic transitions (CINATs), perturbation-facilitated, and perturbation-irrelevant. Collision-induced nonadiabatic transitions between halogen ion-pair states are discussed. A transfer of translational, rotational, and vibrational energies in collisions are examined briefly. Electronic deactivation in which collision-induced nonradiative transition to another electronic state, which occurs with a large energy loss, is also discussed.

Weakly-bound, mainly van der Waals complexes and clusters (Chap. 6). Halogen-containing complexes and clusters are examined in detail.

In the systems described in Chaps. 5 and 6, *intermolecular* perturbations of the states occur in collisions of a species with a partner or optically populated weakly-bound complexes of species. Intermolecular interactions occur in the gas phase, liquids, solids, and polymers. These interactions are of great importance in energy transfer processes in the gas phase, organization of structures, and properties of biomolecules. Understanding the nature of intermolecular interactions is one of the fundamental problems of modern chemical physics and molecular spectroscopy. Van der Waals (vdW) complexes of free molecule excited states are ideal model systems to predict the properties and dynamic behavior of more complex system. The partners building the complex retain their identity, and energy transfers are thus easily identified due to the weakness of the intermolecular noncovalent bond. Investigation of the dependence of the energy redistribution and fragmentation processes on the size of the cluster may help bridge the gap with condensed-phase dynamics.

Collision-induced nonadiabatic transitions in the gas phase, in which vdW complexes are formed in collisions and then decay, play a fundamental role in the kinetics and dynamics of excited molecular electronic states. They are responsible for radiative emission and energy transfer in the atmosphere, energy pooling and conversion in laser media, relaxation phenomena in chemiluminescence processes, plasma formation, and in many other situations where electronically excited states are involved. Description of vdW complex formation and decay is essential for understanding mechanisms of nonadiabatic transitions in weakly-bound complexes of any molecules and clusters. Understanding of these processes is also necessary for the interpretation and modeling of various photoinitiated processes in clusters, liquids, and solids, where the role of intermolecular interaction is greatly magnified through the formation of solvation shell(s) and multiple collisions between a molecule and a solvent.

Mechanism and dynamics of vdW complex formation and decay are governed by multidimensional potential energy surface (*PES*) of intermolecular interaction. These data facilitate understanding the mechanism of interactions between specific reagents and functions of intermolecular interactions in chemical processes. Determination of propensity rules for vdW complex decay is necessary to develop a description of dynamics in weakly-bound complexes.

Photochemical and photophysical investigations of weakly-bound complexes are necessary to understand how the weak intermolecular interaction of partners acts on excitation energy transfer processes. Despite the weak binding of partner molecules in the vdW complex, there are several examples in literature in which vdW complex demonstrates new chemical channels closed in the isolated molecule.

Chemiluminescence (Chap. 7). At the beginning of the chapter, the definitions of various types of chemiluminescence reactions and their rate constants are introduced. The basic patterns of the recombination accompanied by radiation are examined in detail in this chapter.

Chapter 2
General Kinetic Rules for Chemical Reactions, Collisional, and Intramolecular Processes

Abstract This chapter deals with basic information on kinetics for chemical reactions, collisional, and intramolecular processes. Definitions of reaction and process rates, equilibrium constants, partition functions of atoms, and molecules are introduced. Kinetic types of simple reactions, as well as complex reactions and the steady-state method, are discussed.

2.1 Rates of Reaction, Collisional, and Spontaneous Processes. Rate Constants. Kinetic Types of Simple Processes

According to IUPAC Compendium of Chemical Terminology [1], a *chemical reaction is a process that results in the interconversion of chemical species.* A process of interaction of an oxygen atom in the first excited state, $O(^1D)$, with a H_2O molecule in which two OH radicals are formed, i.e., the chemical composition changes:

$$O(^1D) + H_2O(\widetilde{X}^1A_1) \rightarrow 2OH(X^2\Pi, v_X, J_X) \quad (2.1.1a)$$

is an example of chemical reactions.

In principle, electronic deactivation of an $O(^1D)$ atom in a collision with a H_2O molecule

$$O(^1D) + H_2O(\widetilde{X}^1A_1) \rightarrow O(^3P) + H_2O(\widetilde{X}^1A_1, v_X) \quad (2.1.2)$$

can occur. One should note that electronic deactivation of an $O(^1D)$ atom in a collision with an O_2 molecule

$$O(^1D) + O_2\left(X^3\sum\nolimits_g^-, v_X = 0\right) \rightarrow O(^3P) + O_2\left(b^1\sum\nolimits_g^+, v_b\right) \quad (2.1.3)$$

A. Pravilov, *Gas-Phase Photoprocesses*, Springer Series in Chemical Physics 123,
https://doi.org/10.1007/978-3-030-65570-9_2

is the principal channel of the interaction. The chemical compositions do not change in them. Energies and the states of colliding partners vary in these processes. Therefore, they cannot be called as reactions. The 'unreactive process' term is used in [2] (see p. 30, e.g.) for the processes in which chemical composition does not change. The kinetic descriptions of reaction (2.1.1a) and processes of an $O(^1D)$ atom electronic deactivations (2.1.2, 2.1.3) are the same, as the reader will see.

Below, the term *process* is used for the definition of chemical reactions and process leading to the formationes in which chemical compositions do not change. The reader sees that the term *process* is more comprehensive than the term *reaction*.

Let us consider a process of collision of ν_1 species A_1 with ν_2 species A_2 and so on leading to the formation of ν'_1 species, A'_1, ν'_2 species A'_2, and so on. The ν_1, $\nu'_1 = 1, 2, 3\ldots$ numbers called *stoichiometric coefficients* are directly proportional to the numbers of moles of the corresponding reagents.

$$\nu_1 A_1 + \nu_2 A_2 + \ldots = \nu'_1 A'_1 + \nu'_2 A'_2 + \ldots \qquad (2.1.4)$$

or

$$\sum\nolimits_i \nu_i A_i = \sum\nolimits_k \nu'_k A'_k \qquad (2.1.5)$$

The (2.1.4, 2.1.5) are called *stoichiometric equations*.

If ν_1 = 1, 2 and 3, the process has first, second and third order regarding A_1, respectively. For the reaction (2.1.1a), for example, the stoichiometric equations is:

$$O(^1D) + H_2O(\tilde{X}^1A_1) = 2OH(X^2\Pi, v_X, J_X), \qquad (2.1.1b)$$

and the reaction has first order regarding the $O(^1D)$ and $H_2O(\tilde{X}\ ^1A_1)$.

An another example is

$$Cl_2 + 2Na = 2NaCl. \qquad (2.1.6a)$$

It is obvious, that the stoichiometric equation (2.1.6a) does not describe the real reaction mechanism. It states that to produce NaCl molecule it is necessary one Cl_2 molecule and two Na atoms, and two NaCl molecules are produced as a result. In other words, a stoichiometric equation is an 'account' equation.

One should note that the equal signs and arrows are used in the stoichiometric equations and processes, respectively. Below, concentrations of reactants (initial products) and process products (final products) are denoted as $[A_i]$ and $[A'_k]$, respectively.

The rate of a gas-phase process (a chemical reaction, in particular) is determined by the changes of reactant (or reaction product) concentrations with time. Reactant product states are taken into account in this definition. According to IUPAC Compendium of Chemical Terminology [1], 'for the general chemical reaction:

$$aA + bB \longrightarrow pP + qQ + \ldots \quad (2.1.7)$$

occurring under constant-volume conditions, without an appreciable build-up of reaction intermediates, the rate of reaction is defined as:

$$v = -\frac{1}{a}\frac{d[A]}{dt} = -\frac{1}{b}\frac{d[B]}{dt} = \frac{1}{p}\frac{d[P]}{dt} = \frac{1}{q}\frac{d[Q]}{dt} \quad (2.1.8a)$$

or

$$v = -\frac{1}{a}\frac{c_A}{dt} = -\frac{1}{b}\frac{c_B}{dt} = \frac{1}{p}\frac{c_P}{dt} = \frac{1}{q}\frac{c_Q}{dt} \quad (2.1.8b)$$

where symbols placed inside square brackets or c in (2.1.8b) denote amount (or amount of substance) concentrations (conventionally expressed in units of mol dm^{-3}). The symbols R and r are also commonly used in place of v.' The author will use the symbol r.

The author usually follows to symbols used in [2].

– *Amount of substance concentration is the number of moles n divided by volume, for example,*

$$c_B = \frac{n_B}{V} \text{mol/dm}^3 \left(\text{mol/1}, \text{mol/cm}^3\right)$$

The mole is the amount of substance of a system which contains as many elementary entities (atoms, molecules, ions and so on) as there are atoms in 0.012 kg of ^{12}C *atoms* [1] *or amount of elementary entities (species) equal to the Avogadro number,* $N_A = 6.02 \cdot 10^{23}$ *entities/mol.*

Another definition of concentration is preferably for use if one deals with processes in which atoms or radicals participate. It is rather strange to write 'concentration of $O(^1D)$ atoms is equal to n mol/l.' It is much more convenient to measure concentration as *the amount of species N in unit volume*:

$$\frac{N_B}{V} = [B] = c_B \cdot N_A \left(\text{species/cm}^3\right)$$

Below, the author uses the species/cm^3 units of concentration. For these units, the rate of a process is defined as:

$$r = -\frac{1}{v_i}\frac{d[A_i]}{dt} = \frac{1}{v'_k}\frac{d[A'_k]}{dt}, \quad (2.1.8c)$$

if the volume V does not change during the reaction. *The rate of the process is the rate of any reactant concentrations changes with a minus sign divided by the reactant stoichiometric coefficient or the rate of any product concentrations change with a plus sign divided by the product stoichiometric coefficient.*

It is easy to see that the r value does not depend on which reactant or product one uses to calculate the reaction rate, since, as can be seen from the stoichiometric equation, changes in the concentrations of reagents and products are proportional to the corresponding stoichiometric coefficients. The definitions (2.1.8a–c) are not valid for describing processes in an open space such as a gas jet. Nevertheless, they are valid for small degrees of conversion and low-temperature variation even for a flow reactor.

The rate of a simple (see below) process can be written as:

$$r = k \cdot \prod\nolimits_i [\mathrm{A}_i]^{\nu_i}, \tag{2.1.9}$$

(see [2], p. 4). For processes proceeding in one stage through simultaneous collisions of species, for example, processes (2.1.1a, b, 2.1.3), this law seems trivial, since only those species that collide can react, and the number of collisions per unit time of collision of ν_1 species A_1 with ν_2 species A_2 and so on, is directly proportional to $\prod_i \mathrm{A}_i^{\nu_i}$ according to the kinetic theory of gases [3]. *The coefficient k in* (2.1.9), *which dimension is determined by the number of species participating in the process or forming in it, is called the rate constant of the process.*

There are two process orders: stoichiometric, equal to the sum of the stoichiometric coefficients in the stoichiometric equation and kinetic or real order. Reaction kinetic order can be determined using *van't Hoff differential method* [4]. One makes the concentration of one substance much less than the concentrations of others, for example,

$$[\mathrm{A}_1] \ll [\mathrm{A}_2], [\mathrm{A}_3], \tag{2.1.10}$$

and measures the dependence of the reaction rate on the concentration of the A_1 substance. It is obvious, that if condition (2.1.10) is fulfilled, then in reaction describing by the stoichiometric equation (2.1.4), only $[\mathrm{A}_1]$ changes noticeably. Consequently (Fig. 2.1),

$$\prod\nolimits_{i \geq 2} \mathrm{A}_i^{\nu_i} = \mathrm{const}$$

and one can assume in (2.1.9) that

$$k \prod\nolimits_{i \geq 2} \mathrm{A}_i^{\nu_i} = k' = \mathrm{const}.$$

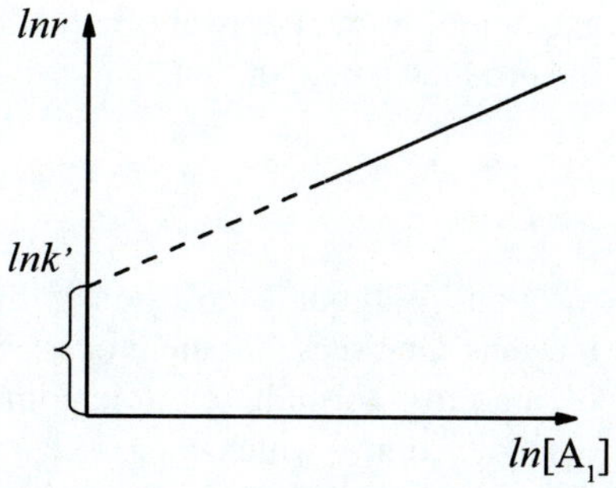

Fig. 2.1 To the van't Hoff method

Therefore,

$$r = k'[A_1]^{\nu_1},$$

$$lnr = lnk' + \nu_1 ln[A_1] \quad (2.1.11)$$

and reaction kinetic order regarding A_1 is

$$\nu_1 = \frac{d\{lnr\}}{d\{ln[A_1]\}} \quad (2.1.12)$$

One can determine reaction kinetic orders regarding A_2, A_3, and so on. *The real reaction order is equal to the sum of reaction kinetic orders regarding A_i.*

As it mentioned above, the stoichiometric equation (2.1.6a) does not describe the real reaction mechanism of the sodium vapor burning in chlorine, and the reaction (2.1.6b)

$$Cl_2 + 2Na \rightarrow 2NaCl \quad (2.1.6b)$$

is a *complex* one. One can suppose that, at small Na and Cl_2 concentrations and [Na] << [Cl_2], this process is described by reactions

$$Cl_2 + Na \rightarrow NaCl + Cl \quad (2.1.13)$$

$$Cl + Na \xrightarrow{M} NaCl, \quad (2.1.14a)$$

reaction (2.1.14a) can be also written as

$$Cl + Na + M \rightarrow NaCl + M. \quad (2.1.14b)$$

The reaction (2.1.14a, b) is a *termolecular reaction*, M is a species, a third body, which stabilizes the (Na … Cl)$^{\#}$ quasi-molecule ($^{\#}$ is vibrational-rotational excitation) by withdrawal of a part of the energy, thus ensuring physical stability of the product. The (Na … Cl)$^{\#}$ quasi-molecule (collision complex) is produced in the

$$Cl + Na \rightarrow (Na\ldots Cl)^{\#} \tag{2.1.15}$$

reaction. Without a collision with M, this complex has to dissociate to initial reactant. One sees that the process of the interaction of Cl_2 and Na atoms, described by the gross formula (or stoichiometric equation) (2.1.6a), can be characterized by other reactions, which more accurately reflect the essence of the process and are themselves stoichiometric equations.

Another example of a complex reaction is a recombination of radicals, methyl radicals, for example:

$$2CH_3 \xrightarrow{M} C_2H_6, \tag{2.1.16a}$$

The stoichiometric equation for the reaction (2.1.16a) is:

$$2CH_3 + M = C_2H_6 + M, \tag{2.1.16b}$$

and stoichiometric order of reaction (2.1.16a) is equal to 3. It would seem that the M species can be thrown out from (2.1.16a, 2.1.16b), but, as the author has shown above, it is impossible. The $(CH_3\ldots CH_3)^{\#}$ complex decays to initial reactant without a collision with a M species.

In fact, the reaction (2.1.16a) can be described as follows:

$$2CH_3 \leftrightarrow (CH_3\ldots CH_3)^{\#} \tag{−2.1.17}$$

$$(CH_3\ldots CH_3)^{\#} \xrightarrow{M} C_2H_6. \tag{2.1.18}$$

The collision complex $(CH_3\ldots CH_3)^{\#}$ (or $C_2H_6{}^{\#}$) which excitation energy is equal to the C_2H_6 dissociation energy dissociate to 2 CH_3 (reaction (−2.1.17)) without collision with a M species. A lifetime of collision complex depends strongly on a number of its vibrational modes, 3 N−6 (N is a number of atoms), since excitation energy is stochastized among them. Therefore, the $(CH_3\ldots CH_3)^{\#}$ lifetime is some orders of magnitude higher than that of for collisions of atoms [2], p. 98. If M concentration is large, so the rate of the process (2.1.18) is much more than that of the $(CH_3\ldots CH_3)^{\#}$ complex dissociation, $k_{1.18}[M] >> k_{-1.17}$, the complex energy is getting less than its dissociation energy, and $(CH_3\ldots CH_3)^{\#}$ dissociation becomes impossible. In this case, the rate of the reaction (2.1.16a) is independent of [M], and the reaction order becomes equal to 2, less than the stoichiometric order 3 (see Sect. 2.4.1).

The processes which rates follow to (2.1.9), *and the kinetic orders coincide with the stoichiometry of the stoichiometric equation, are called simple processes. If they also proceed in one stage, they are called elementary processes*. For example, the reaction (2.1.13) is simple and elementary.

2.1.1 Kinetic Types of Simple Processes

Let us consider a simple process far from equilibrium where the reverse process, in which the reactants of the direct reactions are produced, can be neglected. The reactions (–2.1.13, –2.1.14a) at low temperatures are examples of such cases. The reactions

$$Cl_2 + Na \leftarrow NaCl + Cl \tag{$-2.1.13$}$$

$$Cl + Na \overset{M}{\leftarrow} NaCl, \tag{$-2.1.14a$}$$

reverse to reactions (2.1.13, 2.1.14a) are impossible in these cases. Three types of simple processes are possible:

$$\textit{Unimolecular}, \Sigma_i v_i = 1:$$

$$A \rightarrow \text{products} \tag{2.1.19a}$$

$$\textit{Bimolecular}, \Sigma_i v_i = 2:$$

$$A_1 + A_2 \rightarrow \text{products} \tag{2.1.19b}$$

$$\textit{Termolecular}, \Sigma_i v_i = 3:$$

$$A_1 + A_2 + A_3 \rightarrow \text{products} \tag{2.1.19c}$$

Simultaneous collisions more than three species have a negligible probability (see below). *The value*

$$v = \sum\nolimits_i v_i \tag{2.1.20}$$

is called the overall process order (see [1]).

Let us consider the kinetic features of simple processes.

A unimolecular process.

The rate of a unimolecular process is

$$r_I = -\frac{d[A]}{dt}, \tag{2.1.21}$$

and

$$[A] = [A]_0 \exp(k_I \cdot t) = [A]_0 \exp(-t/\tau), \quad (2.1.22)$$

$\tau = 1/k_I$ [s] and units of k_I is [s^{-1}].

Emission (radiative decay) of excited atoms and molecules are unimolecular processes, also, for example,

$$H(2p^2P) \to H(1s^2S) + hv,$$

$$\tau_H(^2P) = 1.6 \text{ ns}, A_{H(^2P)} = 6.2 \cdot 10^8 s^{-1}.$$

In the case of radiative decay, *k_I is called an Einstein spontaneous emission coefficient or sums of Einstein coefficients if optical transitions occur to several lower states, τ is a radiative lifetime.* (see Sect. 4.3).

An example of unimolecular reactions is predissociation of $Cl_2(B^3\Pi(0_u^+), v_B > 12)$ molecules

$$Cl_2(B^3\Pi(0_u^+), v_B > 12) \to Cl_2(A^3\Pi(1_u), C^1\Pi_u) \to 2Cl(3p^{5\,2}P_{3/2})$$

(see Sect. 4.5). In this case, *k_I is called the predissociation rate constant, and τ is the lifetime related to the predissociation.*

The rate of a bimolecular process is

$$r_{II} = -\frac{d[A_1]}{dt} = -\frac{d[A_2]}{dt} = k_{II} \cdot [A_1] \cdot [A_2] \text{species/cm}^3 \cdot s \quad (2.1.23)$$

Let us get a look how reactant concentration are changed if $[A_1] = [A_2] \equiv [A]$. This is the case of a termolecular recombination at high pressures (see Sect. 2.4.1). One gets after integration of the (2.1.23) with $[A_1] = [A_2] \equiv [A]$ the following:

$$[A] = \left([A_0]^{-1} + k_{II} \cdot t\right)^{-1},$$

i.e., reactant concentration is inversely proportional to a reaction time in the long time limit.

If $[A_1] << [A_2]$, for example, and $[A_2] \approx$ const, the

$$[A_1] = [A_1]_0 \exp(-k_{II} \cdot t)$$

as in the case of a unimolecular process.

Briefly about the dimension of rate constants of bi- and termolecular processes.

It follows from (2.1.23) that

$$k_{\rm II} = \frac{r_{\rm II}\ \text{species/cm}^3\cdot\text{s}}{[\text{A}_1]\cdot[\text{A}_2]\text{species}^2\cdot\text{cm}^6}\left(\text{cm}^3/\text{species}\cdot\text{s}\right) \tag{2.1.24}$$

or simply cm^3/s.

If one measures concentrations in mol/cm^3 or mol/l units, the dimensions of $k_{\rm II}$ are cm^3/mol · s and 1/mol · s, and

$$k_{\rm II}\left(\text{cm}^3/\text{species}\cdot\text{s}\right) = k_{\rm II}\left(\text{cm}^3/\text{mol}\cdot\text{s}\right)/\text{N}_\text{A} \tag{2.1.25}$$

$$k_{\rm II}\left(\text{cm}^3/\text{species}\cdot\text{s}\right) = k_{\rm II}(1/\text{mol}\cdot\text{s})/10^3\cdot\text{N}_\text{A}, \tag{2.1.26}$$

since there are N_A species in mole.

It is easy to understand in what units the concentrations are measured:
the $k_{\rm II}$ value is $\sim\ 10^{-9}$ or less (the exponent is negative) for cm^3/species · s;
the $k_{\rm II}$ value is $\sim\ 10^{14}$ or less (the exponent is positive) for cm^3/mol · s;
the $k_{\rm II}$ value is $\sim\ 10^{17}$ or less (the exponent is positive) for 1/mol · s.

(see Arrhenius law, Sect. 2.3).

Similarly,

$$k_{\rm III} = \frac{r_{\rm III(species/cm^3s)}}{[\text{A}_1]\cdot[\text{A}_2]\cdot[\text{A}_3](\text{species}^3\cdot\text{cm}^9)}(\text{cm}^6/\text{species}^2\cdot\text{s}) \tag{2.1.27}$$

$$k_{\rm III}\left(\text{cm}^6/\text{species}^2\cdot\text{s}\right) = k_{\rm III}\left(\text{cm}^6/\text{mol}^2\cdot\text{s}\right)/\text{N}_\text{A}^2 \tag{2.1.28}$$

$$k_{\rm III}\left(\text{cm}^6/\text{species}^2\cdot\text{s}\right) = k_{\rm III}\left(1^2/\text{mol}^2\cdot\text{s}\right)10^6\cdot/\text{N}_\text{A}^2 \tag{2.1.29}$$

the $k_{\rm III}$ value is $\sim\ 10^{-28}$ or less (the exponent is negative) for (cm^6/species^2 · s);
the $k_{\rm II}$ value is $\sim\ 10^{18}$ or less (the exponent is positive) for cm^6/mol · s;
the $k_{\rm II}$ value is $\sim\ 10^{24}$ or less (the exponent is positive) for 1/mol · s.

2.2 Chemical Equilibrium. Equilibrium Constant

In the previous section, the reactions (–2.1.13, –2.1.14a, b) reverse relative to direct reactions (2.1.13, 2.1.14a, b) have been neglected. In general, for the cases when products of a process do not being left from a reaction zone, both direct, and reverse processes can proceed. Therefore, a process described by the stoichiometric equation (2.1.5) has to be written as follows:

$$\sum\nolimits_i \nu_i \mathrm{A}_i \leftrightarrow \sum\nolimits_k \nu'_k \mathrm{A}'_k. \qquad (-2.2.1)$$

The rate of the direct processes is (2.1.9), and the rate of the reverse process is

$$r' = k' \prod\nolimits_k [\mathrm{A}'_k]^{\nu'_k}. \qquad (2.2.2)$$

The k and k' (or $k_{2.1}$ and $k_{-2.1}$ for the process (2.2.1, −2.2.1)) are rate constants of direct and reverse processes. It is obvious, that what process is direct and what is reverse is determined by which species are written in the left and right parts of an (2.2.1, −2.2.1). In principle, one can measure the rate constants of direct and reverse processes.

Let us analyze the reaction

$$2\mathrm{CH_3} \overset{\mathrm{M}}{\leftrightarrow} \mathrm{C_2H_6}, \qquad (-2.1.16\mathrm{a})$$

as an example.

One can produce CH_3 radicals by photolysis of an unsaturated hydrocarbon at room temperature and measure the rate constant of direct reaction $2CH_3 \overset{M}{\rightarrow} C_2H_6$:

$$r_{1.16\mathrm{a}} = -\frac{1}{2}\frac{d[\mathrm{CH_3}]}{dt} = k_{1.16\mathrm{a}} \cdot [\mathrm{CH_3}]^2[\mathrm{M}]$$

The reverse reaction $2CH_3 \overset{M}{\leftarrow} C_2H_6$ has negligible rate constants at room temperature, even if one utilizes C_2H_6 molecule as unsaturated hydrocarbon: $[C_2H_6] >> [CH_3]$, but C_2H_6 dissociation energy is high, 3.91 eV [5], and scale factor for energy values in molecular-scale systems for room temperature, $k\mathrm{T} \approx 0.03$ eV (k is Boltzmann constant).

On the other hand, one can heat C_2H_6 molecules at very high temperatures at which their dissociation is possible using an explosive driven shock tube. A simple shock tube utilizing in chemical physics is a tube in which a gas or a gas mixture at low pressure and a gas at high pressure are separated using a diaphragm [6], p. 4. The diaphragm suddenly bursts open, for example, due to an explosion of a $H_2 + O_2$ mixture to produce a wave propagating through the low-pressure section. The shock that eventually forms increases the pressure, and temperature, up to tens of thousands of degrees (several eV) of the test gas, and induces a flow in the direction of the shock wave. Observations can be made in the flow behind the incident front or take advantage of the longer testing times and vastly enhanced pressures and temperatures behind the reflected wave.

If one heats C_2H_6 molecules using an explosive driven shock tube, their thermal decomposition: $C_2H_6 \overset{M}{\rightarrow} 2CH_3$ (reaction (−2.1.16a)) occurs only. The CH_3 radical concentration is low just after the explosion. Besides, recombination rate constants decrease with temperature for an overwhelming majority of these reactions.

Therefore, in this case, only the reverse reaction (–2.1.16a)) occurs, and one can measure its rate constant:

$$r_{-1.16\mathrm{a}} = -\frac{d[C_2H_6]}{dt} = k_{-1.16\mathrm{a}} \cdot [C_2H_6][M].$$

Further, as the concentration of CH_3 increases and the concentration of C_2H_6 decreases, the rates of the direct and reverse reactions will converge. If one waits a bit, then some rather high slowly decreasing temperature will be established, each value of which corresponds to its own values $k_{1.16\mathrm{a}}$, $k_{-1.16\mathrm{a}}$, $[CH_3]$, $[C_2H_6]$, and the total reaction rate is:

$$\begin{aligned} r^t_{1.16a} &= r_{1.16a} - r_{-1.16a} = -\frac{1}{2}\frac{d[CH_3]}{dt} + \frac{d[C_2H_6]}{dt} \\ &= k_{1.16\mathrm{a}} \cdot [CH_3]^2[M] - k_{-1.16\mathrm{a}} \cdot [C_2H_6][M]. \end{aligned} \tag{2.2.3}$$

At a stationary case, the constant sufficiently high temperature, then one makes an obvious conclusion that under these conditions the rates of the direct and reverse reactions are equal, the resulting rate is 0, and from (2.2.3) for these conditions (*chemical equilibrium*), one gets:

$$\frac{1}{2}\left(\frac{d[CH_3]}{dt}\right)_{eq} = \left(\frac{d[C_2H_6]}{dt}\right)_{eq} \text{ and } k_{1.16\mathrm{a}} \cdot [CH_3]^2_{eq} = k_{-1.16\mathrm{a}} \cdot [C_2H_6]_{eq}, \tag{2.2.4}$$

and

$$\left(\frac{d[CH_3]}{dt}\right)_{eq} = \left(\frac{d[C_2H_6]}{dt}\right)_{eq} = 0 \tag{2.2.5}$$

In the general case:

$$r_t = r - r' = -\frac{1}{\nu_i}\frac{d[A_i]}{dt} + \frac{1}{\nu'_k}\frac{d[A'_k]}{dt} = k \cdot \prod\nolimits_i [A_i]^{\nu_i} - k' \cdot \prod\nolimits_k [A'_k]^{\nu'_k},$$

and in the case of a chemical equilibrium

$$\frac{1}{\nu_i}\frac{d[A_i]_{eq}}{dt} = \frac{1}{\nu'_k}\frac{d[A'_k]_{eq}}{dt}, \tag{2.2.6}$$

$$k\prod\nolimits_i [A_i]^{\nu_i}_{eq} = k'\prod\nolimits_k [A'_k]^{\nu'_k}_{eq} \tag{2.2.7}$$

and

$$\frac{k}{k'} = \frac{\prod_k [A'_k]_{eq}^{\nu'_k}}{\prod_i [A_i]_{eq}^{\nu_i}} \tag{2.2.8a}$$

This ratio is called the *equilibrium constant*:

$$K_c = \frac{k}{k'} = \frac{\prod_k [A'_k]_{eq}^{\nu'_k}}{\prod_i [A_i]_{eq}^{\nu_i}} = \frac{\prod_k c_{k,eq}^{\nu'_k}}{\prod_i c_{i,eq}^{\nu_i}}, \tag{2.2.8b}$$

(the dimensions of the $[A_i]$, $[A'_k]$ and c are species/cm^3 and mol/cm^3, respectively, see above.

It is evident that the equilibrium constant does not depend on which side we approach equilibrium concentrations, i.e., the concentration of the initial or final (as part of our agreement) products was larger than the equilibrium one at the time $t = 0$. The equilibrium constant depends on temperature only (see later), and, for this reason, it is honored to be listed in reference books.

How can one define this constant experimentally? It follows from (2.2.8b) that this can be done in two ways:

1. Determine the values of k and k' in some way;
2. Determine the equilibrium concentration for a given temperature.

It is clear that if one knows the equilibrium constant and the constant of the direct or reverse process, one can easily determine the constant of the reverse or direct process, respectively.

The rate constants of the direct and reverse processes used to calculate the equilibrium constant has to be determined at the same conditions. If the constant is a tabulated value (not a particular value that one defined for any specific nonequilibrium conditions), then the constant is related to *thermodynamically equilibrium conditions*, that is characterized by the temperature, and all temperatures, of translational motion, rotational, vibrational and electronic excitation are the same.

This book has been written for the readers who research or study in the fields of molecular spectroscopy/molecular physics, as well as chemical physics/physical chemistry and deal with processes, occurred in electronically-excited states of small molecules and complexes. These are purely non-equilibrium systems, in which, at best, there are local thermodynamic equilibria in certain types of motion. For example, in the reaction (2.1.1a) $O(^1D) + H_2O(\tilde{X}^1A_1) \rightarrow 2OH(X^2\Pi, v_X, J_X)$ under photolysis of the O_2 + H_2O mixture at T = 293 K, the electron excitation energy of the oxygen atom is 1.97 eV, and the concept of temperature is not applicable to it, the translational temperature $O(^1D)$ is 293 K, the temperature of H_2O is also 293 K in all degrees of freedom (translational, rotational, vibrational and electronic motion), and the energy distribution in degrees of freedom OH depends on the experimental conditions.

Nevertheless, if one uses the equilibrium constants correctly, this concept and even sometimes tabulated data can be useful not only for 'traditional equilibrium' chemists but also for a researcher who investigates non-equilibrium systems. In 1996, the author discovered the effect of the presence of local thermodynamic equilibrium in covalent states of a chlorine molecule in all degrees of freedom, excluding electronic. The electronic 'temperature', so to speak, of the triplet states, was much higher than the ground singlet state, since these groups of states were not collisionally mixed (see Sect. 7.2.3).

In thermodynamic reference books and tables, equilibrium constants are calculated using partial *pressures* of the species. These constants are given with the index p (pressure) and are called *equilibrium constants at constant pressure* K_p (2.2.8a is *equilibrium constants at constant concentration*).

$$K_p = \frac{\prod_k p_{k,eq}^{\nu'_k}}{\prod_i p_{i,eq}^{\nu_i}} \tag{2.2.9}$$

A partial pressure $p_i = n_i k\mathrm{T} = [\mathrm{A_i}]k\mathrm{T} = c_i R\mathrm{T}$ [2], p. 1 (R is the gas constant). Therefore,

$$K_p = (R\mathrm{T})^{\Delta\nu}\frac{\prod_k c_{k,eq}^{\nu'_k}}{\prod_i c_{i,eq}^{\nu_i}} = (R\mathrm{T})^{\Delta\nu} K_c, \tag{2.2.10}$$

$$\Delta\nu = \sum_i \nu_i - \sum_k \nu'_k$$

Using the apparatus of statistical physics and thermodynamics, one can express the values of K_p and K_c in terms of *partition functions F* of reactants and products [2], p. 12, [3], p. 356.

$$F = \sum_j g_j \exp\left(-\frac{\varepsilon_j}{k\mathrm{T}}\right). \tag{2.2.11}$$

Here, g_j *and* ε_j *are the statistical weight (degeneracy)* and the energy of a j-th species state. *The partition function is equivalent to the number of possible states of the system at the given energy.*

One can determine K_c and K_p values using the partition functions of reactants and products:

$$K_c = \frac{\prod_k \left(F_{k,c}^{\nu'_k}/N_A\right)}{\prod_i \left(F_{i,c}^{\nu_i}/N_A\right)} \cdot \exp\left(\frac{Q_0^0}{RT}\right) \tag{2.2.12a}$$

(concentration units, mol/cm^3)

$$K_c = K = \frac{\prod_k \left(F_{k,c}^{\nu'_k}\right)}{\prod_i \left(F_{i,c}^{\nu_i}\right)} \cdot \exp\left(\frac{Q_0^0}{RT}\right) \tag{2.2.12b}$$

(concentration units, species/cm^3)

$$K_p = \frac{\prod_k \left(F_{k,p}^{\nu'_k}/N_A\right)}{\prod_i \left(F_{i,p}^{\nu_i}/N_A\right)} \cdot \exp\left(\frac{Q_0^0}{RT}\right) \tag{2.2.12c}$$

(concentration units, mol/cm^3).

Here, $F_i^{\nu_i}/N_A$ and $F_k^{\nu'_k}/N_A$ are the partition functions of the i-th and k'-th species, respectively, the indices p and c denote that the corresponding partition functions refer to the standard state at a pressure of 1 atm and 1 mol/cm^3, respectively; Q_0^0 is the thermal effect of the reaction at T = 0 K, equal to the difference between the internal energy of the reactants and products at this temperature $\left(-\Delta U_0^0\right)$ [3], p. 362.

The correctness of (2.2.12a–2.2.12c) seems to be quite understandable:

- in these equations, the Boltzmann factor is present, the need for which is undeniable;
- as regards the partition function, then, as is known from statistical physics, if one analyzes system states with the same energy (the Boltzmann factor is taken into account!), then all these states are equally probable in a thermodynamically equilibrium system, regardless of the degree of freedom (electronic, vibrational, rotational or translational). Naturally, it is necessary to take into account the degeneracies of these states. One can see another manifestation of these laws of nature in Sect. 3.1 in which the detailed balance principle is considered.

The partition function of a species is equal to the product of the functions corresponding to different forms of energy. This statement also seems to be indisputable; indeed, in each molecular electronic state can be many vibrational states corresponding to vibrational degrees of freedom, and in each vibrational state there can be many rotational levels, and each of the species in each rovibronic state moves, etc.

It follows from (2.2.11–2.2.12) that if one knows the spectroscopic constants of the reactants and products, then the equilibrium constants for the process in which they are involved can be calculated.

Let us determine the partition function for the reaction

$$I_2\left(X0_g^+, v_X, J_X\right) \overset{M}{\leftrightarrow} 2I(5p^5\,^2P_J) \tag{2.2.13}$$

J = 3/2 (E = 0), 1/2 (E = 0.9427 eV (7603.1 cm^{-1}) are the components of the spin–orbit splitting (Fig. 2.2) [7], p. 25.

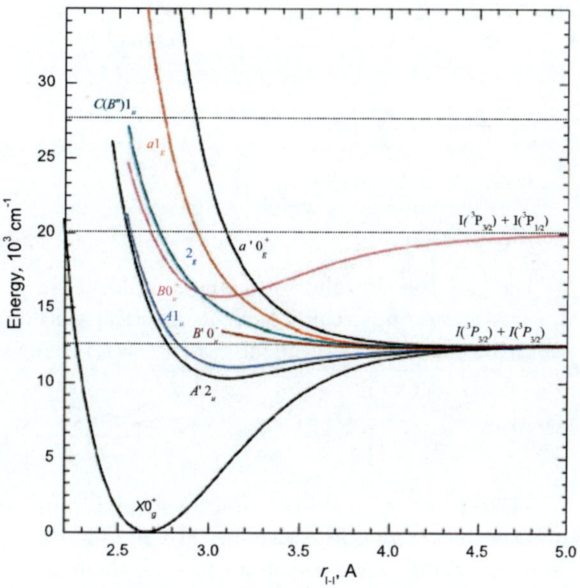

Fig. 2.2 Some potential energy curves of the valence iodine molecule states correlating with the $I(^2P_{3/2}) + I(^2P_{3/2})$ and $I(^2P_{3/2}) + I(^2P_{1/2})$ dissociation limits (see [7], p. 26)

The equilibrium constant K of reaction (2.2.13) is equal to

$$K = \frac{F_{\mathrm{I}}^2}{F_{\mathrm{I}_2}} \exp(-D_0^0/RT) \text{ species/cm}^3. \tag{2.2.14}$$

Here F_{I} is the partition function of the $I(^2P_J)$ atoms, F_{I_2} is the partition function of the $I_2(X0_g^+, v_X, J_X)$ molecule, $D_0^0 = 1.542$ eV is the dissociation energy of the I_2 molecule at T = 0 K, i.e., from the $I_2(X, v_X = 0, J_X = 0)$ state; M partition functions are reduced.

The partition function of an atom is equal to the product of the statistical sums corresponding to the various forms of its energy:

$$F = F_t \cdot F_e \cdot F_n. \tag{2.2.15}$$

Here,

- F_t is the partition function corresponding to the translational motion,
- F_e is the electronic partition function,
- F_n - nuclear partition function.

$$F_t = \frac{(2\pi mk\mathrm{T})^{3/2}}{h^3} \cdot \mathrm{V} \tag{2.2.16a}$$

For one mole of an ideal gas and $p = 1$ (if we need to assume K_p; $V = R\mathrm{T}/p$)

$$F_t = \frac{(2\pi mkT)^{3/2}}{h^3} \cdot RT. \qquad (2.2.16b)$$

for any species. Here

- m is a species mass,
- V is the volume in which the gas is enclosed,
- p is pressure.

The (2.2.16a) is valid for a single molecule in volume V, the presence of other species interacting weakly with it (an ideal gas) does not affect. The (2.2.16b) is valid for one mole of an ideal gas ($pV = RT$). For species/cm^3 units:

$$F_t = \frac{(2\pi mkT)^{3/2}}{h^3} \qquad (2.2.16c)$$

These partition functions can be obtained from the solution of the wave equation of a species moving in a potential box of volume V [3], p. 26. As follows from the solution of this wave equation, the translational motion of a species in a box is quantized, although the energy levels are extremely dense: $\varepsilon_n = n^2h^2/8\ ml$, where l is the side of the box, n is a quantum number, 0 or integer, and $h^2/8\ m = (6.63 \cdot 10^{-27}\ \text{erg s})^2/8 \cdot 16 \cdot 1.67 \cdot 10^{-24}$ g for an oxygen atom, for example, which is $2.05 \cdot 10^{-31}$ erg cm = $1.03 \cdot 10^{-15}\ \text{cm}^{-1}$ for $l = 1$ cm (1 cm^{-1} is $1.98546 \cdot 10^{-16}$ erg/species). Consequently, the average energy corresponding to 300 K, 240 cm^{-1}, with $l = 1$ cm, corresponds

$$n = \frac{\sqrt{240}}{1.03 \cdot 10^{-15}} = 4.8 \cdot 10^8$$

states. Huge value! It follows that the translational partition function, equal to $F_t = \sum_j \exp(\varepsilon_j/kT)$ is also very large, for $m = 16$ amu (oxygen atom) and V = 1 cm^3 $F_t = 6.2 \cdot 10^{25}$. For iodine atoms, $m_{\text{I}} = 127$ amu, $F_t = 1.39 \cdot 10^{27}$.

The electronic partition sum of an atom is equal to its electronic statistical weight (see (2.2.11)). The electronic statistical weight is equal to the degeneracy of the state under consideration; the latter for the angular momentum of (any) S, L, J, I … is equal to $2S + 1$, $2L + 1$, etc. Consequently, for the I($^2P_{3/2}$) state (the spin-orbital degeneracy is removed, and we consider the $J = 3/2$ degeneracy): $F_e = 2 \cdot 3/2 + 1 = 4$. The I($^2P_{1/2}$) state at T $\approx$ 300 K population is exp(−0.94/0.03) = $2.47 \cdot 10^{-14}$. Since the dissociation of the iodine molecule takes place only at high temperatures, for each T, it is necessary to consider the partition function (2.2.11).

The nuclear partition sum of an atom is $F_n = 2I + 1 = 2{\cdot}5/2 + 1 = 6$; ($I = 5/2$ is the nuclear spin of ^{127}I isotope). As one will see later, the nuclear partition functions are reduced.

The partition function of a molecule is equal to the product of the statistical sums corresponding to the various forms of its energy:

$$F = F_t \cdot F_r \cdot F_y F_e \cdot F_n. \qquad (2.2.17)$$

Here,

- F_t is the partition function corresponding to the translational motion,
- F_r is the partition function corresponding to the rotational motion,
- F_v is the partition function corresponding to the vibrational motion,
- F_e is the electronic partition function,
- F_n—nuclear partition function.

At relatively low temperatures and large values of the rotational constants, when one has to distinguish the population of *ortho-* and *para*-isomers of molecules, the nuclear partition function is included in the rotational one.

The translational partition sum of the I_2 molecule differs from the translational partition quantity of iodine atoms only in that the first one has a different mass; in this case, 2 times larger.

$$F_t = \frac{(2\pi mk\mathrm{T})^{3/2}}{h^3} \qquad (2.2.18)$$

(this equation is the same as (2.2.16a). For the iodine molecule, $F_t^{I_2} = 3.93 \cdot 10^{27}\mathrm{cm}^3/\mathrm{species}$.

The rotational partition function of a diatomic molecule is:

$$F_r = \sum\nolimits_J (2J+1)\exp(-E_J/k\mathrm{T}), \qquad (2.2.19)$$

where J is the quantum number of the total angular momentum of the molecule, resulting from the quantum numbers corresponding to the spin and orbital motion of the electrons, the rotation of the molecule (these are the Hund cases *a, b, c, d*) [7], p. 219. For the singlet state with the electron momentum projection $\Lambda = 0$, $^1\Sigma$, J is just a rotational quantum number. It can be shown that if the rotational constant and, as a result, the frequency values of the rotational transition from the first rotational level are not too large (this is true for all molecules except hydrogen and deuterium), and the temperature is not too low, comparable to or greater than room temperature, and the energy of this rotational quantum is much less than kT, then by summing up in (2.2.19) one can get:

$$F_r = \frac{kT}{B_e hc} = \frac{8\pi Ik\mathrm{T}}{\sigma h^2} \qquad (2.2.20)$$

Here, B_e is the rotational constant of the molecule in selected electronic state at the ground vibrational level, $v = 0$, I is the molecule momentum of inertia.

For the iodine molecule, it is equal to $I_{I_2} = \mu_{I_2} r_{I-I}$, where

$$\mu_{I_2} = \frac{m_I m_I}{m_I + m_I} = \frac{m_I}{2} \tag{2.2.21}$$

is the reduced mass, and r_{I-I} is the I-I internuclear distance, σ is the symmetry order; for a homonuclear molecule $\sigma = 2$. For the $I_2(X^1\Sigma_g^+)$ molecule, $F_r^{I_2} = 1556$. A rather large value, since the rotational quantum of the $I_2(X)$ molecule is small ($B_e^X = 0.037\ \text{cm}^{-1}$ [7], p. 192), and a large number of rotational levels are populated at room temperature.

The vibrational partition function of a diatomic molecule is equal to:

$$F_v = \sum\nolimits_v \exp(-E_v/kT), \tag{2.2.22}$$

since the degeneration of each level of a diatomic molecule is equal to 1. If the harmonic approximation is valid for calculation of the partition function, the summation in (2.2.22) gives:

$$F_v = \left[1 - \exp\left(\frac{hc\omega}{kT}\right)\right]^{-1}, \tag{2.2.23}$$

where ω (cm^{-1}) is the energy (wavenumber) of oscillations in the equation for harmonic oscillator levels (see [7], p. 192)

$$G(v) = \omega(v + 1/2)(\text{cm}^{-1}), \nu = \omega c(\text{s}^{-1}), \tag{2.2.24}$$

For the $I_2(X^1\Sigma_g^+)$ molecule, $F_V^{I_2} = 2.68$ due to small vibrational quantum, $\omega_e^X = 214\ \text{cm}^{-1}$.

Electronic partition function of a diatomic molecule. If the spin-orbit interaction is not too large (Hund *a*, *b* cases), the electronic partition function of the molecule electronic state is:

$$F_e = (2 - \delta_{0,\Delta})(2S + 1)\exp(-E_e/kT). \tag{2.2.25a}$$

Here, $\delta_{0,\Lambda}$ is the Kronecker symbol, equal to 1 for Σ states ($\Lambda = 0$) and 0 for all others. It appears due to the so-called Λ—doubling. For the ground electronic states, the term under the exponential is 1. If the spin-orbital degeneracy is removed, one has to treat each Ω-component separately, and

$$F_e = (2 - \delta_{0.\Omega})\exp(-E_d/kT). \tag{2.2.25b}$$

For the $I_2(X0_g^+)$ molecule, $F_e^{I_2(X)} = 1$.

The nuclear partition function of the diatomic molecule. In the general case (AB molecule), it is equal to:

$$F_n = (2I_A + 1)(2I_B + 1)/\sigma, \tag{2.2.26}$$

where I_i are the the nuclei's spins, σ is the order of the axis of symmetry (see above); $\sigma = 1$ (heteronuclear diatomic molecules); for homonuclear molecules, it is equal to 2, because the two positions of the molecule when replacing nuclei are equivalent. It is clear that in the case of a homonuclear molecule, such as I_2, $F_n = (2I + 1)^2/2$. The author deliberately omits here the details associated with different statistics (Fermi–Dirac and Bose–Einstein), which obey homonuclear molecules with different nuclear spins. (the degeneracy of the ortho-isomer is equal to $(I + 1)$ $(2I + 1)$, and para $(2I + 1)$). When summing over all rotational levels, one gets $(2I + 1)^2/2$.

Polyatomic molecules.

The rotational partition function is equal to [3], p. 386

$$F_r = \frac{8\pi^2(8\pi^3 I_A I_B I_C)^{1/2}(kT)^{3/2}}{\sigma h^3}, \tag{2.2.27}$$

I_A, I_B, I_C are the principal moments of inertia.

The vibrational partition function is

$$F_v = \prod_j [1 - \exp(\hbar\omega_j/kT)]^{-1} \tag{2.2.28}$$

ω_j are the normal vibration frequencies.

Using (2.2.14–2.2.25a, b), one can calculate the equilibrium constant of reaction (2.2.13):

$$K^p_{2.13} = \frac{\left[g^e_I \frac{(2\pi m_I kT)^{3/2}}{h^3}\frac{RT}{N_A}\right]^2}{g^e_{I_2}\frac{(2\pi m_{I_2} kT)^{3/2}}{h^3}\frac{RT}{N_A}\cdot\frac{kT}{B^X_e\sigma h}\cdot[1-\exp(-\omega_e/kT]^{-1}}\exp\left(-\frac{D^0_0}{RT}\right) \text{ atm} \tag{2.2.29a}$$

$$K_{2.13} = \frac{\left[g^e_I \frac{(2\pi m_I kT)^{3/2}}{h^3}\right]^2}{g^e_{I_2}\frac{(2\pi m_{I_2} kT)^{3/2}}{h^3}\cdot\frac{kT}{B^X_e\sigma h}\cdot[1-\exp(-\omega_e/kT]^{-1}}\exp\left(-\frac{D^0_0}{RT}\right) \text{cm}^3/\text{species} \tag{2.2.29b}$$

(nuclear partition function are reduced).

One should note that, in *non-equilibrium conditions*, if iodine atoms are *not* produced as a result of heating I_2, as in reaction (2.2.13), but, for example, by photolysis of any iodine-containing substance, and the $I(^2P_{3/2})$ termolecular recombination occurs

$$2I(^2P_{3/2}) \leftrightarrow I_2\left(X0_g^+, A'2_u, A1_u..., v_{max}\right) \tag{2.2.30}$$

$$I_2\left(X0_g^+ A'2_u, A1_u..., v_{max}\right) \overset{M}{\leftrightarrow} I_2\left(X0_g^+, A'2_u, A1_u..., v < v_{max}\right) \tag{2.2.31}$$

The equilibrium constant $K_{2.30} = k_{2.30}/k_{-2.30}$ has nothing in common with the constant calculated by using (2.2.29b) (see the beginning of this Section).

2.3 Arrhenius Equation

The Arrhenius equation describes the dependence of the rate constants of reactions and other collision processes on temperature. The equilibrium constant depends exponentially on temperature (see (2.2.29a, 2.2.29b). The same result can be obtained from the equilibrium constant (2.2.12a, 2.2.12b, 2.2.12c). Indeed, if one takes the derivative of *lnK* with T, then:

$$\frac{dlnK}{dT} = -\frac{Q}{RT^2} \tag{2.3.1}$$

where $Q \equiv \Delta U$ is the heat of reaction equal to the difference of the internal energy of the initial and final products. If one represents Q as the difference of some two energies $Q = E' - E$, then from $\ln K = \ln k - \ln k'$ one has:

$$\frac{dlnk}{dT} = \frac{E}{RT^2} + b \tag{2.3.2}$$

$$\frac{dlnk'}{dT} = \frac{E'}{RT^2} + b \tag{2.3.3}$$

where b is a certain constant. If one assumes that E, b are independent of T then these two equations imply an exponential dependence of k and k' on temperature:

$$k \sim \exp(-E/RT) \tag{2.3.4}$$

Since only species that collide with each other participate in processes other than elementary unimolecular, it can be argued that under conditions of thermodynamic equilibrium for translational degrees of freedom (one says *in cases of local thermodynamic equilibrium for translational degrees of freedom*) the rates of these processes is proportional to the number of collisions:

– for a bimolecular process involving species A and B the number of collisions is [8], p. 606:

$$z_{II} = \overline{V}_{AB} \cdot \pi d_{AB}^2 [A][B] = k_{II}^{gk}[A][B]. \tag{2.3.5}$$

Here $\overline{V}_{AB} = \sqrt{\frac{8RT}{\pi \mu_{AB}}}$ is the average relative velocity of the species A and B, $d_{AB} = (d_A + d_B)/2$, d_A, d_B are the gas kinetic diameters of the species A and B; μ_{AB} is the reduced mass (see (2.2.21)) for identical species $\mu_{A_2} = 1/2m_A$, $k_{II}^{gk} = \overline{V}_{AB} \cdot \pi d_{AB}^2$ is the gas-kinetic rate constant;

– for a termolecular process involving species A, B, and C the number of collisions is are:

$$z_{III} = 8\sqrt{2}\pi^{\frac{3}{2}} d_{AB}^2 d_{BC}^2 \delta \sqrt{RT}\left(\frac{1}{\mu_{AB}} + \frac{1}{\mu_{BC}}\right)[A][B][C] = k_{III}^{gk}[A][B][C]. \tag{2.3.6}$$

Here, δ is the distance that species A and B approach to C; this is a particular value of the order of molecular sizes (0.1 nm). The physical meaning of the gas-kinetic rate constant is clear: if one wishes to count the number of collisions, then one has to multiply this constant by the product of the concentrations of the colliding partners.
Combining (2.3.4–2.3.6), one sees that the rates of bimolecular and termolecular processes are

$$r_{II} \sim k_{II}^{gk}[A][B]\exp(-E_a/RT),$$

$$r_{III} \sim k_{III}^{gk}[A][B][C]\exp(-E_a/RT),$$

respectively (the E and E' are designated here as E_a and called *activation energies*). Clearly, not every the collision can be 'effective', and this circumstance can be taken into account in some approximation by a coefficient, which again in some approximation is independent of temperature. This coefficient is called with the *steric factor P*. Do not let his name mislead a reader, most often the reason for its difference from unit is not some geometric factors, but the forbiddance of the process. We will meet with these things more than once. As part of these approximations and according to (2.1.23):

$$r_{II} = Pk_{II}^{gk}[A][B]\exp(-E_a/RT), \tag{2.3.7}$$

and, similarly,

$$r_{III} = Pk_{III}^{gk}[A][B][C]\exp(-E_a/RT), \tag{2.3.8}$$

(one has to remember that $T^{1/2}$ is 'hidden' in k_{II}^{gk}, k_{III}^{gk}).

The (2.3.7, 2.3.8) are the versions of the Arrhenius equations. The physical meaning of the Arrhenius equation is transparent: the reaction rate constant is equal to the number of collisions, reduced to single concentrations, in which, by energetic, steric, symmetric, and other considerations, an act of process is possible. One sees that the temperature dependence is determined firstly by the $T^{1/2}$ term (the frequency of collisions is proportional to the velocity of the relative motion of the species, and it, in turn, is proportional to $T^{1/2}$), and secondly by the magnitude of the activation energy. The value of E_a determines the rate of the process. If, for example, the temperature doubled, then $T^{1/2}$ increases 1.4 times. The exp(- E_a/RT) relationship can be very weak if E_a is close to 0, or extremely strong.

Measurements and use in publications of the E_a value in units kcal/mol are frequently used, although this unit is nonsystem. The thing in the author opinion is convenience, not just a habit: the gas constant R, if measured in this system of units, is almost exactly 2 cal/mol (more precisely, 1.9858 cal/mol), and the 'usual' value E_a lies within (−5–+40) kcal/mol range (i.e., (−0.2–+1.8) eV, if this value is related to one species. A reader can almost instantly estimate the exponent for E = 1.2 kcal/mol and T = 300 K, for example, exp(−1200/600) = 0.1, if one uses these units.

The author tries to make a reader feels the concepts with which we are familiar. Let us do the same with rate constants.

Gas kinetic constant, what is it equal to? Let us calculate it for species with a gas-kinetic diameter d_A = 0.3 nm, m_A = 16 *amu* and T = 300 K (a collision of two oxygen atoms).

Ad hoc:

$$\overline{V} = \sqrt{\frac{8 \cdot 8.314 \frac{10^7 erg}{mol \cdot K} \cdot 300K}{\pi \cdot 8g/mol}} = 0.9 \cdot 10^5 \mathrm{cm/s}$$

$k_{II}^{gk} = 0.9 \cdot 10^5 \cdot \pi(3 \cdot 10^{-8})^2 \approx 2.5 \cdot 10^{-10} \mathrm{cm}^3/\mathrm{s}$ (see the text following (2.1.26). The reader can use this result as a reference: if the species are 10 times heavier, then the rate, and k_{II}^{gk}, is 3.3 times less (with the same d_A, naturally!).

Now let us estimate k_{III}^{gk} value for termolecular process. Let the values of the masses and gas-kinetic diameters be the same.

$$k_{III}^{gk} = 8\sqrt{2}\pi^{\frac{3}{2}}(3 \cdot 10^{-8} \cdot 3 \cdot 10^{-8})^2 \cdot 1 \cdot 10^{-8} \cdot (8.31 \cdot 10^{-7} \cdot 300)^{1/2} \cdot \left(\frac{1}{8} + \frac{1}{8}\right) \approx 2 \cdot 10^{-32} \mathrm{cm}^6/\mathrm{s},$$

(see the text following (2.1.29).

The author wishes to notice that one does not need to 'pray' to these numbers, because for atoms d_A increases with increasing mass, i.e., the numbers in the periodic table, for two- and polyatomic species, d_A are higher than for atoms, and finally, for electronically-excited or for rovibronically excited species, the values of gas-kinetic diameters can be more than those of unexcited. So it is not so rare to find rate constants of the order of $n \cdot 10^{-9}\,\mathrm{cm}^3/\mathrm{s}$ *(n* < 10). Even more often, constants are many orders of magnitude less than a gas-kinetic one, and not always the activation energy of the process is to blame; the matter is in the steric factor, i.e., as a rule, in the symmetry forbiddance for this process.

As to termolecular processes, sometimes the recombination rate constant can be on the order of $10^{-28}\,\mathrm{cm}^6/\mathrm{s}$, i.e., the value is 4 orders of magnitude higher than mentioned above. But in this case, the main 'fault' does not fall on large gas-kinetic diameters. The reason is that the reaction mechanism cannot be described in any way within the framework of a simultaneous collision of three species (see [9] and references).

Very often, when the activation energy is large, the term ($T^{1/2}$) is neglected, since the dependence of the *k* value on T is determined mainly by the exponent. In these cases, the Arrhenius equation takes the simple form

$$k = A\exp(-E_a/RT), \tag{2.3.9}$$

where $A \equiv Pk_{gk}$ is the so-called *pre-exponential factor*, independent of temperature.

As mentioned above, activation energy can be both positive and negative. It is clear that for endothermic reactions, i.e., reactions in which the internal energy of the products is greater than those of the initial species, the activation energy simply has to be positive. However, it is often positive for exothermic processes, and the matter is that there are potential barriers at the process coordinate, which we will also discuss in Sect. 3.5. Very often (but not always) for the recombination processes, the activation energy is negative, and a decrease of reaction rate with temperature occurs (see below).

To measure the activation energy and the pre-exponential factor *A*, in those cases when one represents *k* in the form of (2.3.9), it is necessary to use the dependence ln $k = f(1/T)$. It is clear from (2.3.9) that this dependence 'should' be direct, and (Fig. 2.3)

$$\frac{d\ln k}{d\left(\frac{1}{T}\right)} = -\frac{E_a}{R} \tag{2.3.10}$$

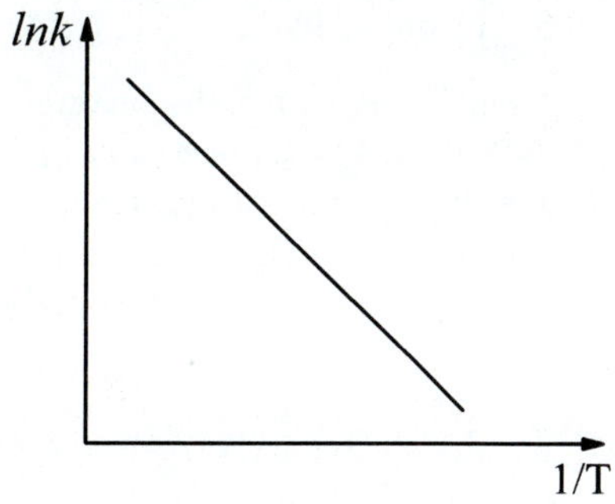

Fig. 2.3 To the activation energy calculation

It is clear that if the temperature interval in which one has measured this dependence is small, then one gets a straight line. It is also clear that in sufficiently large temperature ranges this simple depende once cannot be observed. The reason is that we got this dependence within very simple assumptions. An example would be recombination reactions with a positive activation energy, for example,

$$O(^3P) + CO \xrightarrow{M} CO_2\left(\tilde{X}^1\Sigma_g^+\right). \tag{2.3.11}$$

At temperatures of about 1000 K and below, the activation energy of this reaction, due to the presence of a potential barrier at the reaction coordinate, is positive, ~3 kcal/mol. At very high temperatures, obtained in shock tubes, when the average energy of the species exceeds the height of this potential barrier, this reaction behaves as ordinary termolecular with negative activation energy.

The representation of the Arrhenius equation even in a rather narrow temperature range at high activation energies and often even at small ones is a matter of taste: one writes $k = A \cdot \exp(-E_a/RT)$, $k = A \cdot T^{1/2} \exp(-E_a/RT)$, $k = A \cdot T^n \exp\ (-E_a/RT)$. Getting used to this is necessary.

2.4 Complex Reactions. Consecutive Reactions. Steady-State Method

As mentioned in Sect. 2.1, there are complex, simple processes (reactions) and elementary processes. Let the author reminds that if the stoichiometric order of the process does not coincide with the kinetic, determined in the experiment, then the process is called complex; if it is, then it is simple. If a simple process proceeds in one stage, then this is an elementary process. Most often, simple processes do not proceed in one stage but are a kind of multistage process. The complex reactions are multistage process, indeed.

The author will not discuss in detail the different types of complex chemical reactions, since readers who research or study in the fields of molecular spectroscopy/molecular physics, as well as chemical physics/physical chemistry in the gas-phase they are rarely encountered. They will be mentioned in passing.

So, distinguish:

– *Coupled reactions*, the features of which lie in the fact that one of the reactions $A + B_2$ can take place only in the presence of the second $A + B_1$. The reason is that the product of one reaction is one of the other reactant:

$$A + B_1 \rightarrow X + \text{other products} \tag{2.4.1}$$

$$X + B_2 \rightarrow C + \text{other products} \tag{2.4.2}$$

– *Catalytic reactions* are reactions accelerated in the presence of some foreign substance (a catalyst). Consider homogeneous catalysis when the catalyst and the reactant are in the same phase, gas-phase.

$$A + \underline{K} \rightarrow X + \text{other products} \tag{2.4.3}$$

$$X + B \rightarrow C + \underline{K} + \text{other products}; \tag{2.4.4}$$

catalyst $\underline{K}$ is not consumed. The author gives one example of such processes that he has encountered in his practice. These are processes in the O + NO system: when these species interact, NO_2 molecule is formed, which reacts very fast with O atoms

$$O + \underline{NO} \xrightarrow{M} NO_2$$

$$O + NO_2 \rightarrow \underline{NO} + O_2$$

The rate constant of the first reaction is close to the gas-kinetic one, $7 \cdot 10^{-32}\ \text{cm}^6/\text{s}$, the rate constant of the second one is very high, $6 \cdot 10^{-12}\ \text{cm}^3/\text{s}$, and this reaction, unlike the first, is bimolecular. As a result, atomic oxygen recombines with a high rate in the presence of NO.

– *Autocatalytic reactions* are those in which the catalyst is formed as a product of one of them:

$$A + B \rightarrow \underline{K} + \text{other products} \tag{2.4.5}$$

$$\underline{K} + B \rightarrow C + \underline{K} + \text{other products}; \tag{2.4.6}$$

catalyst $\underline{K}$ is not consumed.

– *Parallel processes.* The simplest examples of these processes, interesting for potential readers of the book, do not differ in kinetics from kinetics simple or even elementary processes. These are optical transitions from an excited state to several lower ones (the most widespread case), predissociation via several channels, see Sect. 4.7, (also quite common), etc. As an example of a parallel *reaction*, these are parallel reactions of recombination and disproportionation:

$$2C_2H_5 \xrightarrow{M} C_4H_{10}$$

$$\rightarrow C_2H_4 + C_2H_6$$

It is obvious that in both cases the rate constant of the decay of the initial products is simply equal to the sum of the constants of parallel processes.

Let us analyze in detail on a widespread type of complex and simple (within the framework of the definitions the author mentioned earlier) processes that are not elementary, but are a sequence of them. These are the so-called *consecutive reactions* in which two or more steps following one another occur:

$$A \rightarrow B; B \rightarrow C, \text{etc.} \tag{2.4.7}$$

As such, one can consider, for example, all the recombination processes described above. The time dependences of the concentrations of the initial, intermediate, and final products of such reactions in the general case can be quite complex. In gas-phase recombination reactions, intermediate products, as a rule, or are very chemically active (atoms, radicals) or short-lived, i.e., they are unstable and decay in high-rate monomolecular processes, or deactivate (relax) in fast collisional processes. In such cases, it is very often possible to use the *steady-state method*, using which one can get rid of systems of differential equations and reduce them to algebraic equations.

2.4.1 Steady-State Method

To understand the essence of the steady-state method, consider one simple example, namely the termolecular recombination of atom A and molecule B described in the framework of so-called *energy transfer mechanism* [9]:

$$A + B + M \rightarrow AB + M. \tag{2.4.8}$$

This reaction is the consecutive one (Fig. 2.4):
After recombination of A and B into a short-lived *intermediate* product (collision complex) $(A\ldots B)^{\#}$, having an excitation energy equal to the dissociation energy AB

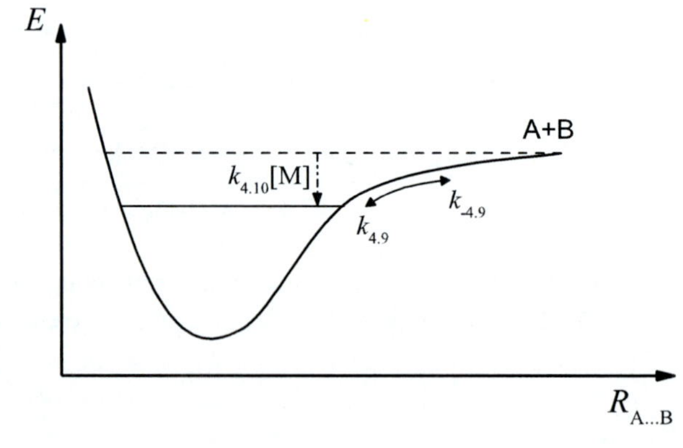

Fig. 2.4 Termolecular recombination of an atom and a diatomic molecule

(without taking into account the kinetic energy of the colliding species), the main channel of which decay is dissociation to the initial products (reactant) (reaction –2.4.9)

$$A + B \leftrightarrow (A\ldots B)^{\#} \tag{−2.4.9}$$

the stabilization of the complex in a collision with the species M, the third body:

$$(A\ldots B)^{\#} + M \rightarrow AB + M \tag{2.4.10}$$

proceed. Suppose that at the time t = 0 one switches on some source of A atoms (for example, one starts photolysis of A containing molecules); let the concentration of B and M species be constant. The equations for the rate of the decay of the A, $(A\ldots B)^{\#}$ species and the formation of AB one are the following:

$$\frac{d[A]}{dt} = r_A - k_{4.9}[A][B] + k_{-4.9}\left[(A\ldots B)^{\#}\right] \tag{2.4.11}$$

$$\frac{d\left[[A\ldots B]^{\#}\right]}{dt} = k_{4.9}[A][B] - k_{-4.9}\left[(A\ldots B)^{\#}\right] - k_{4.10}\left[(A\ldots B)^{\#}\right][M] \tag{2.4.12}$$

$$\frac{d[AB]}{dt} = k_{4.10}\left[(A\ldots B)^{\#}\right][M] \equiv r_{4.8} = k_{4.8}[A][B][M] \tag{2.4.13}$$

(here r_A is a certain constant rate of production of the A atoms).

It is evident that the time after switching on the source of the atoms A, some stationary value $[A]_{ss}$ (ss—steady-state, stationary) will be established, and d[A]/dt will become equal to 0. Concentration $(A\ldots B)^{\#}$ after switching on the pumping has to also increase from 0 to some stationary value $[(A\ldots B)^{\#}]_{ss}$, as well. The concentration of AB species also has to grow, and at a constant $[(A\ldots B)^{\#}]$, its growth rate is constant (see 2.4.13). These are qualitative and, the author hopes, right reasoning; despite the simplicity, they are entirely reliable.

Return to (2.4.11) and see what one can get from it, setting d[A]/dt = 0, i.e., assuming that the steady-state condition is satisfied for the A species.

$$r_A - k_{4.9}[A]_{ss}[B] + k_{4.9}\left[(A\ldots B)^{\#}\right]_{ss} = 0; \tag{2.4.14}$$

and the A steady-state concentration is:

$$[A]_{ss} = \frac{r_A + k_{-4.9}\left[(A\ldots B)^{\#}\right]_{ss}}{k_{4.9}[B]} \tag{2.4.15}$$

Substitute *rhs.* of (2.4.15) in (2.4.12), taking its left side to be 0, also. It is clear that such a ‘beautiful’ time is when d[A]/dt = 0 and d[(A…B)$^{\#}$]/dt = 0 should ever come.

$$\frac{k_{4.9}[\mathrm{B}]\cdot\{r_A + k_{-4.9}\left[(\mathrm{A...B})^{\#}\right]_{\mathrm{ss}}}{k_{4.9}[\mathrm{B}]} = \left[(\mathrm{A...B})^{\#}\right]_{ss}\cdot\{k_{-4.9}+k_{4.10}[\mathrm{M}]\} \quad (2.4.16)$$

Reduce $k_{4.9}$[B] in the *lhp* of (2.4.16), and one gets the obvious result:

$$r_{\mathrm{A}} = k_{4.10}[(\mathrm{A...B})^{*}]_{\mathrm{ss}}[\mathrm{M}] = r_{4.10} = r_{4.8},$$

i.e., the rate of formation of AB in steady-state conditions is equal to the rate of formation of atoms A.

Go back to (2.4.12). Putting d[(A…B)$^{\#}$]/dt = 0 on the *lrp* of (2.4.12), one gets:

$$k_{4.9}[\mathrm{A}]_{\mathrm{ss}}\cdot[\mathrm{B}] = [(\mathrm{A...B})]_{\mathrm{ss}}\cdot\{k_{-4.9}+k_{4.10}[\mathrm{M}]\} \quad (2.4.17)$$

and

$$[(\mathrm{A...B})]_{\mathrm{ss}} = \frac{k_{4.9}[\mathrm{A}]_{\mathrm{ss}}[\mathrm{B}]}{k_{-4.9}+k_{4.10}[\mathrm{M}]} \quad (2.4.18)$$

Equation (2.4.17) *is the steady-state condition for (A…B)$^{\#}$ intermediate.* Beneficial thing.

Let us now see what the steady-state concentrations of atoms A and the (A…B)$^{\#}$ complexes should be at reasonable values of r_{A}, [B], [M], $k_{4.9}$, $k_{-4.9}$ and $k_{4.10}$. Take $r_{\mathrm{A}} = 1\cdot10^{14}$ atoms/cm^3 s (these are normal values that can be obtained by photolysis), [B] = $3.3\cdot10^{16}$ cm^{-3} (p_{B} = 1 Torr), [M] = $3.3\cdot10^{17}$ cm^{-3} (p_{M} = 10 Torr). Let $k_{4.9} = k_{\mathrm{II}}^{gk}$ (this is quite normal), $k_{-4.9} \approx 10^{10}$ s^{-1} (the lifetime of a triatomic molecule with an excitation energy of equal dissociation energy and $D_0^0 = 5$ eV (see [2], p. 98 and references)) and $k_{4.10} = k_{\mathrm{II}}^{gk}$ (this is quite a common picture too). Convert (2.4.17) to the form:

$$\frac{[\mathrm{A}]_{ss}}{\left[(\mathrm{A...B})^{\#}\right]_{ss}} = \frac{k_{-4.9}+k_{4.10}[\mathrm{M}]}{k_{4.9}[\mathrm{B}]} \quad (2.4.19)$$

Compare the values of the two addends in the numerator of the *rhs* of (2.4.19) for the conditions specified above. One sees that the term $k_{4.10}$[M] = $3\cdot10^{-10}\,3\cdot10^{17}$ << $k_{-4.9} = 10^{10}$ s^{-1} can certainly be neglected, i.e., the stabilization rate (A…B)$^{\#}$ at given lifetimes (A…B)$^{\#}$ and pressures M is much less than its decay rate (the usual picture for ‘low-atomic species’). It is also seen from (2.4.19) that [A]$_{\mathrm{ss}}$/[(A…B)$^{\#}$]$_{\mathrm{ss}}$ $\approx k_{-4.9}/k_{4.9}$[B] = $10^{10}/3\cdot10^{-10}\,3.3\cdot10^{16} = 1\cdot10^{3}$ >> 1.

Therefore, the overwhelming majority of atoms A is in the unbound state; it is clear that if the pressure is greatly increased, the picture may become the opposite.

One can estimate the time after which d[A]/dt = 0 and d[$(A\ldots B)^{\#}$]/dt = 0. It has been already shown that in steady-state conditions $r_A = k_{4.10}[(A\ldots B)^{\#}]_{ss}[M]$. This allows to calculate $[(A\ldots B)^{\#}]_{ss} = r_A/k_{4.10}[M] = 10^6\ cm^{-3}$ and $[A]_{ss} = 10^3\cdot[(A\ldots B)^{\#}]_{ss} = 10^9\ cm^{-3}$. This number of atoms can be accumulated within $\sim 10^{-5}$ s. Although this value is not equal to the time of establishment of the steady-state concentrations, it may indicate the order of time for the establishment of it.

The rate at which steady-state conditions are reached can be judged by the magnitudes of the coefficients in the *rfs* of (2.4.11, 2.4.12) for [A] and [$(A\ldots B)^{\#}$]. In the case under discussion, this is $k_{4.9}[B] = 10^7\ s^{-1}$ and $k_{-4.9} = 10^{10}\ s^{-1}$. If they are so large, then it can be assumed that at times not exceeding the time of accumulation of the stationary concentration of active intermediate products (in our case A and $(A\ldots B)^{\#}$) the steady-state conditions have have occurred. Of course, in unobvious cases, it is not bad to solve a system of differential equations, and calculate the time to establish stationary concentrations. Only then the steady-state method can be used.

And this method is very, very useful. Indeed, one replaces the system of differential equations, often non-linear, using a system of algebraic equations. Using this method, one can often easily get the dependences of the rate of a change of the final product concentrations on the experimental conditions and, from the dependencies observed in the experiment, obtain the values of the rate constants or any of their combinations (ratios, etc.). One should only remember the expression of one of the authors of the book [2], Professor V.N. Kondratyev: "Kinetic proof of the mechanism is the weakest proof!".

But where the steady-state method is merely irreplaceable, is in understanding the mechanism of the process. The reader behind these algebraic exercises can practically understand, in the first approximation, the mechanism of atom and molecule recombination. The reader will see below how to get the recombination rate constant using the rate constants of elementary processes and what essential conclusions can be drawn from this.

Using (2.4.13, 2.4.18), one gets:

$$r_{4.8} = \frac{k_{4.9}k_{4.10}[A]\cdot[B]\cdot[M]}{k_{-4.9}+k_{4.10}[M]} \tag{2.4.20}$$

A very simple and elementary equation. Analyze it using the above values of constants and concentrations. The most simple are the limits of small and large pressures, which can be formulated as:

$$k_{-4.9} \gg k_{4.10}[M] \tag{2.4.21}$$

and

$$k_{-4.9} << k_{4.10}[\mathrm{M}] \tag{2.4.22}$$

which is equivalent to [M] << $k_{-4.91}/k_{4.10} = 10^{10}/3 \cdot 10^{-10} = 3 \cdot 10^{19}$ cm^{-3} and [M] >> $3 \cdot 10^{19}$ cm^{-3}, i.e., p_M << 1 atm and p_M >> 1 atm, respectively. In the low-pressure limit, this is a purely termolecular reaction, its constant is equal to:

$$k_{4.8} = \frac{k_{4.9}}{k_{-4.9}} \cdot k_{4.10} = K_{4.9} \cdot k_{4.10}\ \mathrm{cm}^6/\mathrm{s} \tag{2.4.23}$$

and rate of the AB molecule formation

$$r_{4.8} = K_{4.9} \cdot k_{4.10} \cdot [\mathrm{A}] \cdot [\mathrm{B}] \cdot [\mathrm{M}]\ \mathrm{cm}^{-3}\mathrm{s}^{-1} \tag{2.4.24}$$

is proportional to [M]. In the high-pressure limit, the kinetics of the reaction is bimolecular, and its rate constant is simply equal to the bimolecular rate constant of the $(\mathrm{A}\ldots\mathrm{B})^{\#}$ formation. (All $(\mathrm{A}\ldots\mathrm{B})^{\#}$ quasimolecules are stabilized, the reaction (-2.4.9) do not occur). The rate of formation of AB does not depend on [M] under these conditions. The order of reaction (2.4.8) relative to [M] is 0. Under intermediate conditions, the kinetics is intermediate between the termolecular and bimolecular and has to be described using the bimolecular rate constant $k_{4.8}$([M]) depending on [M]. The reaction (2.4.8) order relative to [M] is fractional. The $k_{4.8} = f$([M]) function has the form of a curve with saturation (Fig. 2.5).

That is why the author prefers the way of writing

$$\mathrm{A} + \mathrm{B} \overset{(\mathrm{M})}{\rightarrow} \mathrm{AB}$$

and

$$\mathrm{A} + \mathrm{B} \overset{\mathrm{M}}{\rightarrow} \mathrm{AB}$$

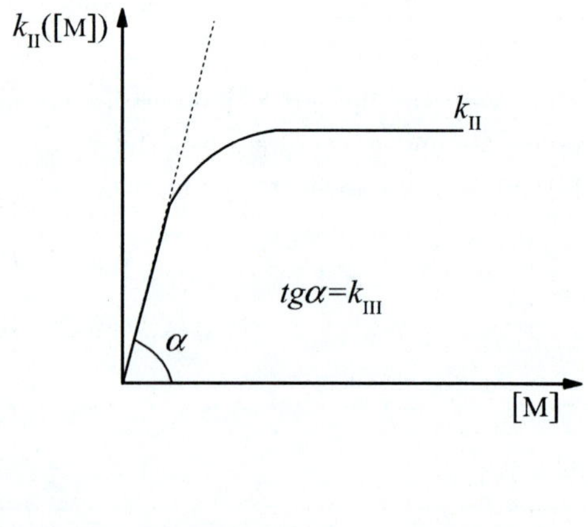

Fig. 2.5 The bimolecular rate constant of a termolecular recombination k_{II}([M]) as a function of M concentration

to the

$$A + B + M \rightarrow AB + M$$

one.

The author wishes to note that this behavior of the dependence of the recombination rate constant is typical for all recombination processes, and the values of the limits of small and large pressures naturally depend on the complex lifetime, that is, mainly on the number of atoms in it, and the magnitude of the stabilization rate constant; however, it is usually close to gas-kinetic. The same behavior is also characteristic of termolecular recombination accompanied by radiation (see Sect. 7.2.3).

Problems

1. What is the physical meaning of the rate constant of a unimolecular process?
2. What is the physical meaning of the rate constant of a bimolecular process?
3. What is the physical meaning of the rate constant of the termolecular process?
4. Can the reaction order and reagent order be zero or fractional? Give examples.
5. The rate constant is $3 \cdot 10^{14}$•exp(−4.8 (KJ/mol)/*R*T) $cm^3/mol \cdot 1$.

   Count the value of the constant in cm^3/species·s at T = 293 K.
6. The rate constant is $8 \cdot 10^{14}$•exp(4.8 (KJ/mol)/RT) $cm^6/mol^2 \cdot s$. Count the value of the constant in cm^6/species^2•s at T = 293 K.
7. What is the difference between the concepts of 'stoihiometric' and 'kinetic (real)' reaction order?
8. Calculate the equilibrium constant of the reaction

$$Cl(^2P_{3/2}) + Cl(^2P_{3/2}) \overset{M}{\leftrightarrow} Cl_2\left(X^1\Sigma_g^+\right) \quad (1)$$

   at T = 293 K if it is known that $Cl_2(X^1\Sigma_g^+)$ dissociation energy is 2.4 eV, $\omega_e^X = 559\ cm^{-1}$, $B_e^X = 0.244\ cm^{-1}$, and the chlorine atom mass is 35 *amu*.
9. Using the steady-state method, write an expression for the reaction rate constant

$$Xe(^3P_1) + Kr(^1S) \rightarrow XeKr(B0^+, v_{max}) \rightarrow XeKr(X0^+) + hv \quad (1)$$

   if it is known that its mechanism is described by elementary processes

$$Xe(^3P_1) + Kr(^1S) \leftrightarrow XeKr(B0^+, v_{max}) \quad (2)$$

$$XeKr(B0^+, v_{max}) \rightarrow XeKr(X0^+) + hv \quad (3a)$$

$$\rightarrow XeKr(X0^+, A1) \quad (3b)$$

$$\mathrm{XeKr}(B0^+, v_{\max}) + \mathrm{Kr}(^1\mathrm{S}) \rightarrow \mathrm{XeKr}(B0^+, v_{\max} - \Delta v) + \mathrm{Kr}(^1\mathrm{S}) \quad (4)$$

$$\rightarrow \mathrm{XeKr}(X0^+, A1) + \mathrm{Kr}(^1\mathrm{S}) \quad (5)$$

$$\rightarrow \mathrm{Xe}(^3\mathrm{P}_1) + 2\mathrm{Kr}(^1\mathrm{S}) \quad (6)$$

10. Calculate the steady-state concentration of the $O(^3P)$ and $(O\ldots O_2)^{\#}$ intermediate species if it is known that the processes

$$\mathrm{O}_2 + h\nu \rightarrow \ldots \rightarrow 2\mathrm{O}(^3\mathrm{P}) \quad (1)$$

$$\mathrm{O}(^3\mathrm{P}) + \mathrm{O}_2 \leftrightarrow (\mathrm{O}\ldots\mathrm{O}_2)^{\#} \quad (2)$$

$$(\mathrm{O}\ldots\mathrm{O}_2)^{\#} \xrightarrow{\mathrm{M}} \mathrm{O}_3 + \mathrm{M} \quad (3)$$

occur in the system under study.

References

1. IUPAC Compendium of Chemical Terminology, 2nd ed. (the Gold book). Complied by A.D. McNaught and A. Wilkinson, Blackwell Scientific Publications, Oxford (1997). https://doi.org/10.1351/goldbook
2. Kondratiev, V.N., Nikitin, E.E.: Gas-phase Reactions: Kinetics and Mechanism. Springer-Verlag, Berlin Heidelberg New York (1981). https://doi.org/10.1007/978-3-642-67608-6
3. Glasstone, S.: Theoretical Chemistry, D. van Nostrand Comp., Inc. (1944)
4. Laidler, K.J.: Chemical Kinetics. McGaw Hill Book Co, New York-London (1965). https://doi.org/10.1002/bbpc.19660700316
5. Blanksby, S.J., Ellison, G.B.: Bond dissociation energies of organic molecules. Acc. Chem. Res. **36**, 255–263 (2003). https://doi.org/10.1021/ar020230d
6. Liberman, M.A., Velikovich, A.L.: Physics of Shock Waves in Gases and Plasmas. Springer (1986). https://doi.org/10.1017/S0022112094211059
7. Herzberg, G.: Molecular Spectra and Molecular Structure, vol. 1. Spectra of Diatomic Molecules, Van Nostrand, New York (1950)
8. Calvert, J.G., Pitts, J.N., Jr.: Photochemistry. John Wiley & Sons; 1st edition (1966)
9. Porter, G.: Mechanism of third-order recombination reactions. Disc. Far. Soc. **33**, 198–204 (1962)

Chapter 3
Theory of Elementary Processes

Abstract The chapter deals with elementary processes, i.e., the simple processes which proceed in one stage. The definitions of differential and total cross-sections, microscopic rate constants, and probabilities of elementary processes are introduced. The detailed balance principle, adiabatic approximations, as well as potential energy curves and surfaces, are discussed. Different types of intermolecular interactions, descriptions of collisional processes and reactions including nonadiabatic transition using potential energy curves and surfaces are examined.

3.1 Cross-Sections, Rate Constants, and Probabilities of Elementary Processes. The Detailed Balance Principle

When species collide (below, in this section, we will consider the collision of two species) two different results are possible (not according to the formal kinetics of the process): either the chemical composition of the colliding species does not change or it changes (i.e., a chemical reaction occurs). In general, the first option can be presented as [1], p. 30:

$$\mathrm{A}(i) + \mathrm{B}(j) \rightarrow \mathrm{A}(l) + \mathrm{B}(m) + \Delta E_{ij,lm}, \tag{3.1.1}$$

and the last one as

$$\mathrm{A}(i) + \mathrm{B}(j) \rightarrow \mathrm{C}(l) + \mathrm{D}(m) + \Delta E_{ij,lm} \tag{3.1.2}$$

where i, j and l, m are some sets of quantum numbers of the states of the species before and after the collision, respectively, $\Delta E_{ij,lm}$ is the change in the internal energy of the colliding species, converting to their kinetic energies.

The above quantum numbers describe the electronic states of atoms and the rovibronic states of two- and polyatomic species. It should be noted that to

A. Pravilov, *Gas-Phase Photoprocesses*, Springer Series in Chemical Physics 123,
https://doi.org/10.1007/978-3-030-65570-9_3

characterize the process fully, it is necessary to indicate the relative velocity of the particles.

What are the possible variants of the process (3.1.1), i.e., a process without changing the chemical composition?

1. *Elastic collision*: $i = l$, $j = m$, $\Delta E_{ij,lm} = 0$.
2. *Inelastic collision*: the sets of quantum numbers of the system before and after the collision are different, and $\Delta E_{ij,lm} \neq 0$. However, the cases, $\Delta E_{ij,lm}$ *is equal to or close to 0*, are also possible. Such processes are referred to as *resonant* or *quasi-resonant*, respectively. That means close to 0 depends on the type of process. This problem will be discussed below.

The number of species scattered per unit volume per unit time in the ϑ, φ direction to the vector of relative motion of particles A(i) and B(j), $\boldsymbol{u}$, in process (3.1.1), for example, is proportional to the product $u \cdot [A(i)] \cdot [B(j)]$ (see (2.1.9) and Sect. 2.1:

$$\frac{d[A(l, u, \vartheta, \varphi)}{dt} \sim u \cdot [A(i)] \cdot [B(j)] \text{ cm}^{-3} \cdot \text{s}^{-1}.$$

The proportionality coefficient, i.e., the characteristic of the processes (3.1.1, 3.1.2) at a fixed velocity of relative motion of species is the differential cross-section of the scattering (reaction) $q_{ij,lm}(\boldsymbol{u}, \vartheta, \varphi)$ (cm^2), defined as the ratio of the concentration of species A(l) or B(m) (C(l) or D(m)) scattered in a specific direction per unit time in a single solid angle (cm$^{-3} \cdot$ s^{-1}), to the flow of species A (i), B(j), $\boldsymbol{u} \cdot [A(i)] \cdot [B(j)]$ (cm$^{-5} \cdot$ s^{-1}) (Fig. 3.1).

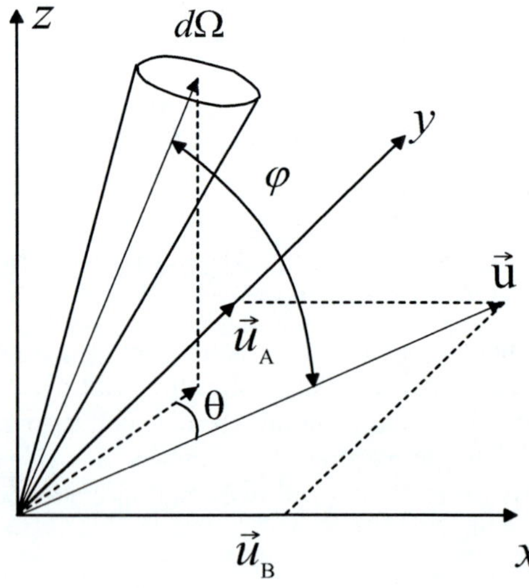

Fig. 3.1 To the definition of the differential cross-section

ϑ, φ are the angles defined relative to the velocity vector $\boldsymbol{u}$ of the relative motion of the reactants A(i), B(j), characterizing the direction of scattering of the interaction products A(l), B(m) (C(l), D(m)). In this way:

$$\frac{d[\mathrm{A}(l,u,\vartheta,\varphi)}{dt} = q_{ij,lm}(\boldsymbol{u},\vartheta,\varphi)\cdot u\cdot[\mathrm{A}(i)][\mathrm{B}(j)]\mathrm{cm}^{-3}\cdot\mathrm{s}^{-1} \quad (3.1.3)$$

If one integrates the right side of the equation under consideration over the solid angle 4π, then one gets the total number of species in a given state, scattered in a unit of volume per unit time in all directions; for the process (3.1.1), for example, this is:

$$\frac{d[A(l,u)}{dt} = \sigma_{ij,lm}(\boldsymbol{u})\cdot u\cdot[\mathrm{A}(i)][\mathrm{B}(j)]\ \mathrm{cm}^{-3}\cdot\mathrm{s}^{-1} \quad (3.1.4)$$

where

$$q_{ij,lm}(\boldsymbol{u}) = \int q_{ij,lm}(\boldsymbol{u},\vartheta,\varphi)d\Omega\cdot\mathrm{cm}^{-3} \quad (3.1.5)$$

is *total scattering cross-section.*

Neither under thermodynamic equilibrium conditions nor even in molecular beams, the velocities of colliding species A and B are constant for all species of the same kind but are described by certain distribution functions $f_{\mathrm{A}}(u_{\mathrm{A}})$, $f_{\mathrm{B}}(u_{\mathrm{B}})$. Taking into account this fact and assuming that $f_{\mathrm{A}}(u)$ and $f_{\mathrm{B}}(u)$ are independent of species concentrations (and this happens very often), (3.1.4) is transformed into:

$$\begin{aligned}\frac{d[A(l)]}{dt} &= [\mathrm{A}(i)][\mathrm{B}(j)]\int\sigma_{ij,lm}(u)f_{\mathrm{A}}(u_A)f_{\mathrm{B}}(u_B)du_A du_B) \\ &= k_{ij,lm}[\mathrm{A}(i)][\mathrm{B}(j)]\ \mathrm{cm}^{-3}\cdot\mathrm{s}^{-1}\end{aligned} \quad (3.1.6)$$

(the latter is valid according to (2.1.9). Here $f_{\mathrm{A}}(u_A)$,$f_{\mathrm{B}}(u_B)$ are the velocity distribution functions of the species normalized to 1, and the integral $k_{ij,lm}$ is the *microscopic process rate constant.* Microscopic means related to a process in which species with states of a set of quantum numbers i, j form species with a set of quantum numbers l, m.

For the reaction (3.1.2), similarly, one has the differential reaction cross-section $q^r_{ij,lm}(\boldsymbol{u},\vartheta,\varphi)$, the total, in the solid angle 4π, reaction cross-section $q^r_{ij,lm}(\boldsymbol{u})$, and microscopic reaction rate

$$\begin{aligned}\frac{d[C(l)]}{dt} &= [\mathrm{A}(i)][\mathrm{B}(j)]\int\sigma^r_{ij,lm}(u)f_{\mathrm{A}}(u_A)f_{\mathrm{B}}(u_B)du_A du_B) \\ &= k^r_{ij,lm}[\mathrm{A}(i)][\mathrm{B}(j)]\ \mathrm{cm}^{-3}\cdot\mathrm{s}^{-1}.\end{aligned} \quad (3.1.7)$$

It is evident that the above reasoning fully applies to the reverse reaction (process), since which process is direct, and which is the reverse, is determined by our arbitrariness.

One must take into account that very often, the state of colliding species is not described by the only sets of *i* and *j*, i.e., there is a certain distribution over quantum numbers or the distribution functions X_i^{A}, X_j^{B} of species A and B over internal degrees of freedom. In this case, the reaction rate for scattering in the 4π angle for all initial and final states with the distribution functions $f_{\mathrm{A}}(u_A),f_{\mathrm{B}}(u_B)$, X_i^{A}, X_j^{B} independent of the concentrations A and B is

$$\frac{d[C]}{dt} = k^r[\mathrm{A}(i)][\mathrm{B}(j)]\ \mathrm{cm}^{-3}\cdot\mathrm{s}^{-1}, \tag{3.1.8}$$

where

$$k^r = \sum\nolimits_{lm}\sum\nolimits_{ij} X_i^{\mathrm{A}}X_j^{\mathrm{B}}\int \sigma_{ij,lm}^r(u)f_{\mathrm{A}}(u_{\mathrm{A}})f_{\mathrm{B}}(u_{\mathrm{B}})du_{\mathrm{A}}du_{\mathrm{B}} \tag{3.1.9}$$

is the ‘chemical’ reaction rate constant so familiar to us.

The author has long argued differential and total scattering cross-sections, microscopic rate constants, and rate constants, because cross-sections are measured in molecular beam experiments, whereas, in bulk conditions or in flow reactors and shock tubes, rate constants are measured. One can determine rate constants using cross-sections (but not vice versa!), if the velocity distributions of the colliding species are known. In particular, in the cases of local thermodynamic equilibrium for translational degrees of freedom

$$k = \sigma\cdot\overline{V}_{AB}. \tag{3.1.10}$$

Here $\overline{V}_{AB} = \sqrt{\frac{8R\mathrm{T}}{\pi\mu_{\mathrm{AB}}}}$ is the average relative velocity of the species A and B. For example, the values $\sigma = 2.5\cdot 10^{-15}\ \mathrm{cm}^2$ ($d_{\mathrm{AB}} = 0.5$ nm) and $\overline{V}_{AB} = 10^5\ \mathrm{cm/s}$ $k \approx 3\cdot 10^{-10}\ \mathrm{cm}^3/\mathrm{s}$ is the gas-kinetic rate constant (see Sect. 2.3).

The definition of a microscopic rate constant is introduced above, using the total cross-section of the process. The author has noted many times already that readers who research or study in the fields of molecular spectroscopy/molecular physics, and chemical physics/physical chemistry deal with thermodynamically nonequilibrium systems. As a rule, only local thermodynamic equilibrium, either in the translational degrees of freedom or the translational and rotational ones. These cases happen because it is installed quickly (see Sects. 5.1, 5.2). Very often, one does not have a thermodynamic equilibrium for vibrational or electronic degrees of freedom. Consequently, as a rule, one deals with microscopic rate constants. It is also clear that the ‘integral, macroscopic’ constant is determined by the distributions $f_{\mathrm{A}}(u_A),f_{\mathrm{B}}(u_B)$, X_i^{A}, X_j^{B} for given microscopic constants.

In principle, differential and total cross-sections of processes can be calculated, especially if this is done in the so-called *semi-classical approximation*. The author will explain some features of the approximation and some results of this approach briefly.

The principal idea of the semi-classical approximation is that one group of degrees of freedom of the system under consideration, for example, the translational one, is described in the framework of classical mechanics, and the rest in the framework of quantum mechanics. They form the classical and quantum subsystems. It is believed that in the classical subsystem the motion occurs along a specific trajectory, and the interaction between the classical and quantum subsystems induces time-dependent transitions between the quantized states of the quantum subsystem.

One of the conditions, the fulfillment of which is necessary for the applicability of the semi-classical approximation, is that the *de Broglie wavelength* of the classical subsystem:

$$\lambda = \hbar \cdot p = \left(\frac{\hbar^2}{2E \cdot m}\right)^{1/2} \tag{3.1.11}$$

(here, p and E are the momentum and energy of the species, m is its mass) has to be much smaller than the characteristic size of the potential action, in which the interaction of species occurs, which is about 10^{-8} cm. This condition is equivalent to the one when discussing the translational partition function: quantum numbers corresponding to the classical motion must be much greater than 1; here, it is equivalent to $n = l/\lambda$, where l is the same size, ~ 1 Å. Let us see if this condition is fulfilled for thermal energies and the motion of an oxygen atom, for example, $p = 2.6 \cdot 10^{-23} \cdot 10^5$ g•cm/s: $\lambda = 1.05 \cdot 10^{-27}/2.6 \cdot 10^{-18}$ erg s/s g $= 3 \cdot 10^{-10}$ cm $<< 10^{-8}$ cm (see Sect. 2.2).

Another condition for the semi-classical approximation applicability is that the quantum subsystem energy change should be small; more precisely, there should be a lot less than the kinetic energy of the classical degrees of freedom. Such an approximation is the first step to the adiabatic approximation, which will be discussed below.

The semi-classical approximation allows to formulate the scattering problem as a problem to solve a time-dependent wave equation (and the solution to this problem is more complicated than a stationary solution), but for a smaller number of degrees of freedom, namely, only for the quantum ones.

Let the quantum subsystem be determined by a set of quantum numbers i, j before a collision. Then the time-dependent wave function $\Phi(t \rightarrow -\infty)$ can be represented as a product of the wave functions of non-interacting species φ_i^{A} and φ_j^{B}

$$\Phi(t \to -\infty) = \varphi_i^{\mathrm{A}} \varphi_j^{\mathrm{B}} \, exp\left\{ -\frac{i}{\hbar} \left[E_i^{\mathrm{A}} + E_j^{\mathrm{B}} \right] t \right\} \tag{3.1.12}$$

(here, E_i^{A} and E_j^{B} are the energy of their internal degrees of freedom). After an *inelastic* collision (3.1.1), for example, the wave function $\Phi(t \to +\infty)$ can be represented as an expansion of non-interacting species functions, as (for the scattering process):

$$\Phi(t \to +\infty) = \sum\nolimits_{l,m} S_{ij,lm} \varphi_i^{\mathrm{A}} \varphi_j^{\mathrm{B}} \, exp\left\{ -\frac{i}{\hbar} \left[E_i^{\mathrm{A}} + E_j^{\mathrm{B}} \right] t \right\}, \tag{3.1.13}$$

where the $S_{ij,lm}$ coefficients are the amplitudes of the probability to detect the system in the final state *l, m,* provided that before the collision, it was in the state *i, j*. The set of $S_{ij,lm}$ coefficients forms the scattering matrix, and *the square of the modulus of the* $S_{ij,lm}$ *coefficients,* $P_{ijlm} = |S_{ijlm}|^2$, *is the probability of the corresponding transition.* Since transition probabilities are determined for specific trajectories of the classical degrees of freedom, the $P_{ij,lm}$ depends on the classical subsystem trajectory parameters. Each trajectory is defined by the initial relative species velocity *u* and by the *impact parameter b* defined as the closest distance between species center of mass (see Sect. 3.5). Then, the outcome of the collision (3.1.1) for a specific trajectory is characterized by the transition probability $P_{jj,lm}(u, b)$, which depends on the quantum numbers of the initial (i, j) and final $(1, m)$ states as well as on *u* and *b*.

The total cross-section is expressed via the transition probability as

$$\sigma_{ij,lm} = 2\pi \int_0^\infty P_{ij,lm}(u, b) b db \tag{3.1.14}$$

where the (3.1.14) *rhs* is equal to the effective target area of the scattering (3.1.1).

The transition probability as well as the differential and total cross sections for reactive collisions (3.1.2) are defined in the same way [1], p. 30.

The transition probability for a fixed trajectory satisfies the normalization condition.

$$\sum\nolimits_{l,m} P_{ij,lm}(u, b) = 1. \tag{3.1.15}$$

Besides, due to the reversibility of the motion equations in time, the probabilities of direct and reverse transitions between states *i, j* and *l, m* are equal:

$$P_{ij,lm}(u, b) = P_{ij,lm}(u', b') \tag{3.1.16}$$

The final relative velocity $u^{'}$ and impact parameter $b^{'}$ differ from the initial *u* and *b* values, but are related to them.

Equation (3.1.16) represents the *detailed balance principle*. It is equivalent to actually asserting that the various degrees of freedom are equally probable, which we have discussed, expressing the equilibrium constant via partition functions (see Sect. 2.2). Using the detailed balance principle, one can obtain expressions for the ratio of the cross-sections of direct and inverse processes or their rate constants, taking into account the Boltzmann factor and statistical weights of the species.

If the species velocity distribution is Maxwellian (this is again a case of local thermodynamic equilibrium over translational degrees of freedom), the ratio of microscopic rate constants of two elementary processes, direct and reverse (microscopic equilibrium constant), is.

$$\frac{k_{ij,lm}}{k_{lm,ij}} = \frac{g_l g_m \mu'^{3/2}}{g_i g_j \mu^{3/2}} \exp\left(-\frac{\Delta E_{ij,lm}}{k\mathrm{T}}\right). \tag{3.1.17}$$

Here μ and μ' are the reduced masses of reactants A, B, and products (A, B for the process (3.1.1), $\mu = \mu'$), and C, D for the reaction (3.1.2)), g_i, g_j, and g_l, g_m—degeneracies of species (see [1], p. 32 for details). This is a very advantageous expression, and the author will actively use it below.

The transition to the rate constants (i.e., to the 'macroscopic' rate constants) is performed by averaging over the initial and summing over the final quantum states. As to the equilibrium constant, one gets it for averaging and summing for a given distribution over the initial degrees of freedom, rather than a tabular, related to 'complete' (not local) thermodynamic equilibrium.

3.2 Adiabatic Approximations. Potential Energy Curves and Surfaces

Approximations that we discussed above often do not allow us to simplify the problem so much so that one can solve it. Therefore, it is necessary to go on to further simplifications, namely *adiabatic approximations* of different levels. These approximations are very advantageous for:

- separation of the motion of electrons and nuclei in bound states of molecules (radicals); it allows to introduce the concept of the potential energy of nuclei in molecules, *potential energy curves*, *PEC*s, of diatomic species and *potential energy surfaces*, *PES*s, of polyatomic species to separate the electronic and vibrational–rotational motion in them;
- use the concept of *PEC, PES* to describe the process of collision of particles in terms of the motion of the *image point* (see below) along them;
- separation of various types of vibrations in polyatomic species.

We consider adiabatic approximations in a fairly general form. They all boil down to the fact that in total Hamiltonian

$$\widehat{\boldsymbol{H}}(\boldsymbol{r},\boldsymbol{Q}) = \widehat{\boldsymbol{T}}_e(\boldsymbol{r}) + \widehat{\boldsymbol{T}}_N(\boldsymbol{Q}) + \widehat{\boldsymbol{V}}(\boldsymbol{r},\boldsymbol{Q})$$

of the stationary Schrödinger equation

$$\widehat{\boldsymbol{H}}(\boldsymbol{r},\boldsymbol{Q})\Psi(\boldsymbol{r},\boldsymbol{Q}) = E\cdot\Psi(\boldsymbol{r},\boldsymbol{Q}) \tag{3.2.1a}$$

or

$$\left[\widehat{\boldsymbol{T}}_e(\boldsymbol{r}) + \widehat{\boldsymbol{T}}_N(\boldsymbol{Q}) + \widehat{\boldsymbol{V}}(\boldsymbol{r},\boldsymbol{Q})\right]\Psi(\boldsymbol{r},\boldsymbol{Q}) = E\cdot\Psi(\boldsymbol{r},\boldsymbol{Q}) \tag{3.2.1b}$$

produce a separation of the variables (here $\boldsymbol{r}(x_e, y_e, z_e)$; $\boldsymbol{Q}(x_N, y_N, z_N)$ are the coordinates of the fast subsystem (for example, electrons) and the slow subsystem (for example, nuclei), respectively, *$\Psi(r,Q)$ and E are eigenfunctions and the eigenvalues of the Hamiltonian*, $\widehat{\boldsymbol{T}}$ and $\widehat{\boldsymbol{V}}$ are the operators of the kinetic and potential energy, respectively).

There are various ways of this separation and, accordingly, various approximations. Often, they all together are not quite accurately are called the adiabatic approximation or the *Born–Oppenheimer approximation*. Their essence boils down to the fact that the wave function of the *k-th* state is written in the form:

$$\Psi_k(\boldsymbol{r},\boldsymbol{Q}) = \Phi_k(\boldsymbol{r},\boldsymbol{Q})\chi(\boldsymbol{Q}), \tag{3.2.2}$$

where $\Phi_k(\boldsymbol{r},\boldsymbol{Q})$ is the wave function of the fast subsystem, and $\chi(\boldsymbol{Q})$ is that of the slow one. In the adiabatic approximation of the 1-st order, the wave functions of the fast subsystem are sought under the assumption that the slow subsystem does not move at all, $\widehat{\boldsymbol{T}}_N(\boldsymbol{Q}) = 0$, and in the coarse adiabatic approximation, its coordinates are fixed in an equilibrium position, i.e., at $\widehat{T}_N(\boldsymbol{Q}_0) = 0$. So, in the adiabatic approximation, the wave functions Ψ_k $(\boldsymbol{r},\boldsymbol{Q})$, called *adiabatic*, are found as eigenfunctions of the Schrödinger equation (3.2.1b) with $T_N(\boldsymbol{Q}) = 0$, i.e., as a solution of the wave equation for an electron moving in the field of immobile nuclei and having potential energy $V(\boldsymbol{r},\boldsymbol{Q})$:

$$\left[\widehat{\boldsymbol{T}}_e(\boldsymbol{r}) + \widehat{\boldsymbol{V}}(\boldsymbol{r},\boldsymbol{Q})\right]\Phi_k(\boldsymbol{r},\boldsymbol{Q}) = \mathrm{U}_k(\boldsymbol{Q})\cdot\Phi_k(\boldsymbol{r},\boldsymbol{Q}) \tag{3.2.3}$$

$\widehat{\boldsymbol{V}}(\boldsymbol{r},\boldsymbol{Q})$ is the function of $\boldsymbol{r}$, and it depends on $\boldsymbol{Q}$ as on a parameter. Accordingly, $\Phi_k(\boldsymbol{r},\boldsymbol{Q})$, and the eigenvalue of (3.2.3) $\mathrm{U}_k(\boldsymbol{Q})$ is the energy of the *k-th* state or the adiabatic terms of the fast subsystem) depend on $\boldsymbol{Q}$ as a parameter too (more deeply the physical meaning of $\mathrm{U}_k(\boldsymbol{Q})$ we will understand below). So each fixed configuration of the slow subsystem corresponds to a set of eigenvalues of the excitation energies of the fast subsystem $U_k(\boldsymbol{Q})$.

The next step of the adiabatic approximation is that the motion of the slow subsystem does not change the wave function of the fast one, i.e., the fast subsystem follows the slow one without any inertia and always 'tracks' the change in $\boldsymbol{Q}$.

In order to find out to which equation the eigenfunctions and eigenvalues of the slow subsystem satisfy, we must substitute the wave function (3.2.2) into the wave equation (3.2.1b):

$$\left[\widehat{\boldsymbol{T}}_e(\boldsymbol{r}) + \widehat{\boldsymbol{T}}_N(\boldsymbol{Q}) + \widehat{\boldsymbol{V}}(\boldsymbol{r},\boldsymbol{Q})\right]\Phi_k(\boldsymbol{r},\boldsymbol{Q})\chi(\boldsymbol{Q}) = E\cdot\Phi_k(\boldsymbol{r},\boldsymbol{Q})\chi(\boldsymbol{Q}). \tag{3.2.4}$$

Let us substitute $U_k(\boldsymbol{Q})\cdot\Phi_k(r,\boldsymbol{Q})\cdot\chi(\boldsymbol{Q})$ into the left and right sides of (3.2.4) and distribute the terms in the left and right sides with the same variables:

$$\begin{aligned}&\underline{\left[\widehat{\boldsymbol{T}}_e(\boldsymbol{r}) + \widehat{\boldsymbol{V}}(\boldsymbol{r},\boldsymbol{Q})\right]\Phi_k(\boldsymbol{r},\boldsymbol{Q})\chi(\boldsymbol{Q})} + \left[\widehat{\boldsymbol{T}}_N(\boldsymbol{Q}) + U_k(\boldsymbol{Q})\right]\Phi_k(\boldsymbol{r},\boldsymbol{Q})\chi(\boldsymbol{Q})\\ &\underline{= U_k(\boldsymbol{Q})\Phi_k(\boldsymbol{r},\boldsymbol{Q})\chi(\boldsymbol{Q})} + E\,\Phi_k(\boldsymbol{r},\boldsymbol{Q})\chi(\boldsymbol{Q})\end{aligned} \tag{3.2.5}$$

The underlined parts of (3.2.5) is the wave equation for the fast subsystem (3.2.3) (the term $\chi(\boldsymbol{Q})$ is canceled here because the operators on the *lhs* do not act on it, and it is a factor), the rest is the wave equation for the slow subsystem similar to (3.2.3); (the term $\Phi(\boldsymbol{r},\boldsymbol{Q})$ here is also canceled for the same reason):

$$\left[\widehat{\boldsymbol{T}}_N(\boldsymbol{Q}) + U_k(\boldsymbol{Q})\right]\chi(\boldsymbol{Q}) = E_k\,\chi(\boldsymbol{Q}) \tag{3.2.6}$$

A comparison of (3.2.6) with the Schrödinger equation (3.2.3) shows that the eigenvalues of the fast subsystem $U_k(\boldsymbol{Q})$ (the *rhs* of 3.2.3) are the potential energy of the slow subsystem for the *k-th* state of the fast subsystem (*lhs* of 3.2.6).

To further understand the physical meaning of $U_k(\boldsymbol{Q})$ and E_k, we represent $\widehat{\boldsymbol{V}}(\boldsymbol{r},\boldsymbol{Q})$ as:

$$\widehat{\boldsymbol{V}}(\boldsymbol{r},\boldsymbol{Q}) = \widehat{\boldsymbol{V}}_e(\boldsymbol{r},\boldsymbol{Q}) + \widehat{\boldsymbol{V}}_N(\boldsymbol{Q}) \tag{3.2.7}$$

(Q is a parameter in $\widehat{\boldsymbol{V}_e}(\boldsymbol{r},\boldsymbol{Q})$!), and

$$E_k = E_i^{el} + E_k^{v}, \tag{3.2.8}$$

then the wave (3.2.1b) can be reduced to the form:

$$\left[\underline{\widehat{\boldsymbol{T}}_e(\boldsymbol{r})} + \widehat{\boldsymbol{T}}_N(\boldsymbol{Q}) + \underline{\widehat{\boldsymbol{V}}_e(\boldsymbol{r},\boldsymbol{Q})} + \widehat{\boldsymbol{V}}_N(\boldsymbol{Q})\right]\Phi_k(\boldsymbol{r},\boldsymbol{Q})\chi(\boldsymbol{Q}) = \left(E_i^{el} + E_k^{v}\right)\cdot\underline{\Phi_k(\boldsymbol{r},\boldsymbol{Q})}\chi(\boldsymbol{Q}) \tag{3.2.9}$$

or:

$$\begin{aligned}&\left[\underline{\widehat{\boldsymbol{T}}_e(\boldsymbol{r})\Phi_k(\boldsymbol{r},\boldsymbol{Q})\chi(\boldsymbol{Q})} + \widehat{\boldsymbol{T}}_N(\boldsymbol{Q})\Phi_k(\boldsymbol{r},\boldsymbol{Q})\chi(\boldsymbol{Q})\right.\\ &\left.\underline{+\widehat{\boldsymbol{V}}_e(\boldsymbol{r},\boldsymbol{Q})\Phi_k(\boldsymbol{r},\boldsymbol{Q})\chi(\boldsymbol{Q})} + \widehat{\boldsymbol{V}}_N(\boldsymbol{Q})\right]\Phi_k(\boldsymbol{r},\boldsymbol{Q})\chi(\boldsymbol{Q})\\ &\underline{= E_k^{el}\cdot\Phi_k(\boldsymbol{r},\boldsymbol{Q})\chi(\boldsymbol{Q})} + E_{k-}^{v}\Phi_k(\boldsymbol{r},\boldsymbol{Q})\chi(\boldsymbol{Q})\end{aligned} \tag{3.2.10}$$

Let us space the terms on the left and right sides of (3.2.10), which are functions of $\boldsymbol{r}$ and $\boldsymbol{Q}$ into separate equations and reduce again wave functions on which operators do not act:

$$\left[\widehat{\boldsymbol{T}}_e(\boldsymbol{r}) + \widehat{\boldsymbol{V}}_e(\boldsymbol{r},\boldsymbol{Q})\right]\Phi_k(\boldsymbol{r},\boldsymbol{Q}) = E_k^{el}\cdot\Phi_k(\boldsymbol{r},\boldsymbol{Q}) \tag{3.2.11}$$

$$\left[\widehat{\boldsymbol{T}}_N(\boldsymbol{Q}) + \widehat{\boldsymbol{V}}_N(\boldsymbol{Q})\right]\chi(\boldsymbol{Q}) = E_k^{v}\chi(\boldsymbol{Q}) \tag{3.2.12}$$

Equation (3.2.11) is entirely identical to (3.2.3), and, there is E_k^{el}, not $U_k(\boldsymbol{Q})$ on (3.2.11) *rhs*, only.

The physical meaning of the obtained results can be illustrated by the example of the *PEC* of a diatomic molecule:

- Equations (3.2.3, 3.2.6): a set of excitation energy values (eigenvalues) $U_k(\boldsymbol{Q})$ (*rhs* of (3.2.3)), having the meaning of the potential energy of the nuclei (*lrs* of (3.2.6)) corresponds to each internuclear distance. If the internuclear distance changes, then the eigenvalues also change for each state according to their way; with a smooth change in Q, we obtain (for a diatomic molecule, for example) a set of curves that have the meaning of potential energy curves (*PEC*s) (Fig. 3.2).
- Equations (3.2.7, 3.2.8 and 3.2.11, 3.2.12): The excitation energy of a molecule can be represented as $E_k = E_k^{el} + E_k^{v}$ (see (3.2.8)), where E_k^{el} is the energy of the k-th electronic state, i.e., energy with internuclear distances fixed in equilibrium

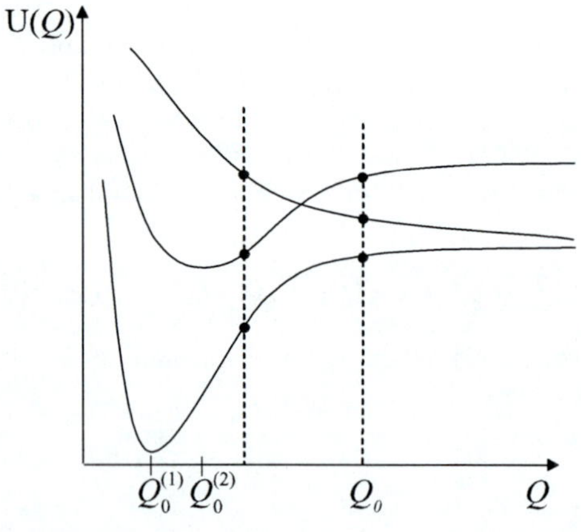

Fig. 3.2 Eigenvalues of the wave equation for a fast subsystem. Potential energy curves

(3.2.11). $V_N(Q)$ is the potential energy of the nuclei (3.2.12), E_k^v is the energy of the oscillations of the *k*-th state described by the wave function $\chi(Q)$ (3.2.12).

If one take rotation into account, the total energy of a molecule in the adiabatic approximation is:

$$E_k = E_k^{el} + E_k^{vr}, \tag{3.2.13a}$$

here, E_k^{vr} is the energy of the rotational-vibrational excitation.
In a very rough approximation

$$E_k = E_k^{el} + E_k^{v} + E_k^{r}, \tag{3.2.13b}$$

and

$$\Psi_{evr} = \Phi \cdot \chi_Q \cdot \chi_r \tag{3.2.14}$$

One should note, that (3.2.13b, 3.2.14) are rough approximations; in particular, they do not suit for a description of degenerate electronic states. However, *symmetry types (species, irreducible representations)*, Γ, of the vibronic, Ψ_{ev}, and rovibronic, Ψ_{evr}, wave functions are equal to the direct products of the Φ,$\cdot\chi_Q$, and χ_r.wave functions:

$$\Gamma(\Psi_{ev}) = \Gamma(\Phi) \times \Gamma(\chi_Q), \tag{3.2.15}$$

$$\Gamma(\Psi_{evr}) = \Gamma(\Phi) \times \Gamma(\chi_Q) \times \Gamma(\chi_r). \tag{3.2.16}$$

If we digress from the PEC of a diatomic molecule and return to the consideration of the general case, we see the following:

The adiabatic approximation allows us to solve the problem of the motion of nuclei, considering the motion of the *slow subsystem*, only but its own for each fast, i.e., motion on *PEC*s, corresponding to fast subsystems. If the slow subsystem has *s* degrees of freedom, then, as we already understand, the function $U_k(Q)$ can be represented by an *s*-dimensional hypersurface in *s* + 1-dimensional space. For one degree of freedom (interatomic distance) using *PEC*s in two-dimensional space.

Under what conditions is the adiabatic approximation applicable? The exact wave function $\Phi(\boldsymbol{r}, t)$, corresponding to the motion of the slow subsystem given by the parameters $Q(t)$, is a solution of the nonstationary Schrödinger equation:

$$i\hbar\frac{d\Phi(r,t)}{dt} = \left[\widehat{\boldsymbol{T}}_e(\boldsymbol{r}) + \widehat{\boldsymbol{V}}_e(\boldsymbol{r},\boldsymbol{Q})\right]\Phi_k(\boldsymbol{r},\boldsymbol{Q}) \tag{3.2.17}$$

This solution can be represented as an expansion in adiabatic wave functions, but the coefficients a_k of the expansion can depend on time.

$$\Phi_k(\boldsymbol{r}, Q) = \sum\nolimits_k a_k(t)\Phi_k^a(t). \tag{3.2.18}$$

Therefore, the probabilities of detecting a system in different adiabatic states of the fast subsystem, i.e., the coefficients $|a_k(t)|^2$ depend on time in this case. It means that the slow subsystem makes transitions between states defined by quantum numbers *k*, i.e., between different *PEC*s. Such transitions are called *nonadiabatic*. To calculate these probabilities is to describe nonadiabatic transitions. If the coefficients $a_k(t)$ = const $\neq f$(t), then the adiabatic approximation is quite applicable.

Let ΔU(*Q*) denotes the difference of any two adiabatic terms, that is, the difference between the energy of any two *PEC*s, at the point *Q* of the configuration space of the slow subsystem. Let *l*(*Q*) is the characteristic length at which the function of the fast subsystem changes significantly. Let *u* be the velocity of the slow subsystem at the point *Q* (Fig. 3.3).

Then the ratio:

$$\zeta = \frac{\Delta U \cdot l}{\hbar \cdot u} = \left(\frac{u}{l}\right) : \left(\frac{\hbar}{\Delta U}\right) \tag{3.2.19}$$

the Massey parameter (a dimensionless quantity) gives the ratio of the transit time of the slow subsystem of the segment *l*, (*l*/*u*) to the characteristic time of motion of the fast subsystem (ħ/ΔU), i.e., the inverse of the frequency of transitions between two adiabatic states. So, suppose the Massey parameter is large. In that case, i.e., the slow subsystem moves rather slowly, and (or) terms are far enough away from each other, the fast subsystem has time to track changes occurring in the slow one, and it remains on the same *PEC*: nonadiabatic transitions are unlikely. In those regions of *PEC*s, where the condition ζ is close to 1, the probability of nonadiabatic transitions is high. In these regions, U(*Q*) loses the meaning of the potential energy of the *k*-th state, and the motion of the slow and fast subsystems cannot be considered independently.

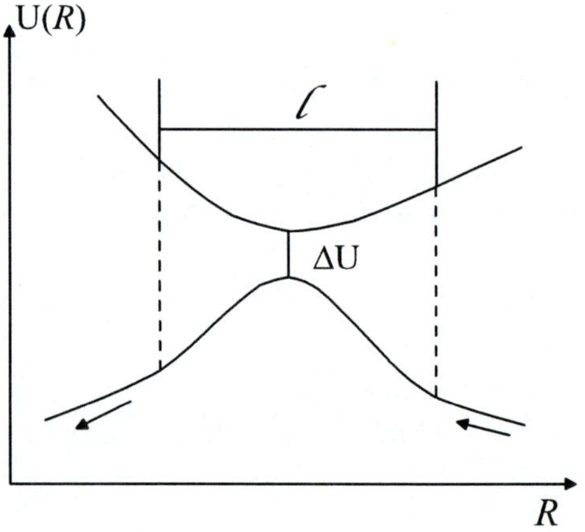

Fig. 3.3 To the definition of the Massey parameter

Before we begin to apply the knowledge we have obtained to the quantum–mechanical description of spontaneous processes, energy exchange, and chemical reactions, we will get acquainted with the modern quantum–mechanical concepts of the nature of intermolecular interactions.

3.3 Intermolecular Interactions. Types of Intermolecular Interactions

A typical interaction potential of two uncharged species (atoms, molecules) that *do not form a chemical bond* (starting now, we will call it the *intermolecular potential*) is shown in Fig. 3.4.

The classification of intermolecular interactions is also given there. The figure identifies three ranges of the distance between the centers of mass of the interacting species A and B, *R*:

I. The range of short distances, in which the potential is repulsive in nature due to the strong Coulomb repulsion between the nuclei; an electronic exchange between species is also very significant due to the strong overlap of their electron shells;
II. The range of intermediate distances with the van der Waals (vdW) minimum, the position of which is determined by the competition between the forces of attraction (see below) and repulsion;

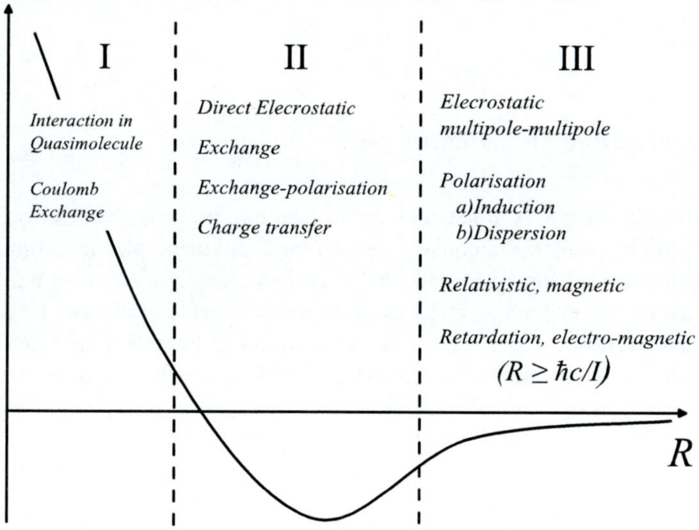

Fig. 3.4 Classification of intermolecular interactions [3], p. 23

III. The range of relatively large R, where the Coulomb and exchange interactions can be neglected, and the intermolecular forces have the character of attraction. For a very large R, the interaction delay must be taken into account.

When species approach each other at a distance of $R \lesssim 2$ Å (range I), the system should be considered as a single quasi-molecule. In the range II, both repulsive and attractive forces exist, and there is a minimum of intermolecular potential energy due to their competition. For distances $R \lesssim 8$ Å (ranges II and I), the exchange interaction must be taken into account. In the range II, it can be considered as low, and *perturbation theory* (PT) (see Sect. 3.6) can be utilized. In the first-order PT, the exchange energy can be separated from the electrostatic energy; in the higher-order, this cannot be done. The terms in the expression for the interaction energy are called, in this case, the exchange-polarization energy. In addition to the exchange in this range, there are interactions due to charge transfer from one species to another.

In the range III, the exchange forces can be neglected. The intermolecular interactions are negligible, and one can use standard PT to describe them. In the first-order PT, it gives the energy of direct electrostatic interaction of species; in higher orders, the polarization energy is obtained, resulting from the mutual polarization of the electronic clouds of species. In the second-order PT, the polarization energy can be divided into induction and dispersion energy; in higher orders, such a division is impossible. (They will be discussed in more detail below). The magnetic interactions (associated with a non-zero total electron spin) are much weaker; they can manifest themselves only at relatively large distances since electrostatic interactions fall off with a distance R quickly (see below).

Consider the nature of the intermolecular interaction forces in somewhat more detail.

3.3.1 Exchange Interaction

At distances less than 8 Å, it is necessary to take into account the electron exchange between particles, which is a consequence of the Pauli principle, according to which the wave function of the system has to be antisymmetric to electron permutations. The energy of the exchange interaction depends exponentially on $1/R$; however, even in the region of the vdW minimum, it must be taken into account (see Fig. 3.4). At short distances, the interacting species must be considered as a single quasi-molecule.

3.3.2 Direct Electrostatic Interactions

Multipole moments. The potential of the field created by the system of charges of the species A at distances larger than the size of the system can be represented as a series of multipole moments. In Cartesian coordinates at a point characterized by a radius-vector $\boldsymbol{R}$, the potential is:

$$\varphi_{\mathrm{A}}(R) = \sum\nolimits_i \frac{e_i}{|\boldsymbol{R} - \boldsymbol{r}_i|} = \frac{q}{R} + \frac{(\boldsymbol{d} \cdot \boldsymbol{R})}{R^3} + \sum\nolimits_{\alpha\beta} Q_{\alpha\beta} \frac{X_\alpha X_\beta}{R^5} + \ldots \tag{3.3.1}$$

Here, e_i and $\boldsymbol{r}_i$ are the charges and coordinates of the nuclei and electrons, X_α are the Cartesian components of the radius vector $\boldsymbol{R}$ (α = 1, 2, 3),

$$q = \sum\nolimits_i e_i$$

is the total charge of the system,

$$\boldsymbol{d} = \sum\nolimits_i e_i \boldsymbol{r}_i$$

is the dipole moment vector

$$Q_{\alpha\beta} = \frac{1}{2} \sum\nolimits_i e_i \left(3x_{i\alpha}x_{i\beta} - \delta_{\alpha\beta} r_i^2\right)$$

is the quadrupole moment tensor ($x_{i\alpha}$ are the Cartesian components of the $\boldsymbol{r}_i$ vector, $\delta_{\alpha\beta}$ is the Kronecker symbol). The multipole moments depend on the arrangement of charges and is independent of the point $\boldsymbol{R}$ in which the potential is determined (see details in [4], p. 26).

The first term in (3.3.1) is non-zero if a species has a non-zero charge, and corresponds to an approximation in which a single point charge replaces all charges at the origin. The second term is determined by the dipole (2 l, l = 1) moment of the species and decreases as $1/R^2$. It is proportional to the cosine of the angle between $\boldsymbol{d}$ and $\boldsymbol{R}$. The simplest example of a molecule with a dipole moment is alkali metal halide molecules, KCl, for example. Due to the low ionization potential of the K atom and the high affinity for the electron of the Cl atom, the electron cloud is drawn to Cl, K^+Cl^- (see Fig. 3.5a), and the dipole moment of this polar molecule is 9.0 D (1 D, Debye, 10^{-18} ESU units). One can show that, for a neutral species, the dipole moment is independent of the choice of the origin of the coordinate system. If a species is charged, then one can always choose a coordinate system such that $d = 0$.

If a molecule A is non-polar, $d = 0$, then the potential $\varphi_{\mathrm{A}}(R)$ is determined by the quadrupole (2 l, l = 2) term in the expansion (3.3.1). The simplest example of the arrangement of charges with a nonvanishing quadrupole moment is shown in Fig. 3.5b; these are equal in magnitude and pairwise opposite in sign charges, located along the vertices of a parallelogram. It can be obtained by moving in the

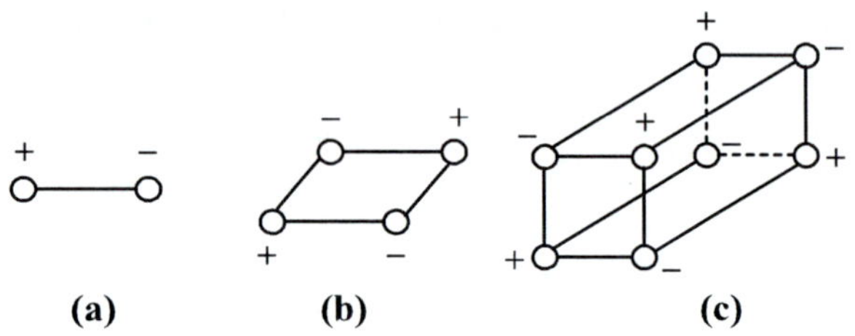

Fig. 3.5 System of point charges possessing dipole **a**, quadrupole **b**, and octopole, **c** moments (see [3], p. 28)

Table 3.1 The values of l for the first nonzero 2^l– pole moment in the ground state of the molecules of different point symmetry groups (see [3], p. 31)

Point group	C_n	C_{nv}	C_s	C_{nh}, $n > 1$	D_{nh}	T_d	O_h	K_h
l	1	1	1	2	2	3	4	Absent

same plane two equal in size and different in the direction of the dipole. Any homonuclear molecule has a nonvanishing quadrupole moment. The quadrupole moment is a symmetric tensor of order 2^2; it is independent of the choice of the coordinate system if the total charge of a species and its dipole moment are equal to 0. The magnitude of the quadrupole moment characterizes the deviation of the charge distribution from the spherical one. If spherical symmetry occurs (point groups T_d, CH_4, and O_h, SF_6), then the system may have higher-order moments, for example, octupole 2^3 (see Fig. 3.5c and Tables 3.1, 3.2).

Table 3.2 Symmetry elements and examples of molecules of some point groups (see [4–7])

Point group	Symmetry elements	Examples[a]
C_1	No symmetry	CHFClBr
C_2	One C_2	H_2O_2, N_2H_4
C_s ($\equiv C_{1s} \equiv C_{1h}$)	One σ	NOCl
C_{2v}	One C_2, two σ_v	H_2O, H_2CO
C_{3v}	One C_3, three σ_v	NH_3, CH_3Cl
C_{4v}	One C_4, one C_2 (coincident with C_4), four σ_v	SF_5Cl
$C_{\infty v}$	One C_∞, any C_p, infinite number of σ_v	ICl, HCN
$D_{\infty h}$	One C_∞, infinite number of C_2 (perpendicular to C_∞, one σ_v, one σ_h, C_p and S_p (coincident with C_∞, i	O_2, CO_2
T_d	Three C_2 (mutually perpendicular), four C_3, six σ_v, three S_4 (coincident with C_2)	CH_4, CCl_4
O_h	Three C_4 (mutually perpendicular), four C_3, i, three S_4 and C_2 (coincident with C_4), six C_2, nine σ, four S_6 (coincident with C_3)	SF_6
K_h	An infinite number of C_∞, σ, i	All atoms

[a] ground states

Table 3.3 The dependences of the energy of the multipole-multipole interactions between species A and B on the distance R between their center of mass (see [3], p. 33)

	Monopole	Dipole	Quadrupole	Octopole
Monopole	$1/R$	$1/R^2$	$1/R^3$	$1/R^4$
Dipole	$1/R^2$	$1/R^3$	$1/R^4$	$1/R^5$
Quadrupole	$1/R^3$	$1/R^4$	$1/R^5$	$1/R^6$
Octopole	$1/R^4$	$1/R^5$	$1/R^6$	$1/R^7$

It can be seen from Tables 3.1 and 3.2, that the dipole moment of a molecule in the ground electronic state equal to 0, if it possesses more than one axis of symmetry, either a mirror-turning axis or a center of symmetry. For excited states of molecules, this statement is false. If the species has spherical symmetry (K_h group, atoms), all multipole moments are equal to 0.

Multipole-multipole interactions. The dependence of the energy of the multipole-multipole interactions between species A and B on the distance R is given in Table 3.3.

One sees, in particular, that the interaction energy of two homonuclear diatomic molecules in the ground states (only the quadrupole moment is nonvanishing) decreases with distance as $1/R^5$.

3.3.3 Polarization Interactions

The forces caused due to the mutual polarization of the species' electron shells as they approach each other are called *polarization*. They are described by the second and higher orders of the PT.

$$E_{pol}^{(2)} = -\sum\nolimits_{n,m} \frac{\left|\left\langle \Psi_n^{\mathrm{A}}\Psi_m^{\mathrm{B}} \middle| \hat{V} \middle| \Psi_0^{\mathrm{A}}\Psi_0^{\mathrm{B}} \right\rangle\right|^2}{\left(E_n^{\mathrm{A}} - E_0^{\mathrm{A}}\right) + \left(E_m^{\mathrm{B}} - E_0^{\mathrm{B}}\right)} = E_{ind}^{(2)} + E_{disp}^{(2)} \tag{3.3.2}$$

the quantum numbers m, n cannot simultaneously take values corresponding to the ground states of the isolated molecules. The sum of n and m can be divided into two parts that have different physical meanings. Consider them separately.

Induction interactions. Induction energy can be obtained from (3.3.3):

$$\begin{aligned} E_{ind}^{(2)} &= -\sum_{m\neq 0} \frac{\left|\left\langle \Psi_0^{\mathrm{A}}\Psi_m^{\mathrm{B}} \middle| \hat{V} \middle| \Psi_0^{\mathrm{A}}\Psi_0^{\mathrm{B}} \right\rangle\right|^2}{E_m^{\mathrm{B}} - E_0^{\mathrm{B}}} - \sum_{n\neq 0} \frac{\left|\left\langle \Psi_n^{\mathrm{A}}\Psi_0^{\mathrm{B}} \middle| \hat{V} \middle| \Psi_0^{\mathrm{A}}\Psi_0^{\mathrm{B}} \right\rangle\right|^2}{E_n^{\mathrm{A}} - E_0^{\mathrm{A}}} \\ &= -\sum\nolimits_{m\neq 0} \frac{V_{0m,00}^2}{E_m^{\mathrm{B}} - E_0^{\mathrm{B}}} - \sum\nolimits_{n\neq 0} \frac{V_{n0,00}^2}{E_n^{\mathrm{A}} - E_0^{\mathrm{A}}} \end{aligned} \tag{3.3.3}$$

The first term in (3.3.3) is the electrostatic interaction of the charges of species A ground state, characterized by electron density

$$\rho_{00}^{A}(i) = N_A \int \left|\Psi_0^A(1\ldots i\ldots N_A\right|^2 dV^{(i)}$$

($dV^{(i)}$ is a configuration space volume element of all electrons of species A, except for the i-th) with species B charges characterized by the function of induced transition electron density:

$$\rho_{00}^{B}(j) = N_B \int \Psi_m^B(1\ldots j\ldots N_B)^* \Psi_0^B(1\ldots j\ldots N_B) dV^{(j)}$$

We are considering the interaction of species A and B in the ground electronic states. The distribution of electron density in species B (transition density) is induced by the field created by species A (ground state). Similarly, the second term in (3.3.3) corresponds to the interaction of species B in its ground state with the induced electron density distribution of species A.

It can be shown that the energy of the induction interaction of species in the ground electronic states is always negative, i.e., corresponds to an attraction. For species in excited states, this may not be the case, and the induction energy may correspond either to attraction or to repulsion.

At large distances R, the energy of the induction interaction can be represented as a multipole series, i.e., represent (3.3.3) in an expansion of $1/R$. The first member of the series corresponds to the interaction of the induced dipole with the field belonging to the inducing species (dipole, quadrupole, etc.). The dependence on $1/R$ is determined by the *square* of the corresponding dipole-multipole interaction, presented in Table 3.3. For example, for the interaction of a dipole with a non-polar molecule, which in this case is an induced dipole, the member most slowly decreasing with distance R, is $1/R^6$. For the interaction of the quadrupole moment of one molecule with an induced dipole of another is decreased as $1/R^8$, etc. The energy of the induction interaction of species depends on their static polarizability. For uncharged species in the ground state, the induction interaction is usually small; some classes make an exception of molecules with a large magnitude of the induced dipole moment, for example, long molecules with conjugated bonds, some biopolymers.

Dispersion interactions. Dispersion interaction can be presented by the rest part of (3.3.2) after subtracting $E_{ind}^{(2)}$

$$\begin{aligned} E_{disp}^{(2)} &= -\sum\nolimits_{m,n\neq 0} \frac{\left|\left\langle \Psi_n^A \Psi_m^A \middle| \hat{V} \middle| \Psi_0^A \Psi_0^B \right\rangle\right|^2}{\left(E_n^A - E_0^A\right) + \left(E_m^B - E_0^B\right)} \\ &= -\sum\nolimits_{m,n\neq 0} \frac{\left|V_{nm,00}\right|^2}{\left(E_n^A - E_0^A\right) + \left(E_m^B - E_0^B\right)} \end{aligned} \tag{3.3.4}$$

The matrix element $V_{nm,00} = \left\langle \Psi_n^A \Psi_m^A \middle| \widehat{V} \middle| \Psi_0^A \Psi_0^B \right\rangle$ included in (3.3.4) corresponds to the electrostatic interaction of two mutually induced electron clouds ρ_{n0}^A and ρ_{n0}^B. Dispersion energy has no classical analogs; it is determined by quantum–mechanical electron density fluctuations. The instantaneous charge distribution corresponding to the instantaneous dipole (and subsequent multipole) moment of one species induces instantaneous multipole moments of another species. The interaction of these moments determines the dispersion energy. For species in the ground electronic states, the dispersion energy is always negative, i.e., corresponds to an attraction.

The multipole decomposition of the dispersion energy is usually written in the form of a series, the coefficients of which C_n are called dispersion constants:

$$E_{disp}^{(2)} = -\sum\nolimits_6^\infty \frac{C_n}{R^n} \tag{3.3.5}$$

In the case of the interaction of atoms (there are no multipole moments, see above), the series (3.3.5) contains only even (due to square $1/R^n$ (Table 3.3)). The first term, proportional to $1/R^6$, corresponds to the dipole–dipole interaction, the second, $1/R^8$, dipole-quadrupole, etc. For spherically symmetric systems or for averaging over the orientations of arbitrary systems, the coefficient C_6 can be expressed in terms of oscillator strengths, f_{n0}^A, f_{m0}^B, and frequencies, ω_{n0}^A, ω_{m0}^B, of transitions in isolated molecules:

$$C_6 = \frac{3}{2}\sum\nolimits_{n,m\neq n} \frac{f_{n0}^A \cdot f_{m0}^B}{\omega_{n0}^A \cdot \omega_{m0}^B \left(\omega_{n0}^A \cdot + \omega_{m0}^B\right)} \tag{3.3.6}$$

If the oscillator strength of one of the transitions in the species significantly exceeds the others, then the summation over the excited states in (3.3.6) can be replaced by one member. It can also be shown [4], p. 49 that the coefficient C_6 can be obtained using the static polarizabilities $\overline{\alpha_0^A}$, $\overline{\alpha_0^B}$:

$$\overline{\alpha}_0 = \frac{\overline{f}_{k0}}{\omega_{k0}^2} = \frac{2}{3}\sum\nolimits_{k\neq 0} \frac{|\mu_{k0}|^2}{E_k - 0}$$

(μ_{k0} is the dipole moment of the k - 0 transition) and the first ionization potentials of I_A, I_B species:

$$C_6 = \frac{3}{2}\overline{\alpha_0^A} \cdot \overline{\alpha_0^B} \frac{I_A I_B}{I_A + I_B} \tag{3.3.7}$$

One can make a qualitative estimation of C_6 if polarizabilities and ionization potentials of colliding particles are known.

If at least one of the interacting species is in an electronically excited state, the coefficient C_6 can increase by orders of magnitude. Let species A is in the excited state n, the density of states i to which a transition can occur is large, and the transition itself has a large probability. In this case, the transition frequency is small, and the polarizability

$$\alpha_A = \frac{2}{3}\sum\nolimits_{n \neq i} \frac{|\mu_{ni}|^2}{|E_n - E_i|}$$

may be orders of magnitude higher than the polarizability of the samespecies in the ground state. It is this case that takes place in the ion-pair states of halogens, which will be discussed in Sect. 5.5.3.

3.3.4 Resonance Interactions

These interactions occur if one of the colliding species is in ground and the other in excited states, and the transition energies to the excited states of both species are the same (in particular, if the same molecules collide).

Let one of the species (A) be in the excited state, and the other (B) in the ground one. In the absence of interaction, the state of such a system is described by the wave function $\Psi_n^A\Psi_0^B$. The same energy corresponds to the state $\Psi_0^A\Psi_m^B$ (resonance!). There is a degeneration, and the wave function of the system is described as symmetric and antisymmetric linear combinations of the initial functions of zero-order:

$$\Psi_{g,u} = \frac{1}{2}\left(\Psi_n^A\Psi_0^B \pm \Psi_0^A\Psi_m^B\right) \tag{3.3.8}$$

The interaction energy in the *first-order* PT is

$$\begin{aligned} E_{g,u}^{(1)} = \frac{1}{2}\Big[\left\langle\Psi_{g,u}\middle|\hat{V}\middle|\Psi_{g,u}\right\rangle = \left\langle\Psi_n^A\Psi_0^B\middle|\hat{V}\middle|\Psi_n^A\Psi_0^B\right\rangle + \left\langle\Psi_0^A\Psi_m^B\middle|\hat{V}\middle|\Psi_n^A\Psi_0^B\right\rangle \\ \pm 2\left\langle\Psi_n^A\Psi_0^B\middle|\hat{V}\middle|\Psi_0^A\Psi_m^B\right\rangle\Big] \end{aligned} \tag{3.3.9}$$

The first two terms in (3.3.9) is the energy of electrostatic interaction (see Sect. 3.3.2) of the species A in the n-th excited state and B in the ground state, and species A in the ground state and B in the m-th excited state. The last term corresponds to the interaction of transition electron densities of species A and B; it is due to the excitation transition from A to B. This term is usually called the matrix element of the excitation transfer or the resonance integral.

At sufficiently large distances, R, (3.3.9) can be expanded into a multipole series. For neutral, even non-polar molecules, the first non-zero term is the term describing the dipole–dipole interaction, decreasing as $1/R^3$, since the interaction energy is obtained in the first order of perturbation theory. Thus, the resonant interaction is more long-range than the polarization, decreasing as $1/R^6$.

The state described by the wave function (3.3.8) is non-stationary, and as a result of the resonant interaction, the species exchange an excitation with a frequency proportional to the resonant integral. If species scatter, or species B energy dissipates in any way (dissociation, predissociation, photon radiation), then, as a result of *resonance detuning*, one-sided energy transfer takes place

$$A^* + B \rightarrow A + B^*.$$

A more detailed description of intermolecular interactions can be found, for example, in the book by I.G. Kaplan [4]. We will also consider such interactions when analyzing collision-induced nonadiabatic transitions (Sect. 5.5.3).

3.4 Semiempirical Model Potentials for Intermolecular Interactions

Model potentials with simple analytical representations, which parameters are determined experimentally, are widely used for describing of intermolecular interactions. They are discussed in detail in [4, 8], for example.

Below, some model potentials that have been extensively used are discussed briefly.

Hard-sphere potentials. This potential function represents solid, impenetrable spheres with a diameter σ:

$$\mathrm{U}(R) = \begin{cases} \infty, & R \le \sigma \\ 0, & R > \sigma \end{cases} \tag{3.4.1}$$

where σ is the diameter of the sphere (Fig. 3.6a).

This potential is widely used for those problems in which a qualitative result is sufficient. An attractive term, a rectangular well with a depth ε and a width $\sigma(a - 1)$ (Fig. 3.6c)

$$\mathrm{U}(R) = \begin{cases} \infty, & R \le \sigma \\ -\varepsilon, & \sigma < R \le a\sigma \\ 0, & R > \sigma \end{cases} \tag{3.4.2}$$

or a potential, which combines the hard-sphere model with an attraction more realistically is the so-called *Sutherland potential* (Fig. 3.6d):

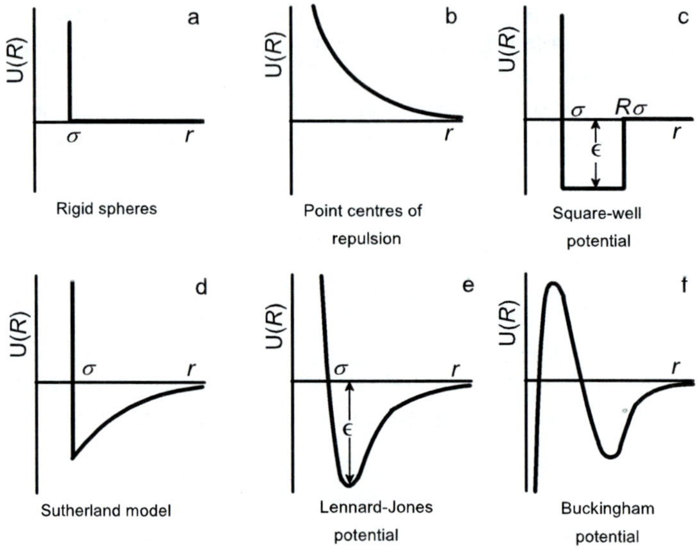

Fig. 3.6 Some semiempirical model potentials widely used for intermolecular interactions (see [8], p. 30)

$$\mathrm{U}(R) = \begin{cases} \infty, & R \le \sigma \\ -\mathrm{u}(\sigma/R)^6, & R > \sigma \end{cases} \tag{3.4.3}$$

is sometimes added to the hard-sphere potential [4] p. 184.

Lennard–Jones potential. The general form of this potential proposed by Jones [9] is:

$$\mathrm{U}(R) = \frac{\lambda_n}{R^n} + \frac{\lambda_m}{R^m} \tag{3.4.4}$$

Below, the so-called Lennard–Jones (6–12) potential:

$$\mathrm{U}_{L-J}(R) = 4\varepsilon[\left(\frac{R_0}{R}\right)^{12} - \left(\frac{R_0}{R}\right)^{6}], \tag{3.4.5}$$

where ε is the depth of the potential well with the minimum at $R_e = R_0\sqrt[6]{2}$ and R_0 is the value of R at $\mathrm{U}(R) = 0$ (gas-kinetic diameter) (Fig. 3.6e). The attractive term corresponds to the dispersion interaction (see Sect. 3.3.3). The repulsion is due to repulsive forces (see Fig. 3.4 and the text below it). The choice of order 12 is due to mathematical convenience, and not the only possible.

Other potential used to describe intermolecular interactions are discussed in detail elsewhere (see [4, 8] and references).

3.5 Descriptions of Collisional Processes Using Potential Energy Curves and Surfaces. Intramolecular Vibrational Relaxation in Polyatomic Molecules

Let us discuss the potential energy of the system of atoms and how to describe the collisional processes by the motion of an image point along the *PES* or *PEC* of the system of atoms.

Let the author reminds once again that the basis for introducing the concepts of *PEC*s, *PES*s is the adiabatic approximation. Within the framework of this approximation, each fixed configuration of atoms defined by the positions of their nuclei corresponds to the set of eigenvalues of the solution of the wave equation for the electron wave function namely adiabatic electronic terms $U_k(Q) = E_k^{el} + U_N(Q)$, where Q and $U_N(Q)$ are coordinates and the potential energy of interaction of nuclei. Each of them corresponds to its electronic state, i.e., its *k*-th wave function. If these terms are sufficiently separated in the entire Q range, then the Massey parameter $\Delta Ul/\hbar u$ is large, and the adiabatic term can be considered as the *PES* (*PEC*) of a system of atoms in the *k*-th electronic state. Within the framework of the adiabatic approximation, the motion of atoms does not cause transitions between *PEC*s, *PES*s, and the elementary collision process (energy redistribution or chemical reaction) is described in terms of the motion of an image point, the point corresponding to the current values of the coordinates of the system of atoms, along this single *PES*, *PEC*. Going beyond the adiabatic approximation involves taking into account the possibility of transitions between electronic states, which is the subject of the theory of nonadiabatic transitions. At low kinetic energies of atoms u (u is in the denominator of the Massey parameter (3.2.19)) adiabatic approximation quite satisfactorily describes the process on a significant part of a *PEC* or *PES*. If the states approach the distance 'dangerous for clean adiabaticity' in any region of a *PEC* or *PES* (again the Massey parameter), then one has to take into account nonadiabatic effects, and in a large number of problems this gives a small correction, which is correctly calculated in the framework of PT. Even if this is not the case, and the 'nonadiabatic effect' is large, most often, the area where this takes place is relatively small, and the description of the process outside it within the framework of the adiabatic approximation is very instrumental. Nonadiabatic processes, on the other hand, can be described as an instantaneous compared with the characteristic times of motion of the nuclei ($\sim 10^{-13}$ s) 'jump' ($\sim 10^{-15}$ s) from the surface to the surface. The author has discussed collision processes and their description in the framework of motion along a *PEC*, *PES*, so far. All of the above applies to the description of spontaneous unimolecular processes, in particular, dissociation and predissociation.

We will consider nonadiabatic processes a little later; now, we will see how to describe an elastic scattering, the energy redistribution processes, and chemical reactions in the framework of the motion of an image point on a *PEC* or *PES*.

It is evident that to solve the problem of collision, dissociation, or predissociation as a motion of an image point on a *PEC*, *PES*, it is necessary to know these curves or surfaces. As for the *PEC*s, i.e., the problems of dissociation, predissociation of a diatomic molecule, or the collision of two atoms, very often, we can get much of the information about the *PEC*s from an experiment. For *PES*s, this happens less often, and usually only after solving the quantum chemical problem of calculating *PES* using some ab initio or semi-empirical method(s), one can begin to solve the dynamic problem of describing the motion of an image point on this *PES*.

Let us now understand how elastic collisions, energy exchange processes during them, and chemical reactions can be described using *PEC*s and *PES*s.

Elastic collisions. In collisions (3.1.1) as we now know, for each pair of species A(i) + B(j), A(l) + B(m), etc. correspond *PEC*, *PES* (Fig. 3.7).

If A and B are molecules or radicals, they have vibrational–rotational excitation; accordingly, the difference in energy, $E_{A(i)}-E_{A(l)}$, and, consequently, the magnitude of the energy gap between the *PES*s of the corresponding pairs of species is small. Consequently, the Massey parameter is low, and nonadiabatic transitions from the surface to the surface are possible, i.e., change of A and B species states, or *inelastic* collisions. Following the above, elastic collisions

$$\left(i = l, j = m, \Delta E_{ij,lm} = 0\right)$$

molecules and radicals are quite rare phenomena. They, as a rule, occur in collisions of atoms. In this case, the Massey parameter should be large, i.e., atomic states should not be degenerate, and the difference in the energies of their terms should be considerable enough.

In the adiabatic approximation for each electronic state of the system under consideration, we can determine the potential energy of the interaction of species, depending on the distance R between them. The types of these interactions and the dependence of their potentials on R we have considered in Sects. 3.3, 3.4. One of the potentials, including dispersion interaction and repulsion, associated with short-range valence forces is the Lennard–Jones (6–12) potential (3.4.5) (see Fig. 3.8):

Fig. 3.7 Collisions involving molecules and radicals

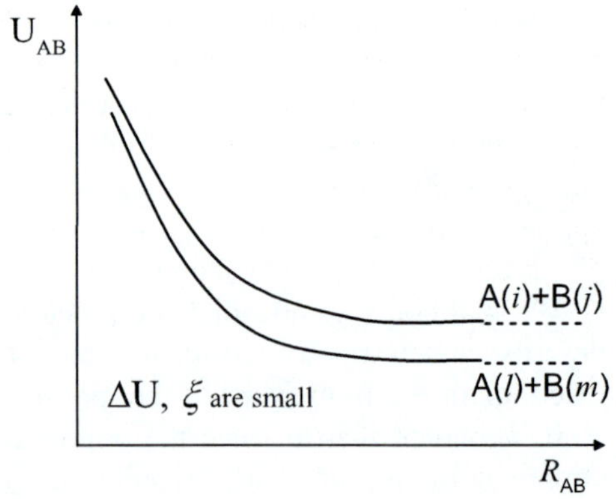

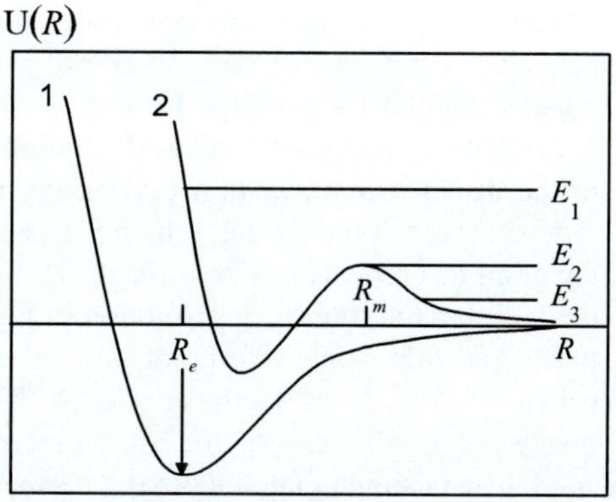

Fig. 3.8 Potential energy of interaction of atoms (1) and effective potential energy (2) (see [2], p. 102)

Lennard–Jones potentials are used to describe the interaction potentials of species that can form only a vdW complex with a well depth ε. A classical motion in the elastic collision of two atoms A, B is equivalent to the problem of the motion of a material point with a mass equal to the reduced mass in the field of the effective potential (Fig. 3.9):

$$\mathrm{U}_{eff}(R) = U_{L-J}(R) + \frac{\boldsymbol{M}^2}{2\mu_{\mathrm{AB}}R^2} \tag{3.5.1}$$

where $\boldsymbol{M} = \mu_{\mathrm{AB}} \cdot \boldsymbol{u} \cdot b$ is the angular momentum vector. A feature of $\mathrm{U}_{eff}(R)$ is the presence of a potential centrifugal barrier for potentials U(R), decreasing faster than $1/R^2$. The magnitude of this barrier is larger, the larger u and impact parameter b. It is somewhat similar to the rotational centrifugal barrier of diatomic molecules.

Now let us discuss the trajectories of relative motion in the elastic collision of species. When the time changes from $-\infty$ to $+\infty$, R decreases to the minimum value of R_t (the so-called turning point) and then increases again to infinity. As a result of motion along such a trajectory, the velocity vector of the relative motion of atoms rotates by the scattering angle. If the impact parameter of a collision is large (the height of the potential barrier $\mathrm{U}_{\mathrm{eff}}$ is large), and the collision energy is small (E_3

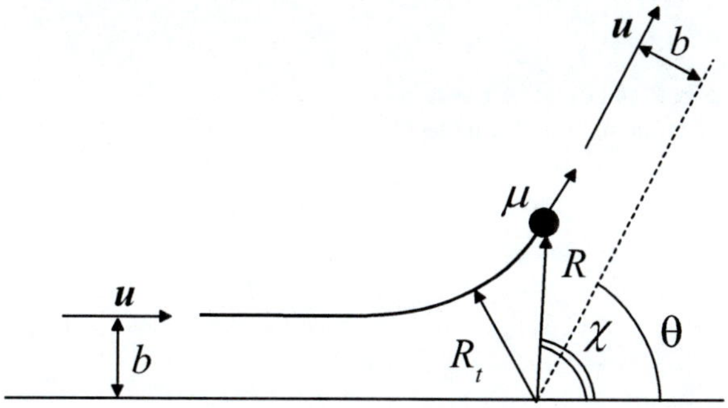

Fig. 3.9 To the motion of a material point with mass μ in the field of the central potential (see [2], p. 102)

in Fig. 3.8), then the image point does not fall over the barrier. It reflects from it, i.e., species moves past each other, not coming closer. If one fixes the value of the relative angular momentum $\boldsymbol{M} = \mu_{AB} \cdot \boldsymbol{u} \cdot b$ and increases the total energy E, so $E > E_2$, the image point 'overruns' through the barrier (i.e., the species approaches to the distance $R < R_m$, Fig. 3.8). Then the image point reaches the point on the repulsive part of the barrier, which is isoenergetic with the energy of the potential at the point R_m, reflects, and goes back. This motion of the image point corresponds to the fact that the trajectory of motion in Fig. 3.9 corresponds to a spiral, which first twists and then spins (the point jumped out of the potential well). This feature is called *orbiting*. If one of the species A, B is a di- or polyatomic molecule, then the image point will 'dangle' for some time in this potential well.

One can show that the total cross-section of a process that corresponds to orbiting, called the capture cross-section is:

$$\sigma_c(E) = \pi b_c^2(E);$$

b_c is the impact parameter corresponding to the orbiting.

It can also be shown that orbiting takes place if

$$\sigma_c(E) = \pi R_0^2,$$

much more than gas-kinetic cross-section. It is also clear that if the collision energy is high (E_1, Fig. 3.8), then orbiting does not take place: the image point reaches the repulsive potential, reflects and quickly fall out of the well.

Potential energy surfaces. Let us consider the *PES* of systems of three atoms, using which one can describe the processes of energy exchange, bimolecular reactions, and spontaneous decay.

The relative position of three atoms A, B, and C, on which the potential energy depends, is described by three coordinates. These can be, for example, the distance between atoms B and C, r_{BC}, the distance R_A from the atom A to the center of mass of the system BC (let this system is a molecule) and the angle γ between the vectors $\boldsymbol{R}$ and $\boldsymbol{r}_{BC}$ (Fig. 3.10).

If the collision is collinear, then $\gamma = 0$, and only two coordinates, R_A and r_{BC}, are required to describe the motion. The *PES* of this system is $(3 \times 3) - 5 = 4$-dimensional

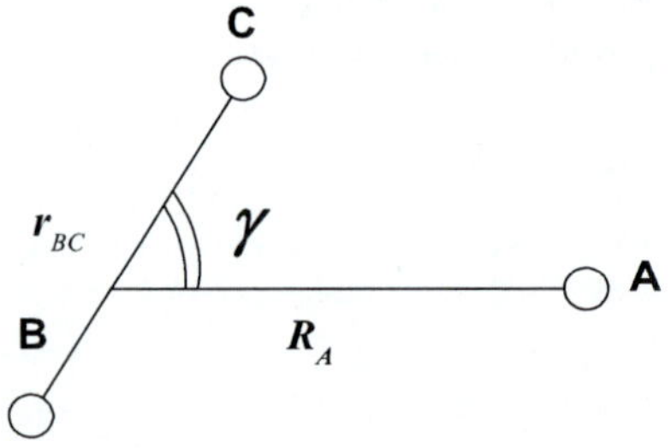

Fig. 3.10 Relative position of an atom A and a molecule BC (see [2], p. 109)

surface in 5-dimensional space. But, if we 'freeze' two bending coordinate, for example, if we make the angle γ equal to 0, then we get a two-dimensional surface in three-dimensional space. This is a completely 'presentable' thing.

It can be 'decorated' using lines corresponding to constant potential energy, *equipotential lines*, and it is quite tangible to imagine the relative motion of atoms as the motion of a point of mass μ, called the image *(representative) point* of this potential well. And not only to present but also to calculate.

To do this, one must transform the R_A, and r_{BC} coordinates to another, an *oblique (skew-angular) coordinate system* ξ_1, ξ_2, with an angle between the axes depending on the masses of the species (Fig. 3.11)

$$\begin{aligned} \xi_1 &= R_A/a, \\ \xi_2 &= r_{BC} \cdot a \end{aligned} \tag{3.5.2}$$

$$\begin{aligned} a &= (\mu_{BC}/\mu)^{1/2} \quad \mu_{BC} = m_B \cdot m_C/(m_B + m_C) \\ \mu &= \left[\left(m_A \cdot m_B \cdot m_C\right)(m_A + m_B + m_C)\right]^{1/2} \end{aligned} \tag{3.5.3}$$

This transformation needs to make the kinetic energy equal to the product of the effective mass (μ/2) by the sum of the squares of velocity components for motions along the ξ_1 and ξ_2, coordinates. The dynamics of the linear system ABC can be simulated by the motion of the heavy mass point μ over the potential surface U(ξ_1, ξ_2). The coordinates ξ_1 and ξ_2 are convenient for describing the relative motion of A and BC; they are generally used in *V-T* energy exchange studies (see below). Such transformations the description kinetic energy of the linear system in the canonical form

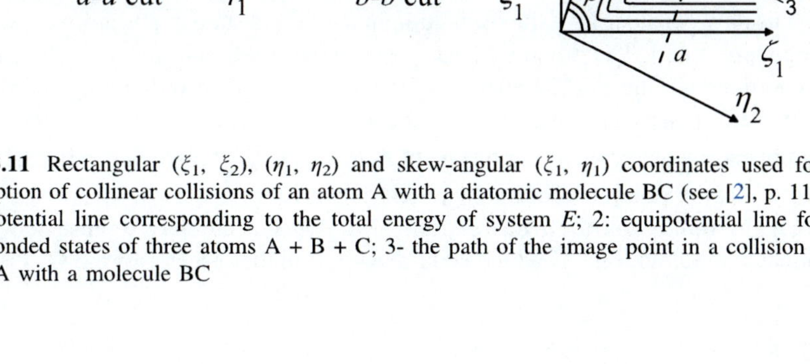

Fig. 3.11 Rectangular (ξ_1, ξ_2), (η_1, η_2) and skew-angular (ξ_1, η_1) coordinates used for the description of collinear collisions of an atom A with a diatomic molecule BC (see [2], p. 110). 1: Equipotential line corresponding to the total energy of system *E*; 2: equipotential line for the non-bonded states of three atoms A + B + C; 3- the path of the image point in a collision of an atom A with a molecule BC

$$T = \mu/2\left[(d\xi_1/dt)^2 + (d\xi_2/dt)^2\right].$$

The energy conservation, $T + V = \text{const}$, is also satisfied.

Let us now consider how a colinear collision of an atom A with a diatomic molecule BC can be described, leading to vibrational excitation of the latter when the atom is approached from side B. The simplest model is a model obtained under the assumption that the interaction potential of particles is the sum of interaction potentials in the BC molecule and, for collinear A + BC configuration, between atoms A and B (Fig. 3.11) (so-called dumb-bell model):

$$\mathrm{U}(R, r) = \mathrm{U}_{\mathrm{BC}}(r_{\mathrm{BC}}) + U_{\mathrm{AB}}(R_{\mathrm{AB}}). \tag{3.5.4}$$

We can take the potential of $\mathrm{U}_{\mathrm{BC}}(r_{\mathrm{BC}})$ equal to the Morse potential

$$U(R) = \varepsilon\left\{1 - \exp\{-\alpha(R - R_e)]\right\}^2 - \varepsilon; \tag{3.5.5}$$

U_{AB} can be an exponent or Lennard–Jones (6–12) potential. This model is widely used in the calculation of the probabilities of vibrational excitation. The *PES* cross-sections corresponding to potential (3.5.4), obtained at $R_{\mathrm{AB}} = \text{const}$, should give $\mathrm{U}_{\mathrm{BC}}(r_{\mathrm{BC}})$, i.e., Morse potential, for example, and that of with $r_{\mathrm{BC}} = \text{const}$, exponent. As a result, the surface should appear as shown in Fig. 3.11.

The trajectory describing the collision of an atom A with an unexcited BC molecule ($r_{\mathrm{BC}} = r^e_{\mathrm{BC}}$), for sufficiently large R_{A}, is a straight line parallel to the R_{A} axis and cutting off a segment on the r_{BC} axis equal to r^e_{BC}. The image point must rest on a dead-end: it cannot rise to a height of U_{BC} higher than the kinetic energy of the relative motion of A and BC at $R_{\mathrm{AB}} \to +\infty$. The trajectory of the image point motion after the reflection is sinusoidal one that corresponds to BC molecule vibrational excitation. If the kinetic energy of relative motion A and BC is large, higher than the BC dissociation energy, then the image point can leave the well and reach the plateau in the figure's upper part. The collision led to dissociation, and instead of an atom and a molecule, one got three atoms. As we understand, if at least one of the species, A or BC, is in a degenerate state, then for large R, we have several *PES*s coinciding in energy, diverging with decreasing R_{AB}. The probability of an image point going on a particular surface depends on many factors, particularly on the degeneration of the state corresponding to the given *PES*.

Let us now consider how the oscillation in the bound state of a triatomic molecule can be represented. The potential energy of a linear triatomic molecule depends on four coordinates. One can visualize this function putting the two bending coordinates equal to zero, so a two-dimensional surface in three-dimensional space represents the potential energy as for the ground state of the $CO_2(\widetilde{X}^1\Sigma_g^+)$ molecule (Fig. 3.12). As in Fig. 3.11, the oblique coordinate system is used.

Near the minimum, the *PES* section can be very accurately represented by an elliptical paraboloid, one of which semi-axes lies on the bisector of the OCO angle,

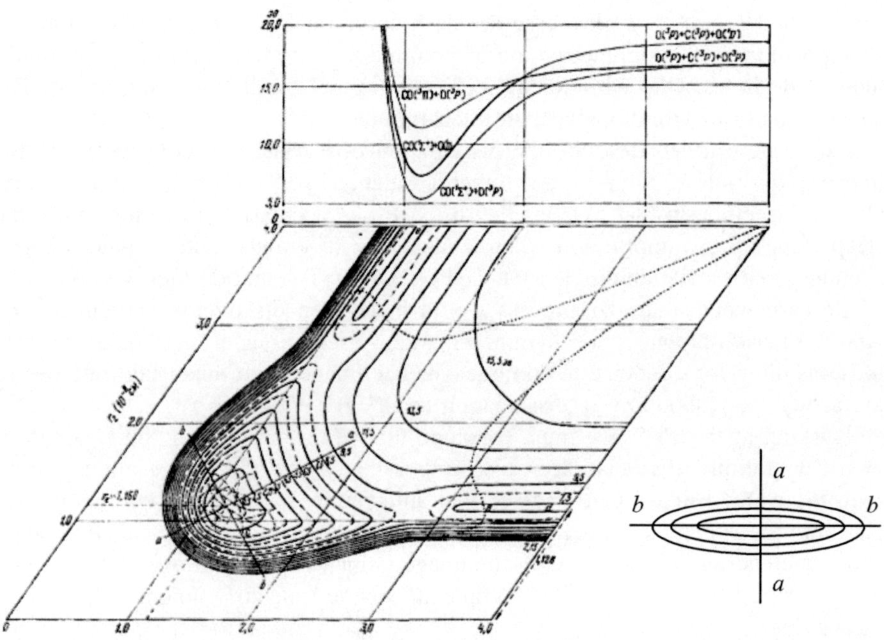

Fig. 3.12 *PES* of the $CO_2(\widetilde{X}^1\Sigma_g^+)$ state as the functions of the OC–O и O–CO distance (see [5], p. 430)

and the other, of course, is perpendicular to it. The *harmonic* approximation describes well the potential energy in this surface region. The study of the dynamics of the motion of such a system shows that it can be described by such a set of variables in which the total vibrational Hamiltonian is represented as a sum of Hamiltonians, each of which depends on this variable and its derivative on time. The total energy of the system in this approximation is represented as the sum of time-independent energies. These coordinates are called *normal*, and the approximation is suitable as long as the *PES* differs little from an elliptic paraboloid. As the amplitude of the oscillations increases, the higher terms of the coordinates, which consider the anharmonicity, appear in the decomposition of the kinetic and potential energies, and an elliptical paraboloid cannot describe the *PES*.

Now let us consider vibrations of the linear triatomic molecule $CO_2(\widetilde{X}^1\Sigma_g^+)$. It is clear that the oscillation described by the motion of an image point along the bisector of the OCO angle (straight line *a-a*) is symmetric valence: the OC and CO distances, in this case, are equal. If the motion of an image point is carried out along the other half-axis of the ellipse (straight line *b-b*), then this is an antisymmetric stretching vibration. It is clear that they can exist independently only as long as the point dangles near the bottom of the potential energy well, where the *PES* is an elliptical paraboloid. If the amplitude of the oscillations is large enough, then the motion cannot be described by straight lines; the image point describes the complicated *Lissajous figures* (see dot-dash curve on Fig. 3.12). This motion is a superposition of the symmetric and antisymmetric vibration with different amplitudes and phases. One vibrational mode, one normal oscillation cannot exist. The anharmonicity of vibrations leads to their mixing.

Let us now try to describe the dissociation of a triatomic molecule using the image point motion. Suppose a symmetric valence oscillation takes place (motion along a bisector). In that case, as the distance r_{OC} decreases, the point climbs the steep slope up to infinity, and when it increases, it 'crawls' onto a plateau corresponding to dissociation on $O(^3P) + C(^3P) + O(^3P)$. The author does not know of a single case when a similar process would have been discovered experimentally. And the reason is that 'pure' symmetric valent oscillations at amplitudes close to those required for dissociation do not exist. The image point must 'fall' into one of the valleys corresponding to dissociation to $O(^1D) + CO(X^1\Sigma^+)$.

Looking at the *PES*, we must conclude that in the process of 'pure' antisymmetric vibrations, dissociation is also impossible, since the image point 'climbs onto the wall'. But we can also see that antisymmetric stretching vibrations will pass into the superposition of oscillations mentioned earlier due to anharmonicity, and if the energy is large enough, the image point will fall into one of the valleys along a complex trajectory. The complexity of the trajectory among other things means that:

- the $CO(X^1\Sigma^+)$ molecule, the product of the dissociation of $CO_2((\widehat{X}^1\Sigma_g^+))$ must be vibrationally excited,
- before dissociation, the molecule will make a large number of vibrations.

The author has mentioned in Sects. 2.1 and 2.4.1 that the lifetime of a triatomic molecule with excitation energy close to the dissociation energy is approximately equal to the time of 10^3 vibrations. Now we have understood the mechanism of this phenomenon on the fingers.

And how can we describe the reaction A + BC → AB + C in terms of the motion of an image point along the *PES*. This reaction can be seen on the example of the reaction $O(^1D) + CO(X^1\Sigma^+) \rightarrow OC(X^1\Sigma^+) + O(^1D)$. Let us accept for definiteness that the horizontal valley (*d*) corresponds to the products $^{18}O(^1D) + C^{16}O(X)$, and the valley *c* to $^{18}OC(X) + {}^{16}O(^1D)$, i.e., we describe isotopic exchange in the CO_2 molecule. It is clear that if the collision of $^{18}O(^1D)$ with the $C^{16}O(X, v_X = 0)$ molecule occurs, and the kinetic energy of the relative motion of O and CO is close to 0, for example, $E = 0.03$ eV, this point will enter *d* valley along its bottom. Then, falling into the wells, it will dangle along it for a long time (various vibrational modes with redistribution of excitation energy will be excited), and finally, fall out or back, or to valley *c*, since its height relative to the bottom of the pit is approximately the same. Apparently, at zero energy ^{18}O and $C^{16}O(X, 0)$, the exit from the well to the valley is possible only when the velocity vector is strictly parallel to the axis of the valley, i.e., all vibrational energy is focused on one of the O – C – O bonds. This is a somewhat rare event, and, therefore, the lifetime of such a complex has to be quite long. However, the decay of the CO_2 molecule will still happen if during the lifetime of this complex it does not collide with another species. It is also evident that if the $C^{16}O_2(X,v)$ molecule is vibrationally excited, or the kinetic energy of the relative motion of O and CO is not equal to 0, the lifetime of the complex has to be less. What is the probability of 'dropping out' of an image point in a given valley at given energies? The answer to this question can be given by solving the problem of the motion of an image point on a given surface and with given initial conditions.

The author has been describing quite a bit about how the oxygen atom in the CO molecule is replaced when it collides with the $O(^1D)$ atom and how the problem of the vibrational excitation of the CO_2 molecule formed can be solved and the probability of the exchange of oxygen atoms. The same process can be described more simply and is done using the *theory of an activated complex.*

The author will not dwell on the description of this theory and mention how to describe the energy of the reaction. Again, as an example, take the same *PES* of the CO_2 molecule. Let us draw a line corresponding to the motion of the image point from valley *d* to valley *c* along a trajectory with the *lowest* CO_2 excitation energy (it is clear that in a real reaction this trajectory does not occur, but in this case we are not interested in the dynamics, but the energy of the reaction). In this case, we will get a line called the *reaction path*. The *PES* cross-section along the reaction path, represented as a one-dimensional line on a plane, is called the *reaction energy profile* (Fig. 3.13).

In the case under discussion, it approximately looks like the right branch of the CO_2 *PEC* obtained with r_{OC} = const, together with its reflection in a mirror.

For a more accurate consideration, it is necessary to take into account the presence of zero oscillations of two- and polyatomic species. Consider, for example, the profile of a reaction path that has a potential barrier, for example, the reaction of thermal decomposition of N_2O to $N_2(X^1\Sigma_g^+) + O(^3P)$. In this case, the profile of the reaction path looks like it is shown in Fig. 3.14a; here, E_Z and $E_Z^{'}$ are the zero-vibration energy $N_2O(\widetilde{X}^1\Sigma^+)$ and $N_2(X^1\Sigma_g^+)$, E_0 is the height of the potential barrier corresponding to the lower intersection point of the *PES* of the lower singlet and triplet states of N_2O, $E_Z^{\neq}$ is the zero-point energy of the N_2O molecule in the position, when it can decay or return to the state of a molecule. This is a position in which the molecule is not like a molecule, is called an *activated complex.*

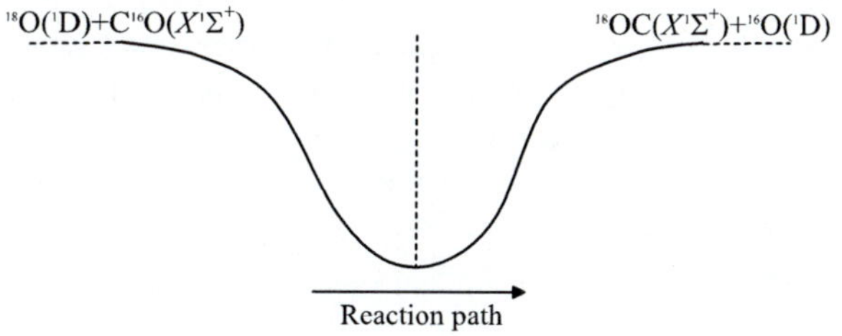

Fig. 3.13 Reaction energy profile for isotope exchange in a $CO_2(\widetilde{X}^1\Sigma_g^+)$ molecule

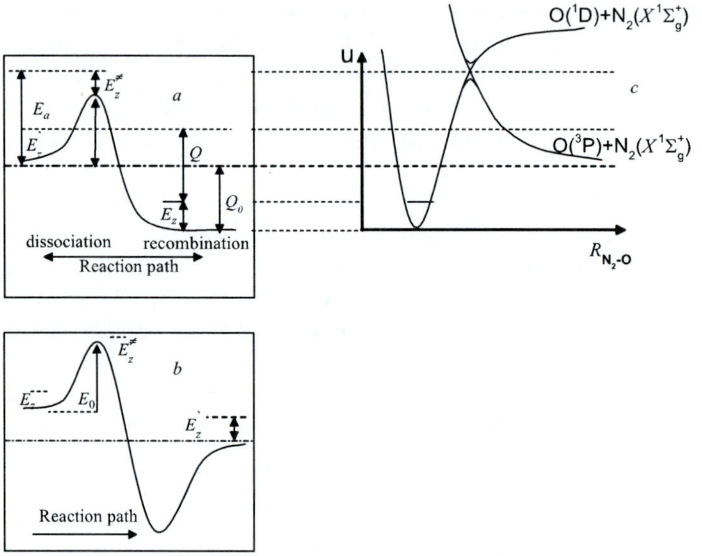

Fig. 3.14 Profile of a reaction proceeding through the potential barrier with (**a**) and without (**b**) an intermediate potential well (see [2], p. 115). **a** without an intermediate potential well; **b** with an intermediate potential well; **c** schematic of the N_2O *PEC*s

Now we can characterize the energy of the reaction completely: $E_a = E_0 + E_Z^{\neq}$, $Q_0 = V - V'$, $Q = Q_0 + \Delta E$, where $\Delta E = E_Z - E_Z'$, Q_0 is the change in potential energy, Q is the heat of reaction. The reaction under consideration is endothermic, i.e., 'absorbing energy'. If we consider the reaction reverse to this one, the reaction of the formation of N_2O

$$\mathrm{O}(^3\mathrm{P}) + \mathrm{N}_2\left(X^1\Sigma_g^+\right) \rightarrow \mathrm{N}_2\mathrm{O}(\tilde{X}^1\Sigma^+)$$

then we will see that there is also a potential barrier in the path of this exothermic reaction. The author deliberately took the thermal decomposition reaction of N_2O known to the reader as an example, although it is not an elementary process (see Sect. 2.4.1). If we consider the exchange reaction

$$\mathrm{H} + \mathrm{Cl}_2 \rightarrow \mathrm{HCl} + \mathrm{Cl},$$

which is an elementary process with positive heat of reaction +45 kcal/mol, the picture is similar. The presence of a potential barrier, as we can see by the example of these two reactions, because, in the H + Cl_2 reaction, for example, the hydrogen atom, when approaching the Cl_2 molecule has to destroy molecular bond Cl – Cl with D_0 = 2.4 eV. The HCl bond is even stronger, D_0 = 4.43 eV, but the former must be destroyed.

The restructuring of the electronic structure of the system during the

$$A + BC \rightarrow AB + C$$

reaction can be very qualitatively described by the superposition of two wave functions $\Psi_{A,BC}$ and $\Psi_{AB,C}$, the first of which corresponds to the system consisting of the atom A and the molecule BC, and the second atom C and molecules AB (Fig. 3.15):

$$\Psi = a\Psi_{A,BC} + c\Psi_{AB,C}. \tag{3.5.6}$$

The coefficients a and c vary along the reaction path in accordance with the continuous transition from state A + BC to state AB + C. Far from the region corresponding to the activated complex of system A + BC, H + Cl_2, O + N_2, a *PES* can be represented as that of corresponding to the interaction of the inert gas atom with the molecule, i.e., pure repulsive potential. If we confine ourselves to exchange reactions, then the potential of AB + C away from the activated complex is the same (see, for example, HCl + Cl). If there is no reaction, then at smaller distances AB − C, HCl − Cl, etc. the potential should be repulsive. In the area near the activated complex, i.e., for identical values of the internal coordinates and the excitation energy of systems A, BC and AB, C, the *PES*s have to intersect. Taking into account the interaction between the states described by $\Psi_{A,BC}$ and $\Psi_{AB,C}$ this intersection 'disappears', and the lower *PES* is formed corresponding to the reaction and upper *PES*. The highest point on the reaction path profile lies lower than the

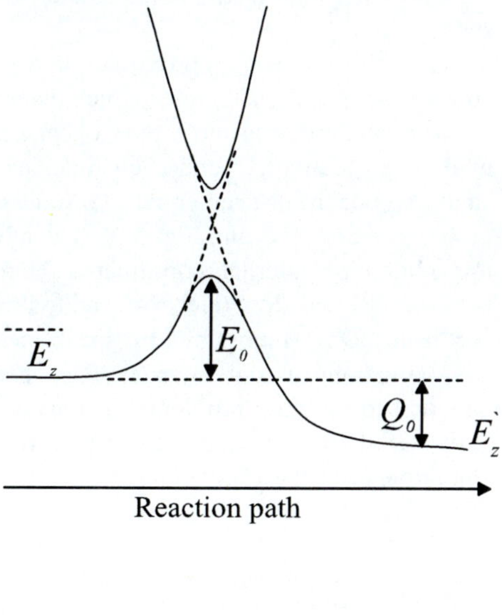

Fig. 3.15 Sections of *PES*s along the reaction path. *Dotted line* represents sections for zero approximation and *full lines* adiabatic surface sections

intersection point of the *PES*s of the zero approximation. The magnitude of the understatement depends on the matrix element of the interaction of the states $\Psi_{A,BC}$ and $\Psi_{AB,C}$

$$\langle \Psi_{A,BC} | \widehat{V} | \Psi_{AB,C} \rangle$$

The form of the operator $\widehat{V}$ is determined by the nature of the interaction of the two states, $\Psi_{A,BC}$ and $\Psi_{AB,C}$.

We have reviewed some processes that can be described within the framework of a single *PEC* (*PES*) i.e., within the framework of the adiabatic approximation. These are elastic scattering of atoms, vibrational excitation of a molecule when colliding with an atom, vibration and dissociation of a triatomic molecule, reaction of an atom and a diatomic molecule, familiarized with the concept of motion of an image point on a *PES*. We now turn to nonadiabatic processes, processes that cannot be described only in the framework of the motion of one *PEC*, *PES*.

3.6 Nonadiabatic Transitions. Perturbation Theory. Probabilities of Adiabatic and Nonadiabatic Transitions

Let us turn to the consideration of nonadiabatic processes, i.e., a transition between *PEC*s, *PES*s, corresponding to different electronic states. The most striking in terms of the consequences of such a transition is predissociation: the molecule ceases to exist, as such, due to the nonadiabatic transition from the bound to the repulsive state.

So, in the adiabatic approximation, each electronic state of the system of atoms corresponds to the *PEC*, *PES*, which determines the motion of the nuclei in it. How we have already seen more than once, *PEC*s, *PES*s obtained in the zero approximation can 'converge' and even 'intersect'. In this case, the value of ΔU may be small (approach) or even equal to 0 (intersection), and the Massey parameter $\xi < 1$ or = 0 (see Sect. 3.2 and Fig. 3.3). The adiabatic approximation is not applicable in this region of internal coordinates. Here, in principle, one cannot distinguish between fast and slow (electronic and vibrational, for example) motion; as we shall see later, it is impossible to speak here about 'pure' adiabatic (in the zero approximation, of course) states; they 'perturb' each other.

Often for solving problems of nonadiabatic interaction, i.e., the mutual perturbation of states, you can use the perturbation theory, that is, to represent the full Hamiltonian in the form:

$$\widehat{\boldsymbol{H}} = \widehat{\boldsymbol{H}}_0 + \widehat{V} \tag{3.6.1}$$

where $\widehat{\boldsymbol{H}}_0$ is the Hamiltonian of the unperturbed state with the eigenvalues E_j^0 and the eigenfunctions Ψ_j^0, and $\widehat{\boldsymbol{V}}$ is the small correction (*perturbation*) of the unperturbed operator $\widehat{\boldsymbol{H}}_0$.

The solutions of the Schrödinger equation with the Hamiltonian $\widehat{\boldsymbol{H}}$ in the second-order of the PT for energy and the first for the wave function have the form:

$$E_j = E_j^0 + V_{jj} + \sum\nolimits_{i \neq j} \frac{|V_{ij}|^2}{E_j^0 - E_i^0} \tag{3.6.2}$$

$$\Psi_j = \Psi_j^0 + \sum\nolimits_{i \neq j} \frac{V_{ij}}{E_j^0 - E_i^0} \tag{3.6.3}$$

Here

$$V_{ij} = \langle \Psi_i^0 | \widehat{\boldsymbol{V}} | \Psi_j^0 \rangle \tag{3.6.4}$$

are matrix elements of the perturbation operator. They should not be equal to 0, otherwise $\Psi_j = \Psi_j^0, E_j = E_j^0$. From here, we get the selection rule for perturbations (nonadiabatic interaction). Since the operator $\widehat{\boldsymbol{V}}$, as well as $\widehat{\boldsymbol{H}}$, is totally symmetric in the point group of the molecule (the exception is the hyperfine and Stark (electrostatic) interaction operators (they are discussed in Sect. 4.6.1.4), then for the perturbation to take place, it is necessary the symmetry types (irreducible representations) of the wave functions of states i and j have to be the same:

$$\Gamma(\Psi_i) = \Gamma(\Psi_j) \tag{3.6.5}$$

i.e., (see Sect. 3.2)

$$\Gamma(\Phi_i) = \Gamma(\Phi_j) \tag{3.6.6a}$$

for electronic interaction,

$$\Gamma\left(\Psi_i^{ev}\right) = \Gamma\left(\Psi_j^{ev}\right) \tag{3.6.6b}$$

for vibronic interaction,

$$\Gamma\left(\Psi_i^{er}\right) = \Gamma\left(\Psi_j^{er}\right) \tag{3.6.6c}$$

for electron-rotational interaction,

$$\Gamma\left(\Psi_i^{evr}\right) = \Gamma\left(\Psi_j^{evr}\right) \tag{3.6.6d}$$

for rovibronic interaction,

$$\Gamma\left(\Psi_i^{es}\right) = \Gamma\left(\Psi_j^{es}\right) \tag{3.6.6e}$$

for the spin–orbit (*S–O*) interaction (Ψ^{es} is the total electron wave function).

These are very significant and completely universal rules for any molecules. According to the type of perturbation operator, nonadiabatic processes (transitions) are also classified.

Approximately, the perturbation operator can be represented as a sum of operators depending only on the electronic and nuclear coordinates.

$$\widehat{V} = \widehat{V}_e + \widehat{V}_Q, \tag{3.6.7}$$

and the wave functions Ψ_i and Ψ_j as the product of the electronic and vibrational wave functions. Then approximate equality will be executed:

$$V_{ij} = \langle \Psi_i^0 | \widehat{V} | \Psi_j^0 \rangle \approx \langle \Phi_i^0 \chi_i^0 | \widehat{V}_e + \widehat{V}_Q | \Phi_j^0 \chi_j^0 \rangle \approx A_{el} \langle \chi_i^0 | \chi_j^0 \rangle, \tag{3.6.8}$$

where A_{el} is the matrix element of the electron interaction of the states i, j, $\langle \chi_i^0 | \chi_i^0 \rangle$ is the integral of the overlap of the vibrational wave functions of these states, equal to the square root of the Frank-Condon factor for these states. Thus, in nonadiabatic processes, a principle similar to the Frank–Condon one for optical transitions should be observed.

Equations (3.6.2, 3.6.3) are valid for the case $\Delta E = E_i^0 - E_i^0 \neq 0$, i.e., for the case when the *PEC*s, *PES*s obtained in the zero approximation, without taking into account the perturbations, do not intersect; if random degeneration occurs ($\Delta E \to 0$), then (3.6.2) should be replaced by [10], p. 303:

$$E = E_i^0 + E_j^0 + V_{ii} + V_{jj} \pm \sqrt{\frac{1}{4}\left(E_i^0 - E_j^0 + V_{ii} - V_{jj}\right)^2 + \left|V_{ij}\right|^2} \tag{3.6.9}$$

and at the point where random degeneration occurs $E_i^0 = E_i^0$ the equality

$$\sqrt{\frac{1}{4}\left(E_i^0 - E_j^0 + V_{ii} - V_{jj}\right)^2 + \left|V_{ij}\right|^2} = 0 \tag{3.6.10}$$

is valid. Since the expression under the root is the sum of two squares, (3.6.9) is valid only if, for the same values of internuclear distances

$$E_i^0 - E_j^0 + V_{ii} - V_{jj} = 0 \text{ and } V_{ij} = 0 \tag{3.6.11}$$

For a diatomic molecule, this is possible if $V_{ij} = 0$, since both values under the root are functions of only one coordinate and cannot be equal to zero at the same time. Therefore, the *PEC*s of states i and j intersect if these states have different types of symmetry of electronic wave functions (the exception is electrostatic interaction). Otherwise, these states are 'repelled'. In polyatomic molecules, (3.6.10) is always valid, but not for all coordinates (*conical intersection*), i.e., *PES*s of these molecules intersect even if the symmetry types of their wave functions coincide.

So, we see that if the Massey parameter for *PEC*, *PES*, obtained in the zero approximation, i.e., without taking into account the perturbations is small, and these *PEC*, *PES* approach or even intersect, then the adiabatic approximation is again not applicable in the zero approximation. In these cases, we must take into account that a jump from surface to surface, so-called *nonadiabatic transition*, can occur with some probability. As we will see later, the probability of these transitions depends on the degree of interaction of these states obtained in the zero approximation; the energy gap between the *PEC*s, *PES*s of the states obtained already taking into account the disturbances, i.e., 'corrected' Massey parameter.

We now consider the question of the factors on which the probabilities of nonadiabatic transitions depend in more detail. For simplicity, consider first the *PEC*s, i.e., cases of predissociation of a diatomic molecule, reversed predissociation, and transitions between their bound states. First, we consider the general laws and then will illustrate them with examples.

Let us first consider the case when the types of symmetry of the electronic wave functions of the two states are different, the perturbation of these states and the value of V_{ij} are small, and their *PEC*s intersect (Fig. 3.16).

The nonadiabatic coupling of these terms occurs near the intersection point, in which the Massey parameter is equal to 0. In this entire small area, the matrix element of the nonadiabatic interaction of two states V_{ij} (3.6.4), whose type and

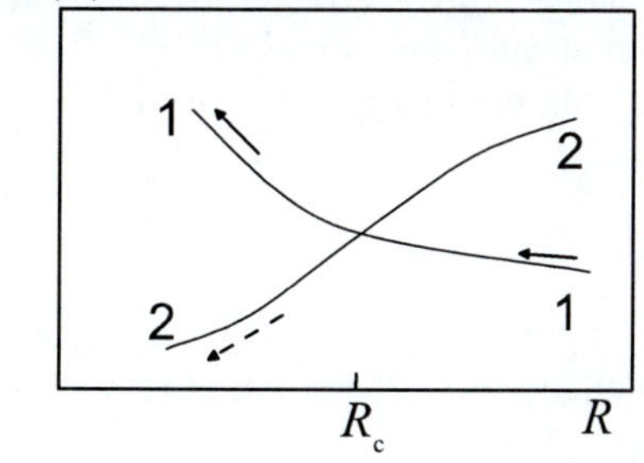

Fig. 3.16 The intersection of adiabatic terms. 1–1, 2–2: adiabatic paths; 1–2: nonadiabatic paths (see [2], p. 120)

magnitude depends on the type of interaction, i.e., the type of the operator, the electronic part of the matrix element, and the overlap integral of vibrational wave functions can be considered as constant. The terms U_1 and U_2 of these states in this small region can be approximated by linear functions with slopes $F_1 = dU_1/dR|_{R_c}$ and $F_2 = dU_2/dR|_{R_c}$, where $R = R_c$ is the coordinate of the crossing point, then the probability $P_{1,2}$ is a nonadiabatic transition from curve 1 to curve 2 is:

$$P_{1,2} = \frac{2\pi V_{1,2}^2}{\hbar \frac{dR}{dt} |F_1 - F_2|}, \tag{3.6.12}$$

where $dR/dt = (R - R_c)/\Delta t$ = const is the radial velocity of the relative motion of the atoms. The probability of motion along a 1–1 curve, i.e., on an adiabatic curve is:

$$P_{1,1} = 1 - P_{1,2} \approx 1. \tag{3.6.13}$$

The author wrote that $P_{1,2} \approx 0$ and $P_{1,1} \approx 1$ since the mutual perturbation of states 1 and 2, as we agreed, is low (the curves intersect), and the matrix element $V_{1,2}$ is small. What else depends on the value of $P_{1,2}$? Note the denominator of the (3.6.12). It 'presents' the velocity of the passage of the image point past the point of intersection of *PEC*s and the difference between the slopes of these *PEC*s: the higher the velocity and the steeper they are located towards each other (the maximum value $|F_1 - F_2|$ is $\pi/2$), the less the probability of the nonadiabatic transition. Qualitatively, this is understandable.

Now we take into account that when $P_{1,1} \approx 1$, the image point passes the intersection point twice, forward and back. Therefore, the probability of a nonadiabatic transition must be equal to:

$$2P_{1,2}(1 - P_{1,2}) \approx 2P_{1,2} \tag{3.6.14}$$

since $P_{1,2} << 1$.

We obtained the probability of a nonadiabatic transition. Just recall the principle of detailed equilibrium, according to which the probabilities of the direct and reverse processes are equal.

And what is the rate constant of the process when moving 'from the side of atoms' (inverse predissociation) and in the case of predissociation for a diatomic molecule, for example?

In the first case, we have:

$$k = k_{gk} \langle P_{1,2} \rangle \exp(-E_0/kT), \tag{3.6.15}$$

where $< P_{1,2} >$ is the average transition probability, E_0 is the energy at the point of intersection of the terms relative to the asymptote. In general, quite a reasonable formula. If we want to obtain a predissociation rate constant, k_{pr}, this can be easily done using the *Fermi Golden Rule* (see [11], p. 535): if for a given vibrational level the matrix element of the interaction of states 1 and 2 (or i and *j*) is known,

$$k_{pr} = 1/\tau_{pr} = \hbar V_{1,2}^2 \text{ s}^{-1} \tag{3.6.16}$$

One can calculate the $V_{1,2}^2$ value if the overlap integral$\langle \chi_i^0 | \chi_i^0 \rangle$, and the A_{el} value are known (3.6.8). How it is interesting: a nonadiabatic transition rate constant is actually calculated in the adiabatic approximation, separating the electron and vibrational motion in the molecule. A little later we will return to this formula.

And now let us consider the case of a strong interaction of states leading to the repulsion of the *PEC*s (Fig. 3.17) which is occurred, for example, for states with the *same symmetry* of the orbital wave function and the *same multiplicity*; $^1\Sigma^+ - ^1\Sigma^+$, $^1\Pi - ^1\Pi$, etc.(see Sect. 4.6.1.1 for details).

In this case, the *PEC*s obtained in the zero approximation without taking into account the interaction, called *diabatic*, intersect, and the *PEC*s obtained concerning the interaction, i.e., *adiabatic*, repels. Again, if we assume that in the *PEC* avoided crossing area, diabatic terms can be represented as straight lines, and near the intersection area by hyperbolas, then, as Landau and Zener showed, the probability of a non-adiabatic transition from curve 1 to curve 2:

$$P_{1,2} = \exp \frac{2\pi a^2}{\hbar \frac{dR}{dt} |F_1 - F_2|} \tag{3.6.17}$$

where $2a$ is the minimal distance between adiabatic terms, and all other notations are the same, as in the formula (3.6.12). Is not it true, (3.6.17 and 3.6.12) are very similar, only in (3.6.17) there is a^2 in the numerator, and the dependence is exponential. This feature, as we shall see a little further, for a reason.

Let us analyze (3.6.17). We see that the probability of a nonadiabatic transition between pieces of the diabatic term 'broken' due to the interaction of the states decreases exponentially with increasing parameter a^2 (which, as we will see below,

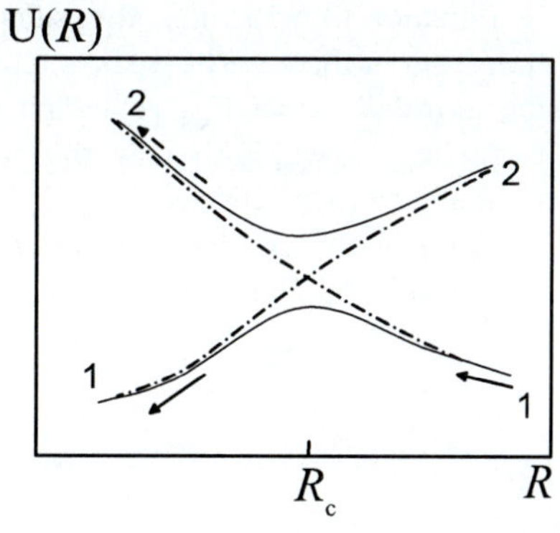

Fig. 3.17 Quasi-intersection of adiabatic terms (see [2], p. 122). 1–1, 2–2: adiabatic paths; 1–2: non-adiabatic paths *(dash-dotted lines* refer to crossing diabatic terms)

is directly related to the degree of interaction of the zero-approximation terms) and the angle between the terms. For a nonadiabatic transition with a weak interaction, the dependence of $P_{1.2}$ on $V_{1.2}$ (3.6.12) is inverse, but the probability of transition 1–1 (3.6.13) (which is nonadiabatic 1–2 path here) also decreases with increasing $V_{1.2}$, decreasing velocity of the motion of the image point and the angle of intersection of the *PEC*s of the zero approximation.

If the velocity of motion of the image point is high, and the splitting of terms $2a$ is small, then the probability of a nonadiabatic transition between *adiabatic* terms can even reach 1. In such a situation, the concept of *diabatic term* is introduced (see Sect. 4.6.1.1). You have already understood that with a not very strong interaction of states, the diabatic terms coincide with the terms of the zero approximation. As you can see, all these definitions of the adiabatic, diabatic term, nonadiabatic transition are very unsteady. Everything is quite clearly defined if the interaction of the zero-approximation states is strong or weak, and the transition probability between the states obtained taking into account the interaction, i.e., nonadiabatic transitions, small. In intermediate cases of interaction, which transitions are considered adiabatic, and which nonadiabatic transitions are not very well understood. We will see this in Sect. 4.7.

Let us return to the discussion of the probability of the adiabatic and nonadiabatic development in the case of a strong interaction of the zero approximation terms. If we denote as $P_{1,2}$ and $P'_{1,2}$ the probabilities of transitions between adiabatic and diabatic terms, respectively, then:

$$P'_{1,2} = 1 - P_{1,2} \equiv P_{1,1} \tag{3.6.18}$$

If the probability of a non-adiabatic transition (1–2), $P_{1,2}$ (Fig. 3.17) is high (the exponent is small, weak interaction), then the probability of an adiabatic transition can be decomposed into a series in terms of the exponent $(1-expx) = x + \ldots$) and one gets that this probability is equal to:

$$P'_{1,2} = \frac{2\pi a^2}{\hbar \frac{dR}{dt} |F_1 - F_2|} \tag{3.6.19}$$

Equation (3.6.19) has the same form as the probability of a non-adiabatic transition in the case of weak interaction (Fig. 3.16), assuming that the splitting of the terms is $2a = 2V_{1,2}$. The first (the identity of the type of probability) is not surprising, since, although the transitions are called differently (nonadiabatic between diabatic curves and adiabatic, that is, passing along the same adiabatic curve), in both cases we are discussing the same transition.

Second, the fact that

$$a = V_{1,2} \tag{3.6.20}$$

enables us to realize the connection between the energy gap a between repulsive states due to mixing and the matrix element of this interaction.

If we take into account the double passage of the quasi-intersection region, then the probability of transition between adiabatic terms (Fig. 3.17)

$$\langle P_{1,2} \rangle = 2\exp(-\gamma)[1 - \exp(-\gamma)], \tag{3.6.21}$$

where $\gamma = P'_{1,2}$ (see (3.6.19)). This is the Landau-Zener formula [1], p. 50. The maximum probability is obtained when $\gamma = 0.5$ and is equal to 0.5.

As we saw above, the probabilities of adiabatic with strong interaction and non-adiabatic with weak one transitions are the same. Just as before, we can get the rate constants of these transitions.

Now about non-adiabatic transitions in polyatomic molecules or complexes. Here, it is necessary to take into account two significant changes. First, the trajectory of a motion of an image point can be arbitrarily oriented relative to the intersection line or the quasi-intersection of the *PES*s. Secondly, an image point intersects the region of non-adiabatic interaction not twice, as in the case of *PEC*s, but, in general, many times. It is not possible to express the probability through the probability of an adiabatic transition with one passage of an image point through this region.

The probability in the case of quasi-intersection can be calculated using (3.6.19), in which a should be understood as the minimum distance between adiabatic terms, and F_1, F_2 derived from diabatic potentials on the line of their intersection in a direction perpendicular to it. In this case, dR/dt is the component of velocity along this direction.

References

1. Kondratiev, V.N., Nikitin, E.E.: Gas-phase Reactions: Kinetics and Mechanism. Springer-Verlag, Berlin Heidelberg New York (1981). https://doi.org/10.1007/978-3-642-67608-6
2. Kondratiev, V.N., Nikitin, E.E.: Kinetika i Mekhanism Gazofaznykh Reaktzij (Kinetics and Mechanism of Gas-phase Reactions) Nauka, Moscow (1974) (in Russian)
3. Kaplan, I.G.: Vvedenie v Teoriyu Mezhmolekulyarnykh Vzaimodejstvii (Introduction to the Theory of Intermolecular Interactions) Nauka, Moscow (1982) (in Russian)
4. Kaplan, I.G.: Intermolecular Interactions: Physical Picture, Computational Methods and Model Potentials. John Wiley & Sons (2006). https://doi.org/10.1002/047086334X.fmatter
5. Herzberg, G.: Molecular Spectra and Molecular Structure III. Electronic Spectra and Electronic Structure of Polyatomic Molecules. Krieger Publishing Company, Second Edition (1966)
6. Herzberg, G.: Molecular Spectra and Molecular Structure II. Infrared and Raman Spectra of Polyatomic Molecules. Krieger Publishing Company, 16-th Edition (1945)
7. Hochstrasser, R.N.: Molecular Aspect of Symmetry. W.A. Benjamin, Inc. (1966)

8. Hirschfelder, J.O., Curtiss, Ch.O., Bird, R.B.: Molecular Theory of Gases and Liquids, John Willey & Sons, Inc., Chapman and Hall, Lim. (1954). https://doi.org/10.1002/pol.1955.120178311
9. Jones, J.E.: On the determination of molecular fields. II. From the equation of state of a gas. Proc. Roy. Soc. (London) **A106**, 463–477 (1924). https://doi.org/10.1098/rspa.1924.0082
10. Landau, L.D., Lifshitz, E.M.: Course of Theoretical Physics. Volume 3. Quantum Mechanics. Nonrelativistic Theory (Third edition, revised and enlarged), Pergamon Press (1977). https://doi.org/10.1016/C2013-0-02793-4
11. Lefebvre-Brion, H., Field, R.W.: The Spectra and Dynamics of Diatomic Molecules, Elsevier (2004)

Chapter 4
Photolysis of Free Molecules

Abstract The chapter deals with radiative electronic transitions (allowed and forbidden as electric dipole ones), intramolecular perturbations in electronically-excited states of di- and polyatomic molecules. The definitions of primary and secondary processes of gas-phase photolysis and quantum yields of various photoprocesses are introduced in the beginning. Then radiative transitions in absorption and emission, Franck-Condon principle for bound–bound and bound-free transitions are discussed. Intramolecular perturbation in di- and polyatomic molecules, including electronic predissociation, and dissociation, are analyzed in detail.

4.1 Primary and Secondary Processes of Gas-Phase Photolysis. Quantum Yields

What are the *primary* and *secondary* processes of the photolysis of molecules in the gas phase? However, these concepts are different in gas-phase photochemistry and photochemistry of condensed systems. Besides, the terminology of this field of science, at least in gas-phase photochemistry, does not settle yet. Definitions for basic terms are either absent or not implicit. The situation may lead to terminological confusion and mistakes. The author gives some definitions that will be used further.

Let us first consider the processes occurring during the photolysis of *free* (*isolated*) molecules by photons with energy $h\nu$ provided that the photon energy is less than the molecule ionization potential. By free molecules, the author means molecules in conditions allowing the free removal of their fragments without a so-called 'cage effect' [1], i.e., the molecules in gas-phase at not-to-high pressures, less than 10 bars.

$$AB + h\nu \rightarrow AB^*(v, J) \rightarrow A_1 + B_1 \tag{4.1.1}$$

$$\rightarrow A_2 + B_2 \tag{4.1.2}$$

A. Pravilov, *Gas-Phase Photoprocesses*, Springer Series in Chemical Physics 123,
https://doi.org/10.1007/978-3-030-65570-9_4

$$\rightarrow A_i + B_i \tag{4.1.3}$$

$$\rightarrow AB^{**} + h\nu \tag{4.1.4}$$

$$\leftrightarrow AB^{**} \tag{4.1.5}$$

$$AB^{*}(v, J) \xrightarrow{M} AB^{**} + \Delta E_1 \tag{4.1.6}$$

$$\xleftrightarrow{M} AB^{**} \tag{4.1.7}$$

$$\xrightarrow{M} AB^{**} + \Delta E_2 (\Delta E_2 < \Delta E_1) \tag{4.1.8}$$

$$\xrightarrow{M} AB^{*}(v - \Delta v, J - \Delta J) + \Delta E \tag{4.1.9}$$

$$\xrightarrow{M} B_i + C_i \tag{4.1.10}$$

$$\xrightarrow{M} \text{products} \tag{4.1.11}$$

$$AB^{*}(v - \Delta v, J - \Delta J), AB^{**} \xrightarrow{M} \text{products} \tag{4.1.12}$$

$$B_k^{*} \rightarrow C_j + D_j \tag{4.1.13}$$

$$B_k^{*} \xrightarrow{M} \text{products} \tag{4.1.14}$$

$$B_k^{*} \rightarrow B_k^{**} + h\nu \tag{4.1.15}$$

$$B_k^{*} \xrightarrow{M} B_k^{**} \tag{4.1.16}$$

$$\xrightarrow{M} B_k^{*}(v - \Delta v, J - \Delta J) \tag{4.1.17}$$

where AB*(v, J) are rovibronic states of the diatomic or polyatomic molecule resulting after absorption by the molecule AB a photon of energy $h\nu$, AB** are all other rovibronic states including the AB ground state, A_i, B_i are both their fragments.

Photolysis processes are categorized as primary and secondary processes.

Primary photolysis process includes photon absorption by the molecule AB and degradation of excited energy for the resultant rovibronic state AB*(v, J), both

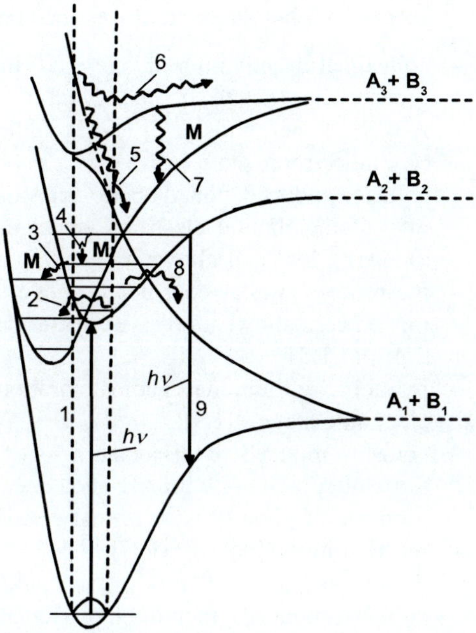

Fig. 4.1 The processes accompanying photon absorption by a free molecule [3], p. 10. 1—photon absorption; 2—nonadiabatic transition into another bound state; 3—collision-induced nonadiabatic transition into another bound state at not-too-high energy loss; 4—rovibrational relaxation within one state; 5—spontaneous or collision-induced nonadiabatic decay of a bound state; 6—dissociation; 7—collisional electronic deactivation (quenching); 8—spontaneous predissociation or collision-induced predissociation; 9—luminescence

proceeding *without* another species M (any medium species, wall) participation. The primary photolysis processes are as follows (Fig. 4.1):

- AB*(*v*,*J*) formation (4.1.1); the optical transition time $\sim 10^{-15}$ s);
- *dissociation* AB*(*v*, *J*), lifetime, $\tau_{AB*} \leq 10^{-13}$ s), predissociation ($\tau_{AB*} \geq 10^{-13}$ s), *nonadibatic decay* (see Sect. 3.6) of AB*(*v*, *J*) bound or repulsive state into A, B products, including those in excited states (4.1.2 and 4.1.3) ($\tau_{AB*} > 10^{-13}$ c) [2, 3];
- luminescence of AB*(*v*, *J*) molecule ($\tau_{AB*} > 10^{-9}$ s, (4.1.4));
- *spontantaneous internal conversion*, a *nonadiabatic* transition into another electronic bound state AB, including *intercombination conversion* [4] ($\tau_{AB*} > 10^{-13}$ s, (4.1.5)).

2. *Photodecay* (*photodecomposition*) of molecule AB is the primary photolysis process reducing the AB* concentration, [AB*], i.e., photon absorption accompanied by dissociation, predissociation, nonadiabatic decay of AB*(*v*, *J*) into fragments of the AB molecule.
3. *Secondary photolysis processes* — processes occurring during a collision of AB*(*v*, *J*) with any medium species M and after this collision, as well as processes occurring with the products of AB photodecay.

Secondary photolysis processes include the follows:

- collisional deactivation of AB*(v, J) into another electronic state, including the ground state (*electronic deactivation*), it is the process proceeding with a large AB*(v, J) energy loss ΔE_1 and excluding the possibility of returning to the AB* (v, J) electronic state (4.1.6);
- collision-induced nonadiabatic transition, CINAT (see [6, 7] and references), into another bound electronic state; this process proceeds with a low AB*(v, J) energy loss and allows a reverse transfer (4.1.7);
- the process similar considered above proceeding with a loss of energy (ΔE_2); this process allows a reverse transfer only to *lower* rovibronic levels AB*(v-Δv, J-ΔJ) (4.1.8);
- rotational, vibrational and rovibrational relaxation of AB*(v, J) within an electronic state (4.1.9);
- collision-induced predissociation of AB*(v, J) into AB fragments, including those of coinciding with AB photodecay products (in other words, with primary products of photolysis or primary photoproducts) (4.1.10);
- reaction involving AB*(v, J) (4.1.11);
- decay processes of AB*(v-Δv, J-ΔJ) and AB** states (spontaneous and collision-induced) including, dissociation, predissociation, and luminescence (4.1.12);
- spontaneous decay of one of the photodecay products of AB*, B_k^*, into fragments, including those coincide with any of the photodecay products of AB (4.1.13), they also include reaction with photodecay products of AB (4.1.10), B_k^* collisional decay (4.1.14), luminescence of these products (4.1.15), electronic deactivation (4.1.16), rotational, vibrational and rovibrational relaxation (4.1.17) and so on.

W.A Noyes, P.A. Leighton [8] and, in the wake of them, H. Okabe [9], proposed to also include as primary processes those occurring with the participation of AB* (v, J) and M (4.1.6—4.1.12). They also proposed to assume by secondary processes those in which photodecay products B_i, C_i participate (4.1.13–4.1.17). Methodologically, in the case of the gas phase, such a definition is wrong as follows. Such absolute quantum yields depend upon of kind and pressure of gases, the ratio of the reactor surface area to its volume, and so on, i.e., they are not *molecular constants describing the molecule features.*

The processes (4.1.6–4.1.12) are usual bi-, and termolecular processes with excited species, which are not in thermodynamic equilibrium with the environment. These processes should be described using rate constants depending on the AB*, AB** states, rather than by the absolute quantum yields.

1. The absolute quantum yield of the product of the AB photolysis product (e.g. AB photodecay product, B_i) is the ratio of the concentration change of photolysis product for a unit time (4.1.18) or for a photon pulse (4.1.19) to the number of absorbed photons for a volume unit, which causes this concentration change:

$$\Phi_{B_i}^{AB}(\lambda) = \frac{d[B_i]/dt}{n_{h\nu}^{a}(\lambda)} \tag{4.1.18}$$

$$\Phi_{B_i}^{AB}(\lambda) = \frac{d[B_i]/dt}{N_{h\nu}^{a}(\lambda)}, \tag{4.1.19}$$

where $n_{h\nu}^{a}(\lambda)$, photon/cm^3 · s, and $N_{h\nu}^{a}(\lambda)$, photon/cm^3 · pulse are absorbed radiation densities. The definition (4.1.18) is suitable for stationary photolysis, and the definition (4.1.19) for pulse photolysis.

2. The absolute quantum yield for the *i*-th AB photodecay process is the ratio of the concentration change taken with the minus sign for photolyzed AB species due to i-th process for unit time or a light pulse interval to the absorbed radiation density:

$$\varphi_i^{AB}(\lambda) = -\frac{\{d[AB]/dt\}_i}{n_{h\nu}^{a}(\lambda)} \tag{4.1.20}$$

$$\varphi_i^{AB}(\lambda) = -\frac{\{d[AB]\}_i}{N_{h\nu}^{a}(\lambda)} \tag{4.1.21}$$

It is obvious that for photodecay processes, the following is true:

$$\Phi_{A_i}^{AB}(\lambda) = \Phi_{B_i}^{AB}(\lambda) = \varphi_i^{AB}(\lambda) \leq 1$$

3. The absolute quantum yield of the AB photodecay:

$$\varphi_{AB}(\lambda) = \sum\nolimits_i \varphi_i^{AB}(\lambda) \tag{4.1.22}$$

Evidently, $\varphi_{AB}(\lambda) \leq 1$. It is less than 1 if, for the given M concentration, the decay rate of AB*(*v*, *J*) is comparable with the rate of AB*(*v*, *J*) collision-induced excitation energy loss to the level lower than the threshold of AB*(*v*, *J*) decay (see 4.1.1)–(4.1.3).

4. The absolute quantum yield of photoluminescence of the AB or AB photodecay products, B_k^*, is the ratio of the number of photons emitted from a volume unit per unit time:

$$n_{h\nu}^{lum} = \int_{\lambda_1}^{\lambda_2} n_{h\nu}^{lum}(\lambda)d\lambda, \text{photon/cm}^3\text{s} \quad (4.1.23)$$

or per photon pulse:

$$N_{h\nu}^{lum} = \int_{\lambda_1}^{\lambda_2} n_{h\nu}^{lum}(\lambda)d\lambda, \text{photon/cm}^3\text{pulse}, \quad (4.1.24)$$

to the density of absorbed radiation calculated to be in the following forms:

$$\Phi_{lum}(\lambda) = n_{h\nu}^{lum}(\lambda)/n_{h\nu}(\lambda) \quad (4.1.25)$$

and

$$\Phi_{lum}(\lambda) = N_{h\nu}^{lum}(\lambda)/N_{h\nu}(\lambda), \quad (4.1.26)$$

respectively, where $n_{h\nu}^{lum}(\lambda)$ and $N_{h\nu}^{lum}(\lambda)$ is the luminescence spectrum in the λ_1 - λ_2 spectral range expressed in photon/cm^3 · s · nm and photon/cm^3 · pulse · nm, respectively. As the $\Phi_{A_i}^{AB}(\lambda)$, $\varphi_i^{AB}(\lambda)$, the absolute quantum yield of luminescence is a function of a wavelength of exciting radiation, in general.
The definitions (4.1.18)–(4.1.26) differ fundamentally from those represented in some monographs. Here, by the absolute quantum yields of primary processes (products) are meant the yields measured at *monochromatic* radiation within the wavelength interval $\lambda \pm \Delta\lambda$, for which quantum yields are independent of photolysis radiation wavelength, and *within volume unit*, where $n_{h\nu}(\lambda)$ = const, and not in the whole irradiated volume as in [9]. The remark above is essential for the study of secondary processes, the rate of which may depend upon the $n_{h\nu}(\lambda)$, $N_{h\nu}(\lambda)$.
5. The absolute integral quantum yield for i-th AB photodecay process within the absorption band of AB is defined as follows. It is equal to the ratio of the area σ_i^{AB} under the spectral curve of AB partial absorption cross-section, which leads to the i-th process, to the total absorption cross-section area $\sigma_{AB}(\lambda)$ under the spectral curve within the AB absorption band:

$$\varphi_i^{AB} = \frac{\int_{\lambda_1}^{\lambda_2} \varphi_i^{AB}(\lambda)\sigma_{AB}(\lambda)d\lambda}{\int_{\lambda_1}^{\lambda_2} \sigma_{AB}(\lambda)d\lambda} = \frac{\int_{\lambda_1}^{\lambda_2} \sigma_i^{AB}(\lambda)d\lambda}{\int_{\lambda_1}^{\lambda_2} \sigma_{AB}(\lambda)d\lambda}. \quad (4.1.27)$$

Virtually, this ratio is the ratio of the oscillator strength of the optical transition corresponding to *i*-th photodecay channel to the sum of the oscillator strengths of all transitions taking place within the band [2, 3, 9]. Its value is characteristic of the optical transition probability for AB photodecay into *i*-th channel, this value is very useful in theoretical evaluations.

6. The absolute integral quantum yield for the formation of the product of AB photodecay, B_i, is found to be as follows:

$$\Phi_i^{AB} = \frac{\int_{\lambda_1}^{\lambda_2} \Phi_{B_i}^{AB}(\lambda)\sigma_{AB}(\lambda)d\lambda}{\int_{\lambda_1}^{\lambda_2} \sigma_{AB}(\lambda)d\lambda} = \frac{\int_{\lambda_1}^{\lambda_2} \sigma_i^{AB}(\lambda)d\lambda}{\int_{\lambda_1}^{\lambda_2} \sigma_{AB}(\lambda)d\lambda}, \quad (4.1.28)$$

where $\sigma_{AB}(\lambda)$ is the AB absorption cross-section.

7. *The convolution quantum yield of i-th primary photoprocess* within the band $\lambda_j \div \lambda_k$ is defined as follows:

$$\varphi_i^{AB,c}(\lambda_j - \lambda_k) = \frac{\int_{\lambda_k}^{\lambda_j} \varphi_i^{AB}(\lambda)\Phi_{hv}(\lambda)\{1 - exp\{\sigma_{AB}(\lambda)[AB]l\}\}d\lambda}{\int_{\lambda_k}^{\lambda_j} \Phi_{hv}(\lambda)\{1 - exp\{\sigma_{AB}(\lambda)[AB]l\}\}d\lambda} \quad (4.1.29)$$

where $\Phi_{hv}(\lambda)$, photon/(s · nm) is the spectral photon flux passing into the photochemical cell, *l* is the cell length.

8. *The convolution quantum yield for the formation of the B photoproduct* is calculated to be as follows:

$$\Phi_i^{AB,c}(\lambda_j - \lambda_k) = \frac{\int_{\lambda_k}^{\lambda_j} \Phi_{B_i}^{AB}(\lambda)\Phi_{hv}(\lambda)\{1 - exp\{\sigma_{AB}(\lambda)[AB]l\}\}d\lambda}{\int_{\lambda_k}^{\lambda_j} \Phi_{hv}(\lambda)\{1 - exp\{\sigma_{AB}(\lambda)[AB]l\}\}d\lambda} \quad (4.1.30)$$

In a typical case, values of the $\varphi_i^{AB,c}(\lambda_j - \lambda_k)$, $\Phi_i^{AB,c}(\lambda_j - \lambda_k)$ are determined by the mutual position of the curves $\Phi_{hv}(\lambda)$, $\sigma_{AB}(\lambda)$, φ_i^{AB}, and the [AB], 1 quantities. Hence, the convolution quantum yields are dependent on experimental conditions, i.e., they all are *incommensurable* quantities. Nevertheless, you may often encounter articles, where authors deal with photolysis processes with very weak intensities in a wide spectral range, wherein the quantum yields of the photoprocesses have not constant values.

4.2 Radiative Electronic Transitions. Selection Rules for Radiative Electronic Transitions. Spin–Orbit Coupling and Spin-Forbidden Radiative Electronic Transitions

Radiative transitions between rovibronic levels of different molecular electronic states are discussed briefly in this section. *Allowed* and *forbidden* transitions are distinguished.

4.2.1 *Allowed Radiative Electronic Transitions*

Radiative electronic transition (transition in the following) is considered as allowed if it occurs as an electric dipole one without taking into account spin–orbit, vibronic and rotational-electronic interactions [10].

4.2.1.1 General Selection Rules for Electric Dipole Transitions

Let Φ_n and Φ_m be the electronic wave functions of the upper and lower states of an electric dipole transition, respectively. The transition is allowed if the *electronic transition matrix element* (transition dipole moment) $R_e^{nm} = \langle \Phi_n | \hat{\mu} | \Phi_m \rangle$

is non-zero

$$R_e^{nm} = \langle \Phi_n | \hat{\mu} | \Phi_m \rangle \neq 0 \tag{4.2.1}$$

for at least one orientation of the transition dipole moment operator

$$\hat{\mu} = (\sum ex_i, \sum ey_i, \sum ez_i). \tag{4.2.2}$$

It is valid if the direct product (see Appendix III in [10]) of the *symmetry types, (species, irreducible representations)*, Γ (see Appendix I in [10]), of the Φ_n, Φ_m wave functions, and $\hat{\mu}$ operator (see Table 9 in [10]) has a totally symmetric component Γ_1,

$$\Gamma(\Phi_n)\,\mathrm{x}\,\Gamma(\Phi_m)\,\mathrm{x}\,\Gamma(\hat{\mu}) = \Gamma_1 + \ldots, \tag{4.2.3}$$

i.e., the direct product of the Φ_n, Φ_m species is the same species as that of one of the $\hat{\mu}$ components (see [10–12]).

Ground states of most stable molecules are totally symmetric. For allowed transitions from these states (absorption), the species of upper states must have the species of a component of the dipole moment operator. If Φ_n and Φ_m wave functions belong to different point groups (for example, CO_2 ground state is linear, CO_2 $(\widetilde{X}^1\Sigma_g^+), D_{\infty h}$ point group, and lower electronically excited states are bent, C_{2v} point group), one has to use species of the lower symmetry point group.

The selection rule (4.2.3) applies strictly for fixed nuclei only. In fact, the nuclei are not fixed, and one has to consider the total wave functions that include nuclear coordinates. Neglecting the rotational motion, one can use adiabatic approximation (see Sect. 3.2) and write

$$\Psi_{ev}(r, Q) = \Phi(r, Q) \cdot \chi(Q), \tag{4.2.4}$$

where $\Phi(r,Q)$ is the electronic wave function (see 3.2.3, 3.2.11), and $\chi(Q)$ is the vibrational wave function, the solution of (3.2.12). In this case matrix element for an electronic transition between vibronic levels is:

$$\mu_{n,m;v',v''} = R_e^{nm}(Q) \cdot \langle \chi_{v'} | \chi_{v''} \rangle, \tag{4.2.5}$$

where $R_e^{nm}(Q)$ is the electric dipole moment of transition for a specific nuclear configuration Q, and $\langle \chi_{v'} | \chi_{v''} \rangle$ is the *overlap integral*. It follows from (4.2.5) that a transition between vibronic states is allowed if their species of the χ_n, χ_m vibrational states are the same (note, that vibrational states of diatomic molecules are totally symmetric).

4.2.1.2 Spin Selection Rules

For weak spin–orbit interaction (see Sect. 4.2.3), the electronic wave function including spin, can be written as a product of an orbital and a spin wave functions

$$\Psi_{es} = \Phi \cdot \sigma, \tag{4.2.6}$$

and the moment of electric dipole allowed transition is:

$$R_{es}^{nm} = \langle \Phi_n | \widehat{\mu} | \Phi_m \rangle \cdot \langle \sigma_n | \sigma_m \rangle. \tag{4.2.7}$$

The $\langle \sigma_n | \sigma_m \rangle$ term vanishes for states of different spin S due to the orthogonality of spin functions corresponding to different S values. Therefore, for weak spin-orbit interaction, the selection rule

$$\Delta S = 0 \tag{4.2.8}$$

is strictly valid [10], p. 131.

4.2.2 Forbidden Electronic Transitions

Transitions which do not comply selection rules discussed in Sect. 4.2.1 are called *electric dipole forbidden*. However, some of these transitions can have appreciable intensities due to perturbations.

4.2.2.1 Transitions in Polyatomic Molecules Allowed Due to Vibronic Interactions

A transition in *polyatomic* molecules forbidden due to the symmetry of electronic wave functions may be allowed when vibronic interactions are taken into account (see 4.2.4). In this case

$$\Gamma(\Psi_{ev}) = \Gamma(\Phi) \times \Gamma(\chi_Q), \tag{4.2.9}$$

(see 3.2.15 and Sect. 3.2). A transition is allowed if direct products of the $\Gamma(\widehat{\boldsymbol{\mu}})$ and $\Gamma(\Psi_{ev})$ species for *both*, lower and upper states, have a totally symmetric component (see Sect. 4.6.2.1).

There are pairs of electronic states in many point groups that cannot be combined in electronic transitions, even when vibronic interaction is taken into account due to *vibrational deficiency*. Vibrationally deficient molecules are defined as molecules that species for the normal vibrational modes of a molecule fail to span all the corresponding point group the irreducible representations. An important consequence of vibrational deficiency is electronic transitions, which are forbidden even if one takes into account a vibronic interaction (*first-order vibronic prohibition* [13]).

An example of vibrational deficiency is transitions

$${}^1A_2({}^1\Sigma_u^-),{}^1B_2({}^1\Delta_u) \leftarrow {}^1A_1\left({}^1\Sigma_g^+\right) \tag{4.2.10}$$

in triatomic molecules of the C_{2v} point group (see Appendix IV in [10]). It is easy to show that any of vibration of triatomic molecules of the $D_{\infty h}$ point group (Σ_g^+, Σ_u^+, Π_u) cannot allow transition (4.2.10) in the $D_{\infty h}$ point group:

$$\Sigma_u^- \times (\Sigma_g^+ + \Sigma_u^+ + \Pi_u) = \Sigma_u^- + \Sigma_g^- + \Pi_g, \tag{4.2.11}$$

$$\Delta_u \times (\Sigma_g^+ + \Sigma_u^+ + \Pi_u) = \Delta_u + \Delta_g + \Pi_g + \Phi_g. \tag{4.2.12}$$

The ${}^1B_2 \leftarrow {}^1A_1$ transition is allowed in the C_{2v} point group (see Table 9 in [10], but electric dipole transitions between Σ_g^+ state and those of *rhp* of (4.2.11, 4.2.12) are forbidden. Authors of [13] believe that weak absorption at the *A*-band of CO_2 (λ

$\approx$ 140–167 nm) is due to this feature. The CO_2 ground state is linear ($^1\Sigma_g^+$, $D_{\infty h}$) and first excited are bent, C_{2v} ($^1A_2$, $^1B_2$ ($^1\Sigma_u^-$ and $^1\Delta_u$ in the linear configuration). Molecule CO_2 'remember' that it is linear in the Franck–Condon zone, and vibronic transitions $^1\Sigma_u^-$, $^1\Delta_u \leftarrow {}^1\Sigma_g^+$ are forbidden. Therefore, the absorption is weak, though the $^1B_2 \leftarrow {}^1A_1$ transition is allowed.

Often, electronic and vibrational wave functions are totally symmetric for ground electronic states, and *vibronic transitions* (transitions allowed due to vibronic interactions) [13] can occur from higher vibrational antisymmetric vibrational states, which are not populated at low temperature.

4.2.2.2 Transitions in Polyatomic Molecules Allowed Due to Electronic-Rotational Interactions

In general, electronic-rotational (Coriolis) perturbation is very weak. However, if no other stronger perturbations exist, the Coriolis interaction can lead to weak transitions. It occurs if a neighboring state has species of rotational wave functions different from those of the states that cannot combine in an allowed transition. This interaction increases with quantum numbers of angular momenta *J*, *K* strongly [10], p. 141.

4.2.2.3 Magnetic-Dipole and Electric-Quadrupole Transitions

Transition probabilities and selection rules for magnetic-dipole and electric-quadrupole transitions are obtained if magnetic dipole moment $\hat{\mu}(R_x, R_y, R_z)$ and electric-quadrupole $\hat{Q}(x^2, y^2, z^2; xy, xz, yz)$ operators are substituted in place of the electric dipole transition operator in (4.2.3). One can find the selection rules of magnetic-dipole and electric-quadrupole transitions for some most important point groups in Table 10 of [10]. Intensities of magnetic-dipole and electric-quadrupole transitions are $\sim 10^5$ and $\sim 10^8$ than those of strong electric dipole transitions in the visible range [10], p. 134.

4.2.3 *Spin–Orbit Coupling and Spin-Forbidden Transitions*

If the spin-orbit interaction and splitting are not negligible, (4.2.6, 4.2.7) are invalid, and the total electronic wave function is more complicated:

$$\Psi_{es} = \Phi \cdot \sigma + \xi_{es}, \tag{4.2.13}$$

where ξ_{es} depends on spin and space coordinates. However, ξ_{es} species has to be the same as that of $\Phi \cdot \sigma$ since Ψ_{es} has to belong to a species of the point group concerned. Therefore, states with total spin quantum numbers different by 1

$$\Delta S = \pm 1 \tag{4.2.14}$$

are mixed, and their 'true' wave functions are superpositions of zero-order wave functions. The 'true' wave function of a triplet (in terms of Λ-*s* coupling) state is:

$$^3\Psi^r_{es} = {}^3\Psi^{0r}_{es} + \sum\nolimits_k \frac{\left\langle {}^1\Psi^0_{esk} | \hat{\boldsymbol{H}}_{SO} | {}^3\Psi^0_{es} \right\rangle}{E^0_{3k} - E^0_{1k}} \cdot {}^1\Psi^0_{esk} \tag{4.2.15}$$

and singlet state is:

$$^1\Psi_{es} = {}^1\Psi^0_{es} + \sum\nolimits_m \sum\nolimits_{r=-1}^{1} \frac{\left\langle {}^3\Psi^{0r}_{esm} | \hat{\boldsymbol{H}}_{SO} | {}^1\Psi^0_{es} \right\rangle}{E^0_1 - E^0_{3m}} \cdot {}^1\Psi^{0r}_{esm}. \tag{4.2.16}$$

Here, ${}^1\Psi^0_{esk}$, ${}^3\Psi^{0r}_{esm}$ are total electronic wave functions of unperturbed singlet and triplet states (r = −1, 0, 1), E^0_{1k} and E^0_{3m} are their energies, $\hat{\boldsymbol{H}}_{SO}$ is the operator of the spin-orbit interaction. The summation is performed on *k*-th singlet and *m*-th triplet states [14], p. 199. Spin–orbit interaction and spin–orbit splitting E_{SO} (denominators in (4.2.15, 4.2.16)) depend strongly on a nuclear charge *Z* and a distance between electrons and heavyweight nucleus. For triplet state hydrogen atom, $E_{SO} \sim Z^4/n^3$ (*n* is the principal quantum number).

It is essential that regardless of the magnitude of the spin–orbit interaction ((4.2.6) or (4.2.13) is valid), states with the same species of the Ψ_{es} are mixed, that is the correction terms in (4.2.15, 4.2.16) are nonzero, if ${}^1\Psi^{0r}_{esk}$ and ${}^3\Psi^{0r}_{es}$ or ${}^3\Psi^{0r}_{esm}$ and ${}^1\Psi^0_{es}$ have the same species of the point group under discussion. For (4.2.15), it means that at least one of the components of the direct product of the species electronic orbital and spin wave functions for the triplet state *r*-component has to be the same as at least one species of the singlet state. This feature means that in the decomposition of the direct product of the species of the total electronic wave function of the triplet and at least one of the singlet states, there must be a totally symmetric irreducible representation:

$$\Gamma\left({}^1\Psi^0_{esk}\right) \times \Gamma\left({}^3\Phi^0\right) \times \Gamma\left({}^3\sigma^0\right) = \Gamma_1 + \ldots \tag{4.2.17}$$

(the species of spin functions of singlet states is totally symmetric, $\Gamma({}^1\sigma^0) = \Gamma_1$). The same result can be obtained for (4.2.16):

$$\Gamma\left({}^{1}\Phi^{0}\right) \times \Gamma\left({}^{3}\Phi_{m}^{0}\right) \times \Gamma\left({}^{3}\sigma^{0}\right) = \Gamma_{1} + \ldots \tag{4.2.18}$$

All this is a consequence of the total symmetry of the $\widehat{\boldsymbol{H}}_{SO}$ operator at symmetry point groups of a molecule, including both spatial and spin coordinates.

The classification of electron motion in a molecule according to symmetry types taking into account the spin–orbit interaction depends on the geometry of the molecule, nuclear charge, etc. Therefore, for theoretical group analysis of (4.2.15–4.2.18), it is necessary to understand the bond classification in Hund's case (*c*). These concepts for diatomic molecules are developed in the Mulliken's works [15–19] (see [20–22] and references, also).

As is known, if there is a spin–orbit interaction in one of the atoms of a diatomic molecule, the axial electric field may not break the connection between the orbital moment of the electron $\boldsymbol{l}$ and its spin moment $\boldsymbol{s}$. In this case, the total electron moment $\boldsymbol{j} = \boldsymbol{l} + \boldsymbol{s}$ precesses around a molecule axis $\boldsymbol{Z}$, and only the quantum number corresponding to the projection of the total moment $\boldsymbol{J} = \sum_i \boldsymbol{j}_i$ on the axis is 'true'. The projections of the total orbital moment $\boldsymbol{L} = \sum_i \boldsymbol{l}_i$, and the spin $\boldsymbol{S} = \sum_i \boldsymbol{s}_i$ of a molecule lose their meaning. The corresponding quantum numbers *L*, *S* remove the meaning, also. This rough description of the coupling of moments in a diatomic molecule in the presence of a strong spin–orbit interaction corresponds to the Hund (*c*) case [23], p. 224.

'Close nuclei' case (c). As a result of the nuclei proximity, the axial component of the electric field is small, the precession of the orbital momentum about the *Z* axis is weak. In this case, *L*, *S*, J_a (*J* for the atom) as well Ω are good quantum numbers (although the first three are not quite so). This is a rarely implemented case (see [15, 17] for details).

Ω-ω coupling. If a molecule can be represented as a charged core characterized by the quantum numbers Λ_c, Σ_c, and Ω_c with a fairly distant electron (so that the interaction is weak), the state of the molecule may be described by analyzing the projections of the orbital and spin momenta of the core and this electron on the internuclear axis. For example, the electron configuration of the *R′*I diatomic molecules, $R' = \mathrm{H}(^2\mathrm{S})$, $\mathrm{Hal}(^2\mathrm{P})$, takes the form [16]:

$$R'\mathrm{I}\left[\left(\sigma^2\pi_{R'}^4\pi_1^2\right)^2\Pi_{3/2}2\sigma^*\right]_{2,1} \tag{4.2.19}$$

$$R'\mathrm{I}\left[\left(\sigma^2\pi_{R'}^4\pi_1^3\right)^2\Pi_{1/2}\sigma^*\right]_{0,1}. \tag{4.2.20}$$

In this case, the iodine molecule electronic configuration is:

$$\mathrm{I}_2\left[\left(\sigma_g^2\pi_u^3\pi_g^4\right)^2\Pi_{3/2}\sigma_g^*\right]_{2,1\mathrm{u}} \tag{4.2.21}$$

$$\mathrm{I}_2\left[\left(\sigma_g^2\pi_u^3\pi_g^4\right)^2\Pi_{1/2}\sigma_g^*\right]_{0,1\mathrm{u}.} \tag{4.2.22}$$

(the configurations of the $R'I^+$ ion is given in parentheses). The $R'I$ state is then characterized by the quantum numbers S_c, Λ_c, Σ_c, Ω_c, $\Omega = \Omega_c \pm 1/2$, the spin quantum number is meaningless. The molecular states having the same quantum numbers $\Lambda = \Lambda_c$, Ω, the same parity, g, u, and the same properties for reflection through a plane containing the molecular axes, + , -, (for $\Omega = 0$ state, only) have the same symmetry types, and are mixed and repelled. In other words, the 'true', i.e., obtained to a reasonably good approximation, wave functions of states belonging to the same species are the linear combination of 'old' unperturbed states. This approximation particularly applies to the $^3\Pi_1$ and $^1\Pi$ states (in terms of Λ-s coupling) of heteroatomic molecules. Thus, compared with the Λ-s coupling, the intensity of the $^1\Pi \leftarrow X^1\Sigma^+$ transition decreases and that of the $^3\Pi_1 \leftarrow X^1\Sigma^+$ transition increases. Another essential factor, in this case, is that the $R'I$ states described by the configuration (4.2.19–4.2.22) form pairs of doublets $^3\Pi_0$, $^1\Pi$ and $^3\Pi_1$, $^3\Pi_2$ (in terms of Λ-s coupling) with an energy difference between these pairs of the order of the spin-orbit interaction energy E_{SO}, which in this case is equal to ΔE ($^2\Pi_{3/2}$-$^2\Pi_{1/2}$). The splitting in the doublets is of the order of $\Delta E(^3\Pi$-$^1\Pi)$, which in this case, is substantially lower than E_{SO} (Fig. 4.2) (see [15, 16, 18, 21] for details).

Mulliken has shown that the coupling type and relative positions of the first excited states of the HI molecule should be described using the Ω-ω coupling, possibly with certain addition of the case (c), type I [17, 18] (see below). Thus, the HI molecule first excited states have to form two groups of doublets, $^3\Pi_0$, $^1\Pi$ and

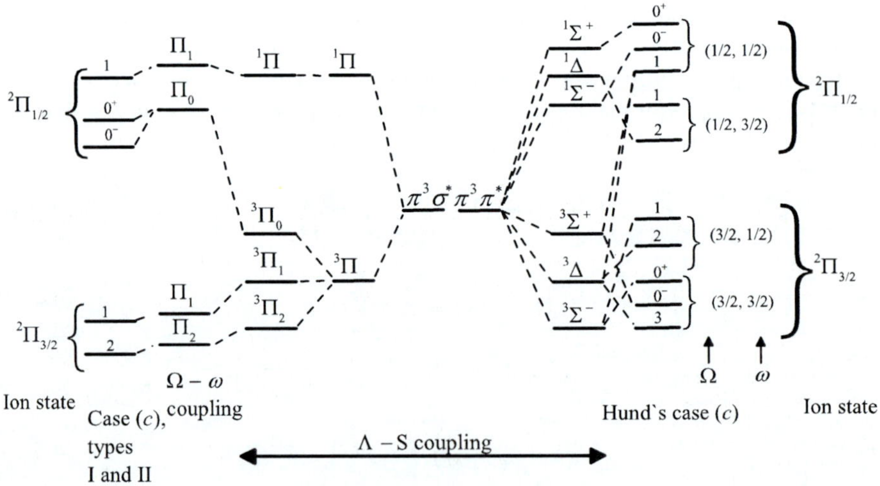

Fig. 4.2 Relative positions of molecular energy levels of the $\pi^3\sigma^*$ and $\pi^3\pi^*$ configurations ($C_{\infty v}$ point group) [3], p. 26 (see [22, 24, 25], also). The positions of the levels are approximate. The spin–orbit interaction increases from the center towards the edges. In the C_s point group, the 0^+ state corresponds to the A' state, and 1, 2, 3 (Π, Δ) states are split into the A', A'' states

$^3\Pi_1$, $^3\Pi_2$ (in terms of Λ-*s* coupling) (see left part of Fig. 4.2). The HI($\sigma^2\pi^3\sigma^{*1}\Pi$, $^3\Pi_1 \leftarrow \sigma^2\pi^4 X^1\Sigma^+$) transitions have to be observed, and the intensity of that of the former state has to be 'transferred' to the ($^3\Pi_1 \leftarrow X^1\Sigma^+$) one. If additional (*c*), type I coupling exist, the $^3\Pi_0^+ \leftarrow X^1\Sigma^+$ ($0^+ \leftarrow 0^+$ in terms of (*c*) coupling) should be observed as a result of the mixing of both states involved in transitions with higher 0^+ states. Mulliken's predictions have been partially confirmed in experiments. All three transitions were discovered, though relative strengths of the transitions and splittings of the $^1\Pi$, $^3\Pi_0^+$ and $^3\Pi_1$ curves do not seem to be consistent with the strong case (*c*), type I effect. The HI($^1\Pi$, $^3\Pi_0^+$ and $^3\Pi_1 \leftarrow X^1\Sigma^+$) transition probabilities are $|\mu_e|^2 = 0.126$, 0.160 and 0.038, inits, D^2 [26]. Note that transition probability to the 'triplet' $^3\Pi_0^+$ state is the largest. (Fig. 4.3).

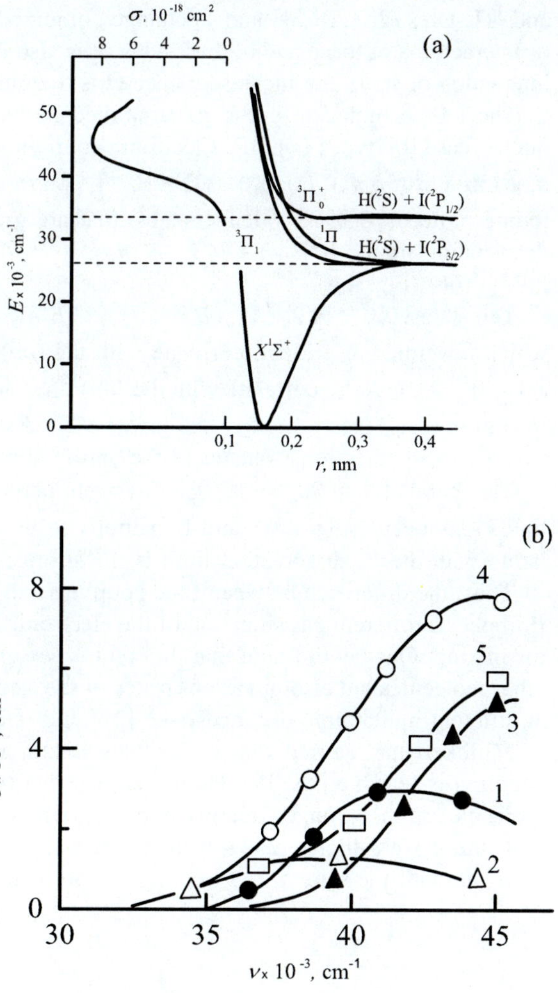

Fig. 4.3 Calculated *PEC*s of some valence state of HI molecule **a**; total (4) and partial cross-sections of HI corresponding to transitions to the $^3\Pi_0^+$ (1), $^3\Pi_1$ (2), $^1\Pi_3$ (3) states, sum of partial cross-sections (1–3) (5) **b** (see [22] and references)

'Far nuclei' case (c), types I and II. If the interatomic distance in a molecule is fairly large, and the dissociation energy is low (iodine molecule, for example), the spin–orbit couplings in the atoms may be conserved despite of axial interatomic field. As a result, the total momentum of an atom J_a does not become meaningless. In this case, the molecule electronic state is characterized by the quantum number Ω, only (Ω is the sum of the J_{a_1}, J_{a_2} projections onto the Z axis). The Λ, S, Σ quantum numbers become meaningless. The symmetry properties (g/u; $\pm$) are conserved. Mulliken subdivides case (c) into (c), type I and (c), type II cases. In the first case, states having the same species belonging to different *molecular* configurations but formed from the same atomic configurations or having the same species in terms of Λ-s coupling, interact and mix. In the second case, states formed from different atomic configurations and having different species in terms of Λ-s coupling may interact. In the case (c), types I and II couplings, the $^3\Pi_0$ state splits to the $^3\Pi_0^+$ and $^3\Pi_0^-$ (0^+ and 0^-). The relative positions of the $^3\Pi_2$, $^3\Pi_1$, $^3\Pi_0^+$, $^3\Pi_0^-$ and $^1\Pi$ states (2, 1, 0^+, 0^- and 1) change compared to the Ω-ω coupling as a result of interactions of these and higher states. It is also essential that possibilities for the interaction of states are increased since Ω is the only meaningful quantum number.

The iodine molecule is the most studied molecule described in the terms 'far nuclei' case (c), types I and II. Electronic configuration of the I_2 ground state is $(5p\,\sigma_g)^2\,(5p\,\pi_u)^4\,(5p\,\pi_g)^4\,(5p\,\sigma_u^*)^0\,X^1\Sigma_g^+\,(0_g^+)$, common designation is 2440 $X0_g^+$. The iodine molecule has 23 valence states, that are grouped by correlation with three dissociation limits: $I(^2P_{3/2}) + I(^2P_{3/2})$ (aa), $I(^2P_{3/2}) + I(^2P_{1/2})$ (ab) and $I(^2P_{1/2}) + I(^2P_{1/2})$ (bb) (Fig. 4.4).

Ten states, $X0_g^+$, $A'2_u$, $A1_u$, $B'0_u^-$, $a1_g$, $C(B'')1_u$, $a'0_g^+$, 2_g, as well as 3_u, $(2)0_u^-$, which unstudied at present, correlate with the (aa) limit. Ten states $B0_u^+$, 0_g^+, $c1_g$, $c'1_g$, 0_g^-, 2_u, $(3,4)1_u$, correlate with the (ab) limit and three, 0_g^+, 1_u and 0_u^-, with the (bb) limit (see Figs. 4.4, 4.5). The lower excited states are populated when one or two σ_g, π_u, π_g electrons transfer to the antibonding σ_u^* orbital.

The parallel I_2 ($B0_u^+ \leftarrow X0_g^+$) as well perpendicular I_2($A1_u$, $C1_u \leftarrow X0_g^+$) transitions occur (Fig. 4.6), and transitions to the 'triplet' 2431 $B0_u^+$ state correlating with the ab dissociation limit is the strongest.

Thus, the difference between Ω-ω coupling and 'far nuclei' case (c), types I and II implies a different classification of the electronic states and different forbiddances for mixing. One should note that different types of coupling can occur not only in other molecules but also in various states of the same molecule or the same state but at different interatomic distances (see [22] and references).

Mulliken has carried out an analysis assuming that case (c), types I and II coupling is feasible [18, 19]. He has shown that transitions to the 2431 state of I_2 molecule are the strongest mainly due to mixing of the 'singlet' ground 2440 state with the 4.1 eV higher 2441 'triplet' state (c), types I mixing since both states correlate with the same $p^5 \cdot p^5$ configuration of iodine atoms. To a certain extent,

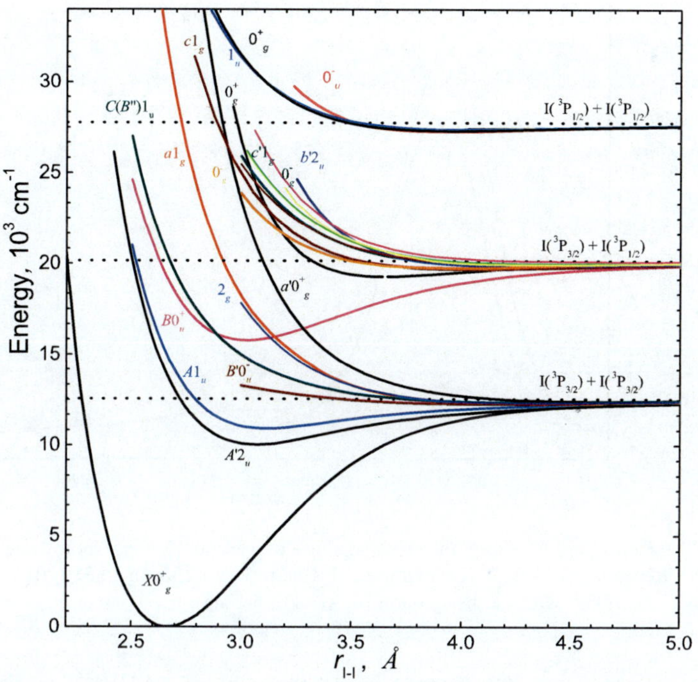

Fig. 4.4 Potential energy curves of the valence iodine molecule states [7], p. 26

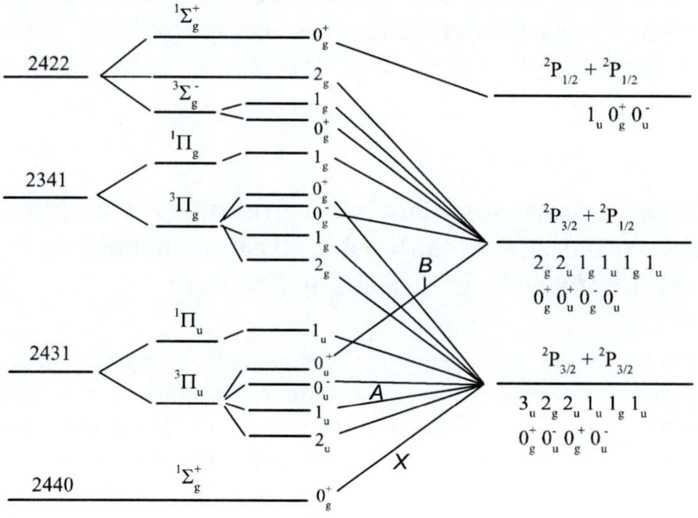

Fig. 4.5 Correlation diagram for the low-lying halogen states with their dissociation products (see [27])

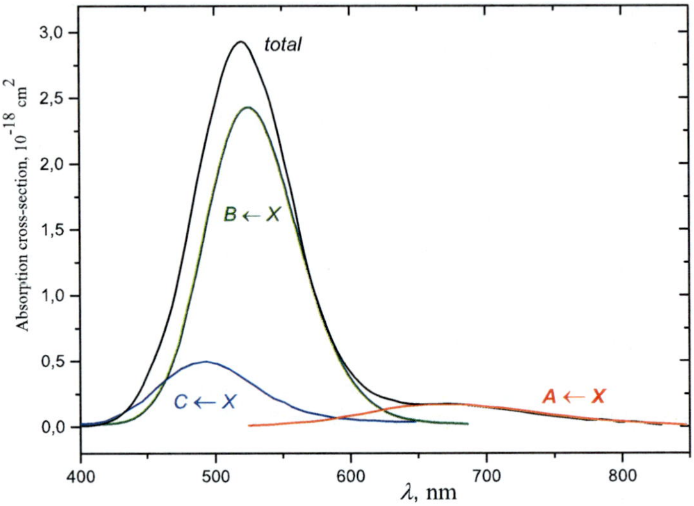

Fig. 4.6 Total and partial absorption cross-sections of gaseous I_2 near room temperature [7], p. 27, [28]. (Reproduced from J. Tellinghuisen, J. Chem. Phys. **135**, 054,301 (2011). https://doi.org/10.1063/1.3616039, with the permission of AIP Publishing).

this effect is also due to the contribution of the 'singlet' 1441 0_u^+ ion-pair state to the 2431 $B0_u^+$ state via 1342 0_u^+ one, i.e., the 1441 0_u^+ 'singlet' state mixes with the 1342 0_u^+ one, and this 'new' state mixes with the 2431 $B0_u^+$ state. This coupling is the (*c*), type II one since the 2431 $B0_u^+$ and 1441 0_u^+ states are formed from different configurations of iodine atoms: the 1342 0_u^+ and 1441 0_u^+ states correlate with the $I^+ + I^-$ ions, $p^4 \cdot p^6$ configuration (see [29, 30] and references, also).

4.3 Absorption. Absorption Band Intensities. Einstein Absorption and Stimulated Emission Coefficients. Beer-Lambert Law. Oscillator Strength

The number of photons with the energy $h\nu$, equal to the energy gap between lower, m, and upper, n, state absorbed per unit time per unit volume (the *absorbed radiation density* $n_{h\nu}^a(\lambda)$, see 4.1.18) in the transition is proportional to the population of the m state, N_m, [cm^{-3}] and the number of photons per unit volume, $\rho_{nm}^{h\nu}$ [cm^{-3}]:

$$n_{h\nu}^a(\lambda) \sim N_m \rho_{nm}^{h\nu}. \tag{4.3.1}$$

The probability of the transition [s^{-1}] is equal to:

$$\frac{8\pi^3}{3h^2c}\left|\mu_e^{nm}\right|^2\rho_{nm} = B_{mn}\rho_{nm}, \tag{4.3.2}$$

μ_e^{nm} is the electronic transition moment (4.2.2), ρ_{nm} [erg · cm^{-2}] is *energy density* (energy per unit volume per unit wavenumber) and B_{mn} [erg^{-1} · cm^2 · s^{-1}] is the *Einstein absorption coefficient* [9], p. 23, [31], p. 349. If one summarizes all rovibronic component of the transitions from a given level $|m, v'', J'\rangle$, and the lower state is degenerate, then

$$B_{mn} = \sum\nolimits_{v',J'} B_{mn,v''v',J''J'} = \frac{8\pi^3}{3h^2cg_m}\left|\mu_e^{nm}\right|^2, \tag{4.3.3}$$

g_m is the lower state degeneracy [10], p. 417. For bound-free transitions (transitions to a continuous range of levels) leading to dissociation, the summation has to be replaced by appropriate integrals.

A transition dipole moment depends on the nuclear coordinates. In diatomic molecules, this dependence is expressed as a power function of the internuclear distance:

$$\mu_e^{nm}(R) = \sum\nolimits_{k=0}^{n} q_k R^k,$$

which, with some approximations, leads to the dependence:

$$\begin{aligned}\left|\mu^{nm,v'v''}\right|^2 &= \langle\chi_{v'}|\chi_{v''}\rangle^2\left(a_0 + a_1\frac{\chi_{v'}|R|\chi_{v''}}{\langle\chi_{v'}|\chi_{v''}\rangle}\right)^2 = \langle\chi_{v'}|\chi_{v''}\rangle^2\left(a_0 + a_1\overline{R}_{v'v''}\right)^2 \\ &= \left|\mu^{nm,v'v''}\left(\overline{R}_{v'v''}\right)\right|^2\langle\chi_{v'}|\chi_{v''}\rangle^2,\end{aligned} \tag{4.3.4}$$

$\overline{R}_{v'v''}$ is *R-centroid* (the *R*-centroid approximation has to be valid, see [31–33] and references), and $\langle\chi_{v'}|\chi_{v''}\rangle^2$ is *Franck-Condon factor*, $q(v', v'')$. The $q(v', v'')$ sum for all v' (absorption) or v'' (spontaneous emission) including continuum, obeys to sum rule

$$\sum\nolimits_{v'} q(v', v'') + \int \left|\langle\varepsilon'|v''\rangle\right|^2 d\varepsilon' = 1 \tag{4.3.5a}$$

$$\sum\nolimits_{v''} q(v', v'') + \int \left|\langle\varepsilon''|v'\rangle\right|^2 d\varepsilon'' = 1, \tag{4.3.5b}$$

here ε', ε'' are *free level energies*, and $|\langle\varepsilon'|v''\rangle|^2$, $|\langle\varepsilon''|v'\rangle|^2$ are *Franck-Condon density* [33] (see Sect. 4.5).

An absorption is characterized by the *absorption coefficient*, $k(\tilde{\nu})$ $(k(\lambda))$ [cm^{-1} · atm^{-1}] at $T = 273$ K, *molar absorption (extinction) coefficient*, $\varepsilon(\tilde{\nu})$ $(\varepsilon(\lambda))$[liter ·

mol^{-1} · cm^{-1}], *absorption cross-section*, $\sigma(\widetilde{\nu})$ $(\sigma(\lambda))$ [cm^2], corresponding to different descriptions of *Beer-Lambert law*:

$$J(\widetilde{\nu},x) = J_0(\widetilde{\nu}) \cdot exp\left\{-\frac{k(\widetilde{\nu}) \cdot x \cdot p \cdot 273}{760 \cdot T}\right\} \tag{4.3.6}$$

$$J(\widetilde{\nu},x) = J_0(\widetilde{\nu})10^{\left\{-\frac{k_{10}(\widetilde{\nu}) \cdot x \cdot p \cdot 273}{760 \cdot T}\right\}} \tag{4.3.7}$$

$$J(\widetilde{\nu},x) = J_0(\widetilde{\nu})10^{\{-\varepsilon(\widetilde{\nu}) \cdot c \cdot x\}} \tag{4.3.8}$$

$$J(\widetilde{\nu},x) = J_0(\widetilde{\nu})exp\{-\sigma(\widetilde{\nu}) \cdot N \cdot x\}. \tag{4.3.9}$$

Here,

$J_0(\widetilde{\nu})$ (or $J_0(\lambda)$ is the incident monochromatic light intensity at wavenumber $\overline{\nu}$ (wavelength λ),

$J(\widetilde{\nu},x)$ $(J(\lambda,x)$ is the intensity after a light beam has traversed a distance x [cm] in the gas medium,

p is the gas pressure [Torr],

$k_{10}(\widetilde{\nu})$ $(k_{10}(\lambda))$ is the absorption coefficient for the decimal base,

N is the concentration (number of molecules per unit volume) [cm^{-3}].

Einstein stimulated emission coefficient, B_{nm} is defined as

$$B_{nm} = |\mu_e^{nm}|^2/6\hbar^2\varepsilon_0 c \tag{4.3.10}$$

where ε_0 is vacuum permittivity, and $B_{nm} \cdot \rho(\widetilde{\nu}_{nm})$ is the $n \rightarrow m$ transition rate for a single molecule stimulated by a radiation field with energy density $\rho(\widetilde{\nu}_{nm})$ with $\widetilde{\nu}_{nm}$ centered at absorption line center, $\widetilde{\nu}^0_{nm}$. The Einstein absorption and stimulated emission coefficients are related as:

$$B_{nm} = \frac{g_n}{g_m}B_{mn}, \tag{4.3.11}$$

where g_n and g_m are total degeneracies of the n and m levels including the 2 J +1 M_J-degeneracy.

Integrated over the whole $\widetilde{\nu}_1 - \widetilde{\nu}_2$ band absorption coefficients [10], p. 418

$$\int_{\widetilde{\nu}_1}^{\widetilde{\nu}_2} k(\widetilde{\nu})d\widetilde{\nu} = N_m \cdot B_{mn} \cdot h\widetilde{\nu}_{nm} = \frac{8\pi^3 \cdot \widetilde{\nu}_{nm}}{3hc} \cdot N_m \cdot |\mu_e^{nm}|^2 \tag{4.3.12}$$

and so-called dimensionless *oscillator strength*

$$f_{nm} = \left|\mu_e^{nm}\right|^2 \cdot \widetilde{\nu}_{nm} \frac{4\pi \cdot m_e \cdot c}{3\hbar e^2} \tag{4.3.13}$$

(m_e and e are the mass and charge of the electron) are also used to characterize intensities of transitions.

The relations between quantities under discussion are as follows [9], p. 26, [10], p. 418, [31], p. 350]:

$$f_{nm} = 4.20 \cdot 10^{-8} \int_{\widetilde{\nu}_1}^{\widetilde{\nu}_2} k(\widetilde{\nu}) d\widetilde{\nu} \tag{4.3.14}$$

$$f_{nm} = \frac{m_e \cdot h \cdot c^2 \cdot \widetilde{\nu}_{nm}}{\pi \cdot e^2} \cdot B_{mn} \tag{4.3.15}$$

$$f_{nm} = 4.703 \cdot 10^{-7} \cdot \widetilde{\nu}_{nm} \cdot \left|\mu_e^{nm}\right|^2 \tag{4.3.16}$$

for μ_e^{nm} in Debye.

There are relations between cross-section for the absorption line center $\sigma_{nm}\left(\widetilde{\nu}_{nm}^0\right)$ [cm^2], selected wavenumber $\widetilde{\nu}$ remote the absorption line center, $\sigma_{nm}(\widetilde{\nu}_{nm})$ [cm^2] and cross-section integrated over the absorption line $\int_{\nu_1}^{\nu_2} \sigma_{nm}\left(\widetilde{\nu}_{nm} - \widetilde{\nu}_{nm}^0\right) d\widetilde{\nu}$ [cm^2 · cm^{-1}] [31], p. 351. They are different for Lorentzian

$$L\left(\widetilde{\nu}_{nm}, \widetilde{\nu}_{nm}^0, \Delta\widetilde{\nu}_{nm}\right) = \frac{1}{\pi} \cdot \frac{(\Delta\widetilde{\nu}_{nm}/2)}{(\Delta\widetilde{\nu}_{nm}/2)^2 + \left(\widetilde{\nu}_{nm} - \widetilde{\nu}_{nm}^0\right)^2} \tag{4.3.17}$$

and Gaussian

$$G\left(\widetilde{\nu}_{nm}, \widetilde{\nu}_{nm}^0, \Delta\widetilde{\nu}_{nm}\right) = \left(\frac{4ln2}{\pi}\right)^{\frac{1}{2}} \cdot \frac{1}{\Delta\widetilde{\nu}_{nm}} \cdot \exp\left[-\left(\widetilde{\nu}_{nm} - \widetilde{\nu}_{nm}^0\right)^2 \cdot \frac{4ln2}{\left(\Delta\widetilde{\nu}_{nm}\right)^2}\right] \tag{4.3.18}$$

lineshapes ($\Delta\widetilde{\nu}_{nm}$ is FWHM of the line). Both $L\left(\widetilde{\nu}_{nm}, \widetilde{\nu}_{nm}^0, \Delta\widetilde{\nu}_{nm}\right)$, $G\left(\widetilde{\nu}_{nm}, \widetilde{\nu}_{nm}^0, \Delta\widetilde{\nu}_{nm}\right)$ are normalized to 1:

$$\int_{\nu_1}^{\nu_2} L\left(\widetilde{\nu}_{nm}, \widetilde{\nu}_{nm}^0, \Delta\widetilde{\nu}_{nm}\right) d\widetilde{\nu} = 1 \tag{4.3.19}$$

$$\int_{\nu_1}^{\nu_2} G\left(\widetilde{\nu}_{nm}, \widetilde{\nu}_{nm}^0, \Delta\widetilde{\nu}_{nm}\right) d\widetilde{\nu} = 1. \tag{4.3.20}$$

The cross-section for a selected wavenumber $\sigma_{nm}(\widetilde{\nu}_{nm})$ and the absorption line center $\sigma_{nm}\left(\widetilde{\nu}^0_{nm}\right)$ [cm^2] are

$$\sigma_{nm}(\widetilde{\nu}_{nm}) = \int_{\nu_1}^{\nu_2} \sigma_{nm}(\widetilde{\nu}_{nm})d\widetilde{\nu} \cdot L\left(\widetilde{\nu}_{nm}, \widetilde{\nu}^0_{nm}, \Delta\widetilde{\nu}_{nm}\right) \tag{4.3.21}$$

$$\sigma_{nm}\left(\widetilde{\nu}^0_{nm}\right) = \int_{\nu_1}^{\nu_2} \sigma_{nm}(\widetilde{\nu}_{nm})d\widetilde{\nu} \cdot \frac{2}{\pi \cdot \Delta\widetilde{\nu}_{nm}} \tag{4.3.22}$$

for a Lorentzian lineshape, and

$$\sigma_{nm}(\widetilde{\nu}_{nm}) = \int_{\nu_1}^{\nu_2} \sigma_{nm}(\widetilde{\nu}_{nm})d\widetilde{\nu} \cdot G\left(\widetilde{\nu}_{nm}, \widetilde{\nu}^0_{nm}, \Delta\widetilde{\nu}_{nm}\right) \tag{4.3.23}$$

$$\sigma_{nm}\left(\widetilde{\nu}^0_{nm}\right) = \int_{\nu_1}^{\nu_2} \sigma_{nm}(\widetilde{\nu}_{nm})d\widetilde{\nu} \cdot \left(\frac{4ln2}{\pi}\right)^{\frac{1}{2}} \cdot \frac{1}{\Delta\widetilde{\nu}_{nm}} \tag{4.3.24}$$

for a Gaussian lineshape.

Since cross-section integrated over the absorption line is [31], p. 351:

$$\int_{\nu_1}^{\nu_2} \sigma_{nm}\left(\widetilde{\nu}_{nm} - \widetilde{\nu}^0_{nm}\right)d\widetilde{\nu} = 4.166 \cdot 10^{-19} \cdot \widetilde{\nu}_{nm} \cdot \left|\mu_e^{nm}\right|^2 \tag{4.3.25}$$

for R_e^{nm} in Debye, the $\sigma_{nm}\left(\widetilde{\nu}^0_{nm}\right)$ is

$$\sigma_{nm}\left(\widetilde{\nu}^0_{nm}\right) = 2.652 \cdot 10^{-19} \cdot \frac{\widetilde{\nu}_{nm}}{\Delta\widetilde{\nu}_{nm}} \cdot \left|\mu_e^{nm}\right|^2 \tag{4.3.26}$$

for a Lorentzian lineshape, and

$$\sigma_{nm}\left(\widetilde{\nu}^0_{nm}\right) = 3.914 \cdot 10^{-19} \cdot \frac{\widetilde{\nu}_{nm}}{\Delta\widetilde{\nu}_{nm}} \cdot \left|\mu_e^{nm}\right|^2. \tag{4.3.27}$$

for a Gaussian lineshape. A Lorentzian lineshape is used when collision broadening occurs. A Lorentzian FWHM is equal to [9], p. 30

$$\Delta\widetilde{\nu}_L = \frac{Z_L}{c \cdot \pi}, \tag{4.3.28}$$

$Z_L = 3.54 \cdot 10^{16} \cdot p_M$ (Torr) is a collision number with species M at T = 273 K.

A Gaussian lineshape is used for the description of the Doppler contour, and Doppler FWHM is

$$\Delta\widetilde{\nu}_D = 7.16 \cdot 10^{-7} \cdot \widetilde{\nu}_0 \cdot \sqrt{\frac{T}{m}}. \tag{4.3.29}$$

(m is the mass of the species in *amu*).

4.4 Luminescence. Radiative Lifetime. Einstein Spontaneous Emission Coefficient

If in an electronically excited state n there are N_n species in a unit volume, then the intensity of the transition to a lower state m is

$$N_n\{A_{nm} + B_{nm} \cdot \rho_{nm}\}, \tag{4.4.1}$$

where

$$A_{nm} = \frac{64\pi^4 \widetilde{\nu}^3_{nm}}{3hg_n} \cdot \left|\mu_e^{nm}\right|^2 = 8\pi hc\widetilde{\nu}^3_{nm} \cdot B_{nm}. \tag{4.4.2}$$

is *Einstein spontaneous emission coefficient* [s^{-1}], and B_{nm} is Einstein stimulated emission coefficient (see (4.3.11 and [9], p. 24). For R_e^{nm} in Debye [31], p. 349,

$$A_{nm} = 3.137 \cdot 10^{-7}\left|\mu_e^{nm}\right|^2 \cdot \widetilde{\nu}^3_{nm} \tag{4.4.3}$$

If the molecules are not oriented in space, then the spontaneous emission is completely spatially isotropic; stimulated radiation has the same direction as that of the photon flux, which excites it. If $N_n << N_m$, stimulated emission can be neglected, and the N_n, A_{nm} quantities characterize the $n \rightarrow$ m transition intensity. The sum of Einstein spontaneous emission coefficient for the transitions from the n state to all lower states m is equal to the reverse n state radiative lifetime:

$$A_n = \sum\nolimits_m A_{nm} = 1/\tau^n_{rad}. \tag{4.4.4}$$

For a molecule, a level n is vibronic or rovibronic one. The concept of radiative lifetime is frequently used for an electronically excited state, i.e., for all rovibronic levels, although A_{nm} value depends on vibrational quantum numbers (see 4.3.4).

Generally, the kinetics and absolute quantum yield of luminescence (see Sect. 4.1) depend on the type and concentration of species M (the third bodies), the presence of walls, and the interaction of the radiating state with others, isoenergetic with it. In the absence of collisional processes, dissociation and predissociation, as well as internal conversion, the luminescence quantum yield of a small, 2–3-atom molecule AB in the state AB*, $\Phi_{lum}^{AB^*}$, is always equal to unity, and the luminescence intensity is described by the formula:

$$J_{lum}^{AB^*} = J_0 \exp\left(-\frac{t}{\tau_{rad}^{AB^*}}\right) \quad (4.4.5)$$

If the luminescence competes with dissociation or/and predissociation, then

$$J_{lum}^{AB^*} = J_0 \exp\left(-\frac{t}{\tau_{AB^*}}\right), \quad (4.4.6)$$

and

$$\Phi_{lum}^{AB^*} = \tau_{AB^*}/\tau_{rad}^{AB^*} < 1, \quad (4.4.7)$$

where reverse AB* lifetime is

$$\tau_{AB^*}^{-1} = \left(\tau_{rad}^{AB^*}\right)^{-1} + \left(\tau_{diss}^{AB^*}\right)^{-1} + \left(\tau_{pred}^{AB^*}\right)^{-1} \quad (4.4.8)$$

(here, $\tau_{diss}^{AB^*}$ and $\tau_{pred}^{AB^*}$ are the lifetimes corresponding to AB* dissociation and predissociation). The internal conversion can change luminescence kinetics very strongly and lead to the quantum beat phenomenon, in particular [34]. Nevertheless, the internal conversion cannot change the luminescence quantum yield of small molecules.

One determines the oscillator strength of the transition using Einstein spontaneous emission coefficient and (4.4.9)

$$f_{nm} = \frac{m_e \cdot c}{8\pi^2 e^2 \widetilde{v}_{nm}^2} \cdot \frac{1}{\tau_{nm}} \cdot \frac{g_n}{g_m} = \frac{1.5}{\widetilde{v}_{nm}^2} \cdot \frac{1}{\tau_{nm}} \cdot \frac{g_n}{g_m} \quad (4.4.9)$$

And finally, one determines $k\left(\widetilde{v}_{nm}^0\right)$

$$k\left(\widetilde{v}_{nm}^0\right) = \frac{\int_{\widetilde{v}_1}^{\widetilde{v}_2} k(\widetilde{v}) d\widetilde{v}}{\Delta\widetilde{v}_D \cdot \frac{1}{2}\sqrt{\frac{\pi}{ln2}}}, \quad (4.4.10)$$

and

$$\sigma\left(\widetilde{v}_{nm}^0\right) = 3.72 \cdot 10^{-20} k\left(\widetilde{v}_{nm}^0\right) \quad (4.4.11)$$

using integral absorption coefficient $\int_{\widetilde{v}_1}^{\widetilde{v}_2} k(\widetilde{v}) d\widetilde{v}$ (4.3.14), and Doppler FWHM, $\Delta\widetilde{v}_D$, (4.3.29).

4.5 Franck–Condon Principle for Bound–Bound and Bound-Free Transitions

The Frank-Condon principle in a semiclassical form, as it was formulated by Franck (see [23] p. 194), contends that those transitions are most likely, under which neither the relative position of the nuclei nor their velocity change. In the quantum–mechanical formulation, the Franck-Condon principle for radiative transitions follows from (4.3.4): transitions with maximum values of integral of vibrational wave function overlap and, therefore, Frank–Condon factors $\langle \chi_{v'} | \chi_{v''} \rangle^2$, are most likely. As a result, vertical transitions are most likely (constant relative position of nuclei according to Franck), as a rule, between classical *PEC* (*PES*) turning points. Near these points, the velocity of the nuclei is close to 0 (constant kinetic energy of the nuclei), the probability of a molecule location in these regions is the largest, and therefore the transition is most likely. It is also evident that for vertical transitions between turning points, the overlap integral is also the largest since $\chi_{v'}$, $\chi_{v''}$ wave functions oscillate between turning points and, as a rule, are not in phase. The Franck–Condon principle is also valid for nonadiabatic transitions (see Sect. 3.6 and (3.6.8).).

R.S. Mulliken has shown that in transitions to or from high vibrational levels of a diatomic molecule bound state, a strong maximum of transition probability can occur at r values where the nuclei in the initial state have maximum kinetic energy if the internuclear equilibrium distance of these electronic states differs significantly [35].

Let us consider, for example, luminescence from a high level of a bound state to vibrational levels of a bound state of a diatomic molecule (Fig. 4.7).

For simplicity, let us assume that the *PEC* of the upper state corresponds to a harmonic oscillator. According to the quasi-classical interpretation of the Franck-Condon principle, a molecule in this state should behave like a harmonic oscillator with energy $E_{AB^*(v')} - E_{AB^*}(r_e)$ oscillating on the *PEC* of this state and emitting photons with energy in the range $hv = E_{AB^*}(r) - E_{AB}(r)$ in the $hv_{min} - hv_{max}$ range (Fig. 4.7a). The energy of these photons can be defined as the length of the vertical segments connecting these two *PEC*s or, equivalently, the length of the vertical dashed segments connecting the straight line with the curve $E_{AB}(r) + E_{AB^*(v')} - E_{AB^*}(r)$ (dashed curve) called *Mulliken difference potential* (see [35], e.g.). Let us forget for a while about the dependence of the spontaneous emission coefficient (4.4.2) on the frequency factor $\tilde{v}^3$ and on the internuclear distance, i.e., assume that $|R_e^{nm}|^2$ = const. Then, according to Franck's 'main idea' (see [23] p.194), the probability of a vertical transition (r = const) should be maximal one at the r_1, r_2 points, and *in the vicinity of the point corresponding to* r_3. Indeed, in the r_3 vicinity, the oscillatory motion kinetic energy in the state AB*(v'), $T'(r'_e)$ is maximal and equal to the potential energy stored at the points R_1, R_2. The total energy of the motion of nuclei during the transition from the state to a point lying on the state *PEC* (r_1, r_2 turning point) is equal to the sum of the potential energy stored in the state AB*(v') and the kinetic energy at the point R'_e, $T'(r'_e)$:

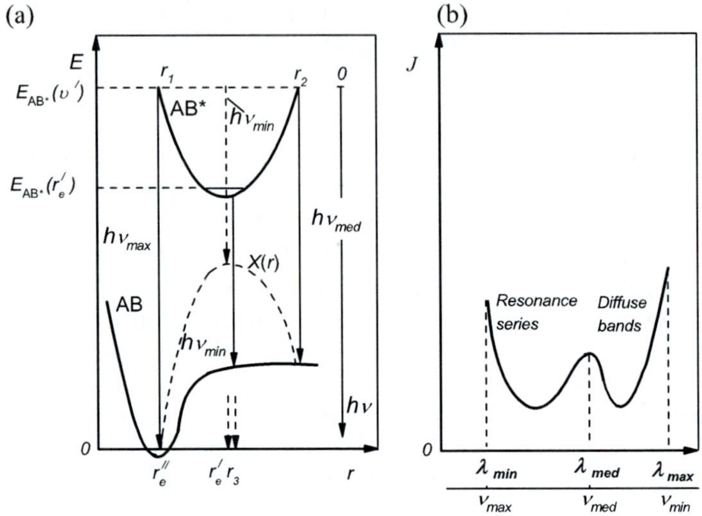

Fig. 4.7 To the Frank–Condon principle for transitions between high vibrational levels of a diatomic molecule [3], p. 37. r_1, r_2 are the classical turning point for the AB*(v') level. $X(r)$ is Mulliken difference potential

$$T''_{nucl} = E_{AB^*(v')} - hv = E_{AB}(r) + T'(r'_e), \tag{4.5.1}$$

$r = r_1$, r_2. If, in a vicinity of the r'_e point, the transition occurs on a flat part of the AB *PEC* (in this case, the points r_3, r'_e coincide)

$$E_{AB}(r) = 0, T''_{nucl} = T''(r) = T'(r'_e), \tag{4.5.2}$$

and the principle of the constancy of the kinetic energy during the radiative transition is retained. Consequently, in the spectral region corresponding to the ν_{min}, the third maximum of the luminescence intensity should be observed (Fig. 4.7b).

One can understand that the similar intensity distribution in the luminescence spectrum should occur if the transition with frequency ν_{min} (λ_{max}) occurs at the right *PEC* part of the lower state close to its dissociation limit. In this case, the luminescence spectrum should look as follows. Long-wavelength bands corresponding to a transition to a flat *PEC* are diffuse, possibly poorly resolvable bands (transitions to the r_1^*, r_2^* points on Mulliken difference potential lying higher than the horizontal dotted line in Fig. 4.7a). Sufficiently clear resonant series correspond to transition to the points on Mulliken difference potential lying lower than the horizontal dotted line in Fig. 4.7a). The following circumstance is distinctive for the considered *PEC* location: the spectrum shape in the ν_{min} (λ_{max}) region has to depend weakly on the vibrational quantum number; a v' decrease up to not too small values leads to a decrease $T'(r'_e)$, $T''(r''_e)$, but not a ν_{min} change. The presence of a 'fixed' long-wavelength band is explained by the fact that a contribution to it comes for transitions from vibrational levels with an extensive range of internuclear distances falling on a small range of change of the transition frequency.

The features described above occur if the overlap integral has a maximum near ν_{min}. Vibrational wave functions of the states AB*(v'), AB*(v''), which at some distance from the classical turning points are the de Broglie wavelengths $\lambda' = (\frac{\hbar^2}{2T'(r)\mu})^{1/2}$, $\lambda'' = (\frac{\hbar^2}{2T''(r)\mu})^{1/2}$ (see (3.1.11) must have the same values $\lambda' = \lambda''$ and be in phase or antiphase. Mulliken has shown [35] that for heavy atoms and high values v', v'' for which there is great anharmonicity and, consequently, the density of vibrational levels, these conditions can be met, since in a small frequency and energy range one can find regions, where the $\chi_{v'}$, $\chi_{v''}$ wave functions, depending on v', v'', coincide. Therefore, $\lambda' = \lambda''$ since $T' = T''$ at $R = R_3$. The effects described above occur if the *PEC*s of the states combined in the transition are sufficiently far apart, and the $\widetilde{\nu}^3$ value does not depend too much on v', v''.

There is the extremum at the Mulliken difference potential in Fig. 4.7a, and there are couple of points r_1^*, r_2^* for each energy values lying higher than the horizontal dotted line on the potential and corresponding to the same $\widetilde{\nu}$ value. It is obvious that the *R*-centroid approximation is invalid in this energy range, and one has to use the trial-and-error procedure to simulate a spectrum (see [36] and references). J. Tellinghuisen has called a band structure in this energy range as interference structure since contributions of the integrand $\chi_{v'} \cdot \chi_{v''}$ near r_1^*, r_2^* to the accumulated overlap integral $\int_0^r \chi_{v'} \cdot \chi_{v''} dr$ 'can add constructively or destructively' [33, 37].

The *PEC* positions similar to shown in Fig. 4.7a are typical for the halogen ion-pair and valence states (see Fig. 4.8 as an example).

4.6 Intramolecular Perturbations. Selection Rules and Franck–Condon Principle for Intramolecular Perturbation. Perturbations Between Bound States

Perturbations in diatomic molecules are described in detail in monographs [31, 38], and the author has used them for writing this section.

Unlike diatomic molecules, an interaction of the states of different species of a polyatomic molecule can occur due to electronic and vibrational motion coupling. Therefore, we consider perturbations in di- and polyatomic molecules separately.

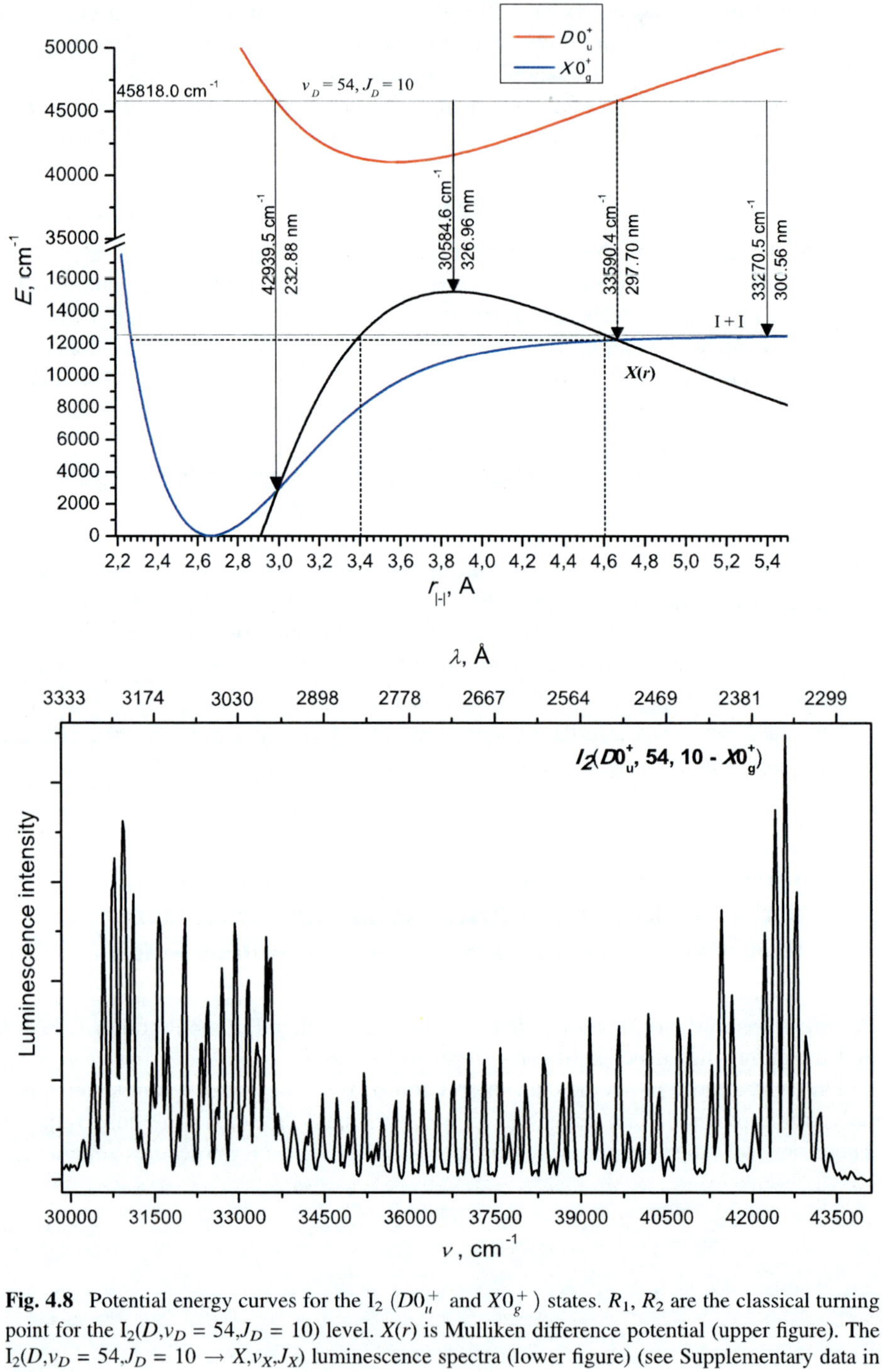

Fig. 4.8 Potential energy curves for the I_2 ($D0_u^+$ and $X0_g^+$) states. R_1, R_2 are the classical turning point for the $I_2(D,v_D = 54,J_D = 10)$ level. $X(r)$ is Mulliken difference potential (upper figure). The $I_2(D,v_D = 54,J_D = 10 \rightarrow X,v_X,J_X)$ luminescence spectra (lower figure) (see Supplementary data in [36])

4.6.1 Perturbations Between Bound States of Diatomic Molecules

In this section, the perturbation examined in [31], which lead to large mixing of states are discussed, namely:

1. Perturbations between states of the same symmetry;
2. Perturbations between electronic states of different symmetry due to interactions between spin and angular momenta of the electrons ($\widehat{\boldsymbol{H}}_{SO}$, spin–orbit operator) as well as rotational perturbations;

Besides, hyperfine interaction, which has not been discussed in [31], will be analyzed. Interactions between bound state rovibronic levels and the vibrational continuum of another electronic state leading to predissociation of the former are discussed in Sect. 4.7.

Selection rules for perturbations are as follows (see [23], p. 285 and Tables 4.1, 4.2):

1. Both states must have the same total angular momentum J; that is $\Delta J = 0$.

Table 4.1 Selection rules for the Λ, S, Σ, Ω and J angular momenta and various nature of the coupling operator [39] (reproduced by permission of EDP Science)

Operator[a]		$\lvert\Delta\Lambda\rvert$	$\lvert\Delta S\rvert$	$\lvert\Delta\Sigma\rvert$	$\lvert\Delta\Omega\rvert$	$\lvert\Delta J\rvert$	Z, μ, J variation[b]	Order of magnitude for I_2, cm^{-1} [c]
$\widehat{\boldsymbol{T}}_N$		0	0	0	0	0	$\sim\mu^{-0,75}$	100
$\widehat{\boldsymbol{H}}_{SO}$		≤ 1	≤ 1	≤ 1	0	0	$\sim Z^4$	$7 \cdot 10^3$
$\widehat{\boldsymbol{H}}_{SS}$		≤ 2	≤ 2	≤ 2	0	0	–	≤ 1
$\widehat{\boldsymbol{H}}_{OR}$		≤ 1	0	≤ 1	≤ 1	0	$\sim J$	$10^{-4} \cdot J$
$\widehat{\boldsymbol{H}}_{SR}$		≤ 1	≤ 1	≤ 1	≤ 1	0	$\sim J$	$10^{-4} \cdot J$
$\widehat{\boldsymbol{H}}_G$		≤ 1	0	≤ 1	≤ 1	0	$\sim\mu^{-1} \cdot Z$	$0.06 \cdot J$
$\widehat{\boldsymbol{H}}_{HFS}$	$\widehat{\boldsymbol{H}}_{MD}$	≤ 1	≤ 1	≤ 1	≤ 1	≤ 1	$\sim Z^3$	0.3
	$\widehat{\boldsymbol{H}}_{EQ}$	≤ 2	0	0	≤ 2	≤ 2	$\sim Z^3$	0.1
$\widehat{\boldsymbol{H}}_Z$		≤ 1	0	≤ 1	≤ 1	≤ 1	–	0.5
$\widehat{\boldsymbol{H}}_S$		≤ 1	0	0	≤ 1	≤ 1	–	1

[a] $\widehat{\boldsymbol{T}}_N$ is the nuclear kinetic energy operator [31] p. 168, $\widehat{\boldsymbol{H}}_{SO}$ is the spin–orbit operator, $\widehat{\boldsymbol{H}}_{SS}$ is the spin–spin operator, $\widehat{\boldsymbol{H}}_{OR}$ is the orbit-rotational operator, $\widehat{\boldsymbol{H}}_{SR}$ is the spin-rotational operator, $\widehat{\boldsymbol{H}}_G$ is the gyroscopic operator, $\widehat{\boldsymbol{H}}_{HFS} = \widehat{\boldsymbol{H}}_{MD} + \widehat{\boldsymbol{H}}_{EQ}$ is the hyperfine operator, $\widehat{\boldsymbol{H}}_{MD}$ is the magnetic-dipole operator, $\widehat{\boldsymbol{H}}_{EQ}$ is the electric-quadrupole operator, $\widehat{\boldsymbol{H}}_Z$ is Zeeman operator, $\widehat{\boldsymbol{H}}_S$ is Stark operator

[b] The dependence of matrix elements on nuclear charge Z, reduced mass μ and total angular momentum J.

[c] A rough estimate of the order of magnitude for iodine molecules for the magnetic field 1 T, the electric field 10^6 V/cm and dipole 1 Debye.

Table 4.2 Parity selection rules. The + (−) sign indicates that the operator is even (odd) [39] (reproduced by permission of EDP Science)

Operator	Total parity	*u*-*g* parity	Krönig parity for $\Omega = 0$
$\widehat{\boldsymbol{T}}_N$	+	+	+
$\widehat{\boldsymbol{H}}_{FS}$ [a]	+	+	+
$\widehat{\boldsymbol{H}}_G$	+	+	+
$\widehat{\boldsymbol{H}}_{HFS}$	+	No rule	No rule
$\widehat{\boldsymbol{H}}_Z$	+	+	−
$\widehat{\boldsymbol{H}}_S$	−	−	+

[a] $\widehat{\boldsymbol{H}}_{FS} = \widehat{\boldsymbol{H}}_{SO} + \widehat{\boldsymbol{H}}_{SS} + \widehat{\boldsymbol{H}}_{SR} + \widehat{\boldsymbol{H}}_{OR}$ is fine-structure operator

2. Both states must have the same multiplicity if the spin-orbit interaction and splitting are negligible (see Sect. 4.2.1), that is $\Delta S = 0$. If it is not the case, $\Delta S = \pm 1$ (see Sect. 4.2.3).
3. Both Λ values of the states may differ only 0 or $\pm$ 1 for Hund's case (*a*, *b*), $\Delta\Lambda = 0, \pm 1$. For Hund's case (*c*), $\Delta\Omega = 0, \pm 1$.
4. Rovibronic levels of both states must have the same parity: + $\leftrightarrow$ + , − $\leftrightarrow$ - (see [23], p. 128)
5. For molecules with identical nuclei, *u*, *g* parity for the permutation of nuclei has to be the same: $g \leftrightarrow g$, $u \leftrightarrow u$. The gerade and ungerade electronic state may interact (mix) due to hyperfine or Stark perturbations, only. Besides, symmetric (*s*) rotational level may interact with symmetric one, and antisymmetric (*a*) rotational level may interact with antisymmetric one, that is $s \not\leftrightarrow a$ (see [23], p.130 and 285).

One has to note that there are two $\boldsymbol{\sigma}_v$-related symmetry classifications relating to the topic under discussion, namely [31], p. 138 ($\boldsymbol{\sigma}_v$ is the operator of a reflection in the molecule-fixed coordinate system through a plane containing the internuclear axis; it specifies the $\pm$ parities [31], p. 70):

Total parity (the effect of the $\boldsymbol{\sigma}_v$ operator on the complete rovibronic wave function ($\pm$);

Total parity of the complete rovibronic wave function exclusive of a $(-1)^J$ or $(-1)^{J-1/2}$ rotational factor for molecules with even or odd number of electrons, respectively (*e*/*f* symmetry) (see below).

Selection rules for perturbation become more straightforward if one uses the *e/f labeling* [40].

For molecules with even number of electrons (integer *total* angular momentum *J*, omitting nuclear spin)

$$\begin{aligned} &\text{levels with parity} + (-1)^J \text{ is called } e \text{ levels, and} \\ &\text{levels with parity} - (-1)^J \text{ is called } f \text{ levels.} \end{aligned} \tag{4.6.1}$$

For molecules with an odd number of electrons (half-integer *total* angular momentum *J*, omitting nuclear spin)

$$\begin{aligned}&\text{levels with parity} + (-1)^{J-1/2} \text{ is called } e \text{ levels, and}\\&\text{levels with parity} - (-1)^{J-1/2} \text{ is called } f \text{ levels.}\end{aligned} \quad (4.6.2)$$

The *e*/*f* labels are independent of the coupling case; they depend on the total angular momentum and the state parity.

If one uses *e*/*f* labels [31], p. 145, [40], perturbations are governed by the

$$\Delta J = 0, e \leftrightarrow e, f \leftrightarrow f, g \leftrightarrow g, u \leftrightarrow u \quad (4.6.3)$$

selection rules. The $g \leftrightarrow g$, $u \leftrightarrow u$ selection rules are valid for all types of perturbation except hyperfine and Stark perturbations.

Rovibronic wave functions of the isolated (free) diatomic molecules are described in a Hund's (*a*) case as follows [41]:

$$|JM\Omega\varepsilon\Lambda v\rangle = 2^{-\frac{1}{2}}[|JM\Omega\rangle|\Lambda\Sigma\rangle + \varepsilon|J, M, -\Omega\rangle|-\Lambda, -\Sigma\rangle|v\rangle. \quad (4.6.4)$$

Here:

J denotes the total angular momentum with projection along space- and molecule-fixed axes of M and Ω, respectively;

$|\Lambda, \Sigma\rangle$ denotes the electronic wave function with projections Λ and Σ, respectively, of the electronic orbital and spin angular momenta along the molecular axis;

$|v\rangle$ denotes the vibrational wave function with quantum number v.

The symmetry index is $\varepsilon = \pm 1$, and the total parity I of the molecular wave function of the ${}^2\Sigma$ states is $\varepsilon(-1)^{J-1/2-s}$ where $s = 0$ and 1 for the Σ^+ and Σ^- states, respectively, and zero otherwise. For Π and Σ^+ states, the $\varepsilon = +1$ levels are labeled *e*, and the $\varepsilon = -1$ levels are labeled *f*. This assignment is reversed for Σ^- states [41].

The assignment of quantum numbers, indices, labels, and parity to the Π and Σ molecular states in (4.6.2) are clarified in Table 4.3.

For the gerade (g) states

$$\varepsilon = (-1)^J \quad (4.6.5)$$

For the ungerade (u) states

$$\varepsilon = (-1)^{J+1} \quad (4.6.6)$$

For the Σ^+, 0_g^+ and $\Lambda \neq 0, \Omega \neq 0$ states, the $\varepsilon = +1$ levels are labeled *e* and the $\varepsilon = -1$ levels, labeled *f*. This assignment is reversed for Σ^- states.

Table 4.3 Assignment of quantum numbers, indices, and labels of rotational levels of a diatomic molecule in $^2\Pi$ and $^2\Sigma$ states [41] (Reproduced from M.A. Alexander, G.C. Corey, J. Chem. Phys. **84**, 100–113 (1986). https://doi.org/10.1063/1.450831, with the permission of AIP Publishing)

		Π states[a]								
Total angular momentum[b]		ε[c]	*e/f*	Total parity	gerade *s/a*	ungerade *s/a*				
$J+1$		+1	*e*	−	*a*	*s*				
		−1	*f*	+	*s*	*a*				
J		+1	*e*	+	*s*	*a*				
		−1	*f*	−	*a*	*s*				
		Σ^+ states					Σ^- states			
	N[d]	ε	*e/f*	Total parity	gerade *s/a*	ungerade *s/a*	*e/f*	Total parity	gerade *s/a*	ungerade *s/a*
$J+1$	$J+1/2$	+1	*e*	−	*a*	*s*	*f*	+	*s*	*a*
	$J+3/2$	−1	*f*	+	*s*	*a*	*e*	−	*a*	*s*
J	$J-1/2$	+1	*e*	+	*s*	*a*	*f*	−	*a*	*s*
	$J+1/2$	−1	*f*	−	*a*	*s*	*e*	+	*s*	*a*

[a]One assumes a pure Hund's case (*a*) representation for a Π state. The assignments of the labels are identical for the $\Omega = 1/2$ and $\Omega = 3/2$;
[b]It is assumed that the $(J-1/2)$ level is even;
[c]Symmetry index;
[d]Nuclear rotational angular momentum, the Hund's case (*b*) label.

4.6.1.1 Perturbations Between States of the Same Symmetry

These perturbations are discussed in Chap. 3 of [31] in detail. In this book, they are examined briefly. The following off-diagonal matrix elements

$$\left\langle \Phi_{i,\Lambda,S,\Sigma,\chi_{v_i,J}} \middle| \hat{\boldsymbol{H}} \middle| \Phi_{j,\Lambda',S',\Sigma',\chi_{v'_j,J}} \right\rangle \tag{4.6.7}$$

of the Hamiltonian

$$\hat{\boldsymbol{H}} = \hat{\boldsymbol{H}}_{el} + \hat{\boldsymbol{T}}_N(R) + \hat{\boldsymbol{H}}_{OR} + \hat{\boldsymbol{H}}_{SO} \tag{4.6.8}$$

cause perturbations. The off-diagonal matrix elements of $\hat{\boldsymbol{H}}_{el}$, $\hat{\boldsymbol{T}}_N(r)$, $\hat{\boldsymbol{H}}_{OR}$ and $\hat{\boldsymbol{H}}_{SO}$ cause *electrostatic perturbations*, *nonadiabatic interactions*, *rotational perturbations*, and *spin-orbit perturbation*, respectively. If a crossing *PEC*s representation is used, off-diagonal matrix elements of the $\hat{\boldsymbol{H}}_{el}$ operator between the states appear. If a noncrossing *PEC*s representation is used, the $\hat{\boldsymbol{T}}_N(r)$ operator becomes responsible for perturbations.

Electrostatic and nonadiabatic perturbations. Electrostatic perturbations occur between electronic states of identical electronic symmetry, i.e., those with identical values of Λ, Σ and S.

Two possible types of *zeroth-order* electronic functions may be defined:

– *Diabatic* electronic wave functions are those for which the equations.

$$\left\langle \Phi_1^d \middle| \hat{\boldsymbol{T}}_N(r) \middle| \Phi_2^d \right\rangle_r = 0, \tag{4.6.9}$$

and

$$\left\langle \Phi_1^d \middle| \hat{\boldsymbol{H}} \middle|_{el} \Phi_2^d \right\rangle_r = H_{12}^{el}(r) \neq 0, \tag{4.6.10}$$

are valid. It follows from (4.6.10) that the diabatic *PEC*s corresponding to Φ_i^d functions *can cross*. Usually, they belong to different electronic configurations (see [31], p. 172, and Sect. 3.6).

– *Adiabatic* electronic wave functions are those for which the equations.

$$\left\langle \Phi_1^{ad} \middle| \hat{\boldsymbol{H}} \middle|_{el} \Phi_2^{ad} \right\rangle_r = 0, \tag{4.6.11}$$

and

$$\left\langle \Phi_1^{ad} \middle| \widehat{\boldsymbol{T}}_N \middle| \Phi_2^{ad} \right\rangle_r \neq 0, \qquad (4.6.12)$$

are valid.

The vibronic interaction between the level v_1 of the *diabatic PEC* $V_1^d(R)$ and the level v_2 of the *diabatic PEC* $V_2^d(R)$ is determined by the matrix element

$$H_{1,v_1;\,2,v_2;} = \left\langle \Phi_1^d \chi_{v_1}^d \middle| \widehat{\boldsymbol{H}}_{el} \middle| \Phi_2^d \chi_{v_2}^d \right\rangle \qquad (4.6.13)$$

($\chi_{v_i}^d$ is vibrational wave function of the $i = 1, 2$ states). After integration over the electronic coordinates, one obtains.

$$H_{1,v_1;\,2,v_2;} = H_{el}^{12}(r) \left\langle \chi_{v_1}^d | \chi_{v_2}^d \right\rangle, \qquad (4.6.14)$$

(see 4.6.10), where $\left\langle \chi_{v_1}^d | \chi_{v_2}^d \right\rangle$ is the overlap integral (see 3.6.8). The $H_{12}^{el}(r)$ value depends on R weakly and can be described using R-centroid approximation see [31], p. 163).

If the approximate deperturbed *PEC*s *do not cross*, they are *adiabatic* curves. The adiabatic curves result from an avoided crossing or correspond to configurations, which differ by only one orbital (see [31], p. 173 and Sect. 3.6). The matrix element for the interaction between the level v_1 of the adiabatic *PEC* $V_1^{ad}(r)$ and the level v_2 of the adiabatic *PEC* $V_2^{ad}(r)$ is

$$H_{1,v_1;\,2,v_2;} = \left\langle \Phi_1^{ad} \chi_{v_1}^{ad} \middle| \widehat{\boldsymbol{T}}_N \middle| \Phi_2^{ad} \chi_{v_2}^{ad} \right\rangle. \qquad (4.6.15)$$

$\widehat{\boldsymbol{T}}^N$ is the nuclear kinetic energy operator.

Diabatic and adiabatic *PEC*s are shown schematically in Fig. 4.9.

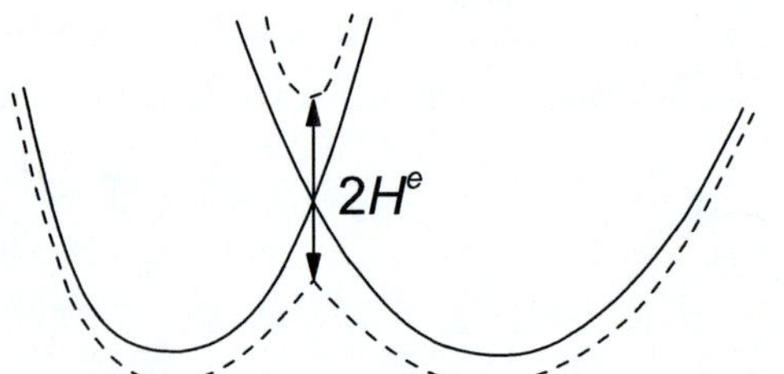

Fig. 4.9 Diabatic (solid lines) and adiabatic (dotted lines) *PEC*s (see [31], p. 164)

The minimal energy gaps between adiabatic *PEC*s occurs at the crossing point of diabatic *PEC*s and equal to $2H_{12}^{el}(r)$. Similar results have been obtained in Sect. 3.6 (see 3.6.20).

4.6.1.2 Perturbations Between Electronic States of Different Symmetry

The $\widehat{\boldsymbol{H}}_{SO}$ and $\widehat{\boldsymbol{H}}_{OR}$ operators cause the largest perturbation between the states of different symmetry.

Spin-orbit perturbation. The form of the spin–orbit Hamiltonian is [31], p. 181

$$\widehat{\boldsymbol{H}}_{SO} = \frac{\alpha^2}{2}\sum_i \left\{ \frac{Z_A e}{r_{iA}^3}\widehat{\boldsymbol{l}}_{iA}\cdot \boldsymbol{s}_i + \frac{Z_B e}{r_{iB}^3}\widehat{\boldsymbol{l}}_{iB}\cdot \widehat{\boldsymbol{s}}_i \right\} - \frac{\alpha^2}{2}\sum_{i\neq j}\frac{1}{r_{ij}^3}(\boldsymbol{r}_{ij}\times \boldsymbol{p}_i)(\boldsymbol{s}_i + 2\boldsymbol{s}_j). \tag{4.6.16}$$

Here,

$\alpha = e^2/\hbar c = 1/137.036$ is the fine-structure constant,
Z_i is the charge of the nucleus,
$\hat{\boldsymbol{l}}_{iA} = \left(\frac{1}{\hbar}\right)(\boldsymbol{r}_{iA}\times \boldsymbol{p}_i)$, $\hat{\boldsymbol{l}}_{iB} = \left(\frac{1}{\hbar}\right)(\boldsymbol{r}_{iB}\times \boldsymbol{p}_i)$ are angular momentum operators.
$\boldsymbol{p}_i$—is the momentum of the electron measured in the molecule fixed coordinate frames.

The first part of (4.6.16) is *S–O* coupling of each electron in the field of two nuclei with charges Z_A, Z_B. The second part is due to interelectronic interactions.

The selection rules for *S–O* interactions are as follows [31], p. 182:

$$\begin{aligned} &\Delta J = 0, \Delta\Omega = 0, g \not\leftrightarrow u, e \not\leftrightarrow f, \Sigma^+ \leftrightarrow \Sigma^- \\ &\Delta S = 0 \text{ or } \Delta S = \pm 1, \Delta\Lambda = \Delta\Sigma = 0 \text{ or } \Delta\Lambda = -\Delta\Sigma = \pm 1 \end{aligned} \tag{4.6.17}$$

(see Tables 4.1, 4.2 and Sect. 4.2.3, also).

Rotational perturbations. There are three terms in the off-diagonal matrix element of $\widehat{\boldsymbol{H}}_{OR}$, which give rise to rotational perturbation, namely [31], p. 210:

$$(1/2\mu R^2)\left(\widehat{\boldsymbol{L}}^+\widehat{\boldsymbol{S}}^- + \widehat{\boldsymbol{L}}^-\widehat{\boldsymbol{S}}^+\right) \tag{4.6.18}$$

(*spin-electronic homogeneous* ($\Delta\Omega = 0$) *perturbation*),

$$-(1/2\mu R^2)\left(\widehat{\boldsymbol{J}}^+\widehat{\boldsymbol{S}}^- + \widehat{\boldsymbol{J}}^-\widehat{\boldsymbol{S}}^+\right) \tag{4.6.19}$$

($\widehat{\boldsymbol{S}}$-*uncoupling operator*), and

$$-\left(1/2\mu R^2\right)\left(\hat{J}^+\hat{L}^- + \hat{J}^-\hat{L}^+\right). \quad (4.6.20)$$

($\hat{L}$-*uncoupling operator*). The (4.6.19 and 4.6.20) terms give rise to heterogeneous ($\Delta\Omega = \pm 1$) perturbation. Here

$$\hat{L}^{\pm} = \hat{L}_x \pm i\hat{L}_y, \hat{S}^{\pm} = \hat{S}_x \pm i\hat{S}_y, \hat{J}^{\pm} = \hat{J}_x \pm i\hat{J}_y \quad (4.6.21)$$

are molecule-fixed raising (+) and lowering (−) operators [31], p. 74.

The selection rules for the spin-electronic homogeneous perturbation are (see general selection rules 1, 4, 5 in the beginning of Sect. 4.6.1, also):

$$\Delta\Omega = 0, \Delta\Lambda = -\Delta\Sigma = \pm 1, \Delta S = 0 \quad (4.6.22)$$

This type of perturbation is much weaker than the spin–orbit one. One can consider the interaction of the $^3\Pi_1$ and $^3\Sigma_1^+$ states, as an example.

The selection rules for the heterogeneous perturbation causes by the $\hat{S}$-uncoupling operator are:

$$\Delta S = 0, \Delta\Omega = \Delta\Sigma = \pm 1 \quad (4.6.23)$$

Generally, this operator mixes component of the same multiplet electronic state, $^2\Pi_{1/2}$ and $^2\Pi_{3/2}$ states, for example, and causes Λ- and Ω-doubling.

The selection rules for the heterogeneous perturbation causes by the $\hat{L}$-uncoupling operator are:

$$\Delta S = 0, \Delta\Omega = \Delta\Lambda = \pm 1 \quad (4.6.24)$$

This perturbation causes Λ- and Ω-doubling. One can consider the interaction of the $^2\Pi_{3/2}$ and $^2\Sigma$ states, as an example.

4.6.1.3 Matrix Elements for Different Types of Perturbation

The magnitude of the perturbation matrix element, $H_{1,v_1;2,v_2;}$, is determined by the product of the electronic, $H_{el}^{12}(r)$, and vibrational, $\langle\chi_{v_1}|\chi_{v_2}\rangle$, factors, both of which can depend on *J* (see 3.6.8 and 4.6.14, for example). A perturbation occurs if *both* electronic and vibrational factors are nonzero. An electronic part of the matrix element depends on the type of interaction, i.e., the type of the operator. An overlap integral of vibrational wave functions, $\langle\chi_{v_1}|\chi_{v_2}\rangle$, is high when *PEC*s approach or cross (see Sect. 4.7).

Electrostatic perturbation. In light molecules, homogeneous perturbations due to electrostatic interaction are often very strong, 400 $cm^{-1} < H_{el}^{12} < 10^4$ cm^{-1} (see Table 3.5 in [31]).

Most frequently, these interactions occur between Rydberg and valence states [31], p. 307 as well as Rydberg and ion pair state in dihalogens [7], p. 82. In the latter cases, interactions are weaker.

Spin–orbit perturbation. Spin-orbit perturbation (interaction) causes the mixing of electronic states according to selection rules (4.6.17) and splitting of the fine-structure components, corresponding to different Ω values. Spin–orbit interaction and splitting between the states belonging to the same configuration are described in Sect. 4.2.3.

Rotational perturbations. The spin-electronic homogeneous perturbation is much weaker than the *S–O* one (see above). Two other types of rotational perturbation mentioned above (heterogeneous interactions) cause Λ- and Ω-doubling. They have been studied in the iodine molecule in details (see Sect. 4.3 in [7]).

In principle, the $B'0_u^- \sim B''1_u$, $a1_g \sim 0_g^+$, $B0_u^+ \sim 1_u(ab)$, and $0_u^- \sim 1_u(bb)$ heterogeneous interactions in the iodine molecule valence states can occur (see Fig. 4.4). To the best of our knowledge, the $0_u^- \sim 1_u(bb)$ one has been observed, only [42] (see Sects. 4.1 and 4.3.1.2 in [7], also).

Possibilities for heterogeneous interaction in iodine molecule ion-pair (IP) states are much more. Six first tier IP states correlate with the $I^-(^1S) + I^+(^3P_2)$ dissociation limit (Fig. 4.10).

There are four pairs of states, $E0_g^+ \sim \beta 1_g$, $D0_u^+ \sim \gamma 1_u$, $\beta 1_g \sim D'2_g$ and $\gamma 1_u \sim \delta 2_u$, which can be coupled due to heterogeneous interaction according to the

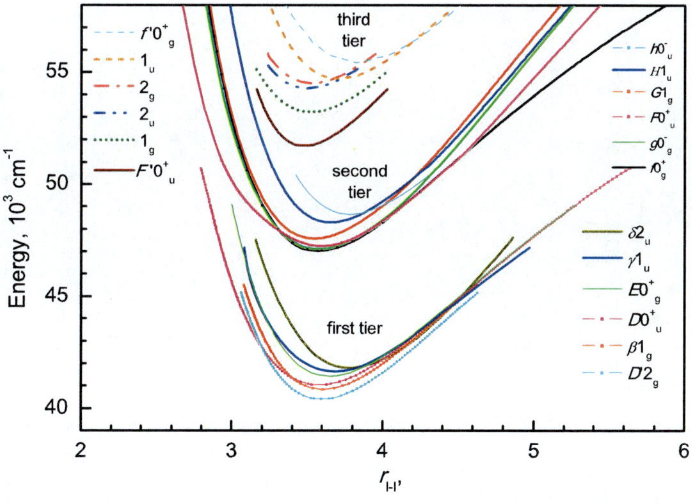

Fig. 4.10 Potential energy curves of the iodine molecule ion-pair states [7], p. 32

$\Delta\Omega = \pm 1$ selection rule. The $E0_g^+ \sim \beta 1_g$ [43] and $\gamma 1_u \sim \delta 2_u$ [44] coupling have been observed and studied.

J. Perrot et al. determined the empirical formula of the $\beta 1_g$ *e*/*f* splitting [43] as follows:

$$\Delta E_\beta(v, J) = -\left[7.11\,10^{-6} + 1.68\,10^{-7}(v + 1/2) - 5.35\,10^{-9}(v + 1/2)^2\right] \\ J(J+1) + 2.91\,10^{-11} J^2(J+1)^2 \quad (4.6.25)$$

Assuming that the interaction between $E0_g^+$ and $\beta 1_g$ states is responsible for the splitting of rovibrational levels, the second order of the perturbation theory can be applied [7], p. 73:

$$\Delta E_\beta(v_\beta, J) = -\left[6.8\,10^{-7} + 3.3\,10^{-9}\left(v_\beta + \frac{1}{2}\right)\right] J(J+1) \left|\left\langle E0_g^+ \middle| J_e^\pm \middle| \beta 1_g \right\rangle\right|^2 \quad (4.6.26)$$

The comparison of (4.6.25 and 4.6.26) gives the electronic matrix element that is in good agreement with the predictions made in the pure precession model [43].

T. Ishiwata's research group carried out analogous experiments. The Ω-splitting of the $\gamma 1_u$ state was observed (Fig. 4.11), and proposed due to interaction with the $D0_u^+$ state. Direct coupling of the other, $\delta 2_u$ state with the $D0_u^+$ one is forbidden as heterogeneous interaction in the first-order perturbation theory due to $\Delta\Omega = \pm 1$ selection rule. The Ω-splitting of the $\delta 2_u$ state occurs through the intermediate $\gamma 1_u$

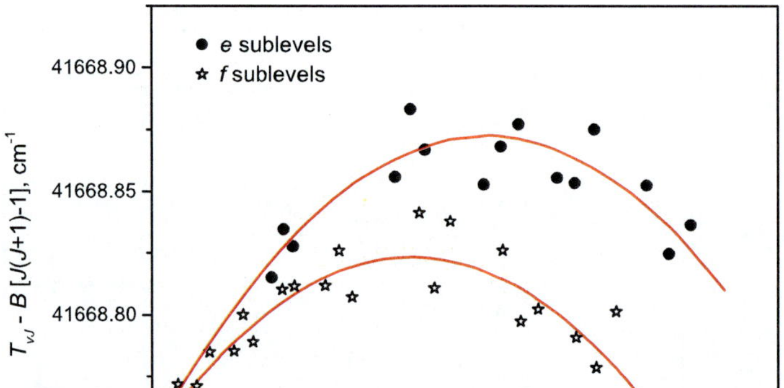

Fig. 4.11 Energy level diagram of the $\gamma 1_u, v_\gamma = 0$ state (see [7] p.75, [44])

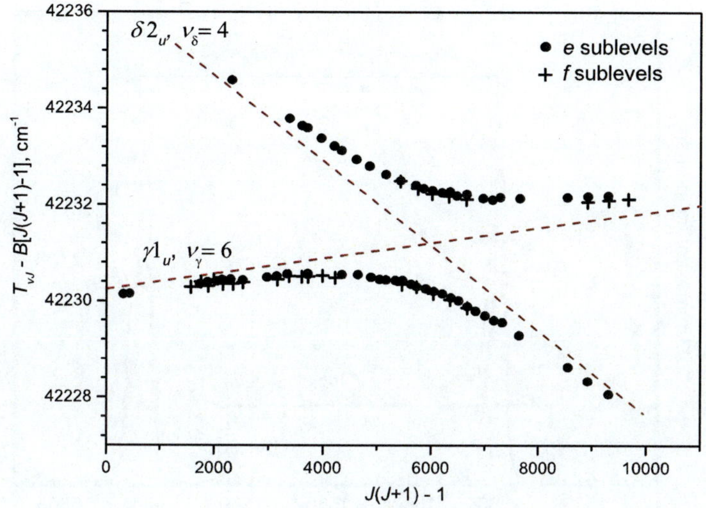

Fig. 4.12 Energy level diagram of the $\gamma 1_u, v_\gamma = 6$ and $\delta 2_u, v_\delta = 4$ states (see [7] p.75, [44])

state, describing in the second-order perturbation theory (see below). This effect is much less than that of for the $\gamma 1_u$ state and was not observed.

However, another effect of interaction, namely, mixing of the $\gamma 1_u$ and $\delta 2_u$ wavefunctions was observed. Due to the admixture of the $\gamma 1_u$ rovibrational levels to the $\delta 2_u$ ones, the latter can also be populated from the $A1_u$ intermediate state in two-step transition [44]. The extra lines in the excitation spectra of the $\gamma \rightarrow A$ luminescence were observed due to 'intensity borrowing' from the γ state and assigned to transition to the $\delta 2_u$ perturbed state. The energy diagram in Fig. 4.12 shows two series of levels corresponding to the γ and δ states. One sees that the heterogeneous interaction leads to avoiding crossing between coupled states. Interaction matrix elements between γ and δ IP states were determined for v_γ = 2–11 vibrational levels [44].

The second tier consists of 6 IP states corresponding to the two dissociation limits: $I^-(^1S) + I^+(^3P_1)$ ($G1_g$, $g0_g^-$, $H1_u$, $h0_u^-$) and $I^-(^1S) + I^+(^3P_0)$ ($f0_g^+$, $F0_u^+$). Therefore, the $G1_g \sim g0_g^-$ and $H1_u \sim h0_u^-$ heterogeneous interaction only can be possible. The Ω-doubling of the $G1_g$ and $H1u$ states was found for a wide range of rovibrational levels [45, 46]. The $1_u(^1D) \sim 2_u(^1D)$ heterogeneous interaction was studied by these authors, also [47].

Indirect heterogeneous perturbation. The more complex 'indirect' $D0_u^+ \sim \delta 2_u$ coupling was observed in [48]. The D state optical population by three-step laser excitation from the ground state via the valence $B0_u^+$ and 0_g^+ (bb) intermediate states was performed. In the excitation spectra near the $D, v_D = 22, J_D$ rotational lines, some extra lines were observed, which were assigned to the population of the $\delta 2_u, v_\delta = 13$ state after luminescence spectra simulation (Fig. 4.13).

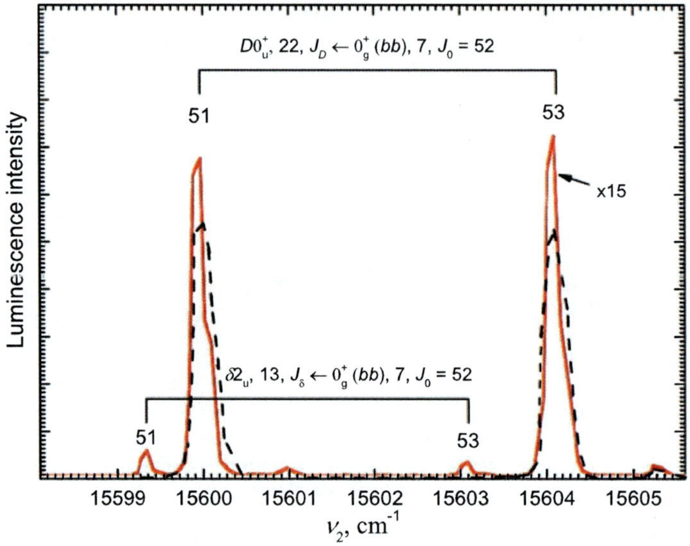

Fig. 4.13 Excitation spectra of the δ,13,J_δ (solid line) and D,22, J_D (broken line) states luminescence [7] p.76, [48] (Reprinted from Chemical Physics Letters, Vol. 638, V.V. Baturo, S.S. Lukashov, S.A. Poretsky, A.M. Pravilov, O.S. Vasyutinskii, Mixing of the $D0_u^+$ and $\delta 2_u$ ion-pair states of iodine molecule. 244–248 (2015) with permission from Elsevier)

Possible hyperfine interaction was excluded due to the lack of coupling between rovibrational levels with $J_D \neq J_\delta$. Therefore, the only explanation of the $\delta 2_u$ population is the heterogeneous coupling of these states occurring through the intermediate $\gamma 1_u$ state. The $D0_u^+ \sim \delta 2_u$ coupling was described in the second-order PT, including $\gamma 1_{u,v_\gamma,}J_\gamma$, $J_\gamma = J_D = J_\delta$ rovibrational states (see details in [48] and Sect. 4.3.2.1 in [7]).

4.6.1.4 Hyperfine Interaction

For a Hund's case (*c*) the ket is $|v, \Omega, v, J, M_J\rangle$ (v stands for all omitted electronic quantum numbers). If the nuclear spins are non-zero, a molecule is homonuclear, $^{127}I_2$, for example, and one uses the symmetrical coupling scheme $|J(I_a, I_b)IFM_F\rangle$ [39] (the total angular momentum $\boldsymbol{F} = \boldsymbol{I} + \boldsymbol{J}$ is strictly conserved in this case), the wave function is

$$|J\varepsilon FM_F\rangle = \sum\nolimits_I \alpha(I\varepsilon JF)|JIFM_F\rangle \tag{4.6.27}$$

(α = 1/137.036, see the previous section).

Due to the Pauli Exclusion Principle, the total wave function of a homonuclear diatomic molecule with half-integer nuclear spin is antisymmetric for the exchange of the nuclei. In a state of even (gerade) electronic symmetry, such as the $^{127}I_2(X\ 0_g^+)$, only states of odd total nuclear spin (I = 1, 3, 5) can combine with odd-J states, and states of even nuclear spin (I = 0, 2, 4) can combine only with even-J states. The total number of hyperfine states with the nuclear spin I is $2I + 1$. For the 0_g^+ states of the $^{127}I_2$ molecule, rotational levels with odd J correspond to I = 1, 3, 5 (*ortho*-state) and split into 21 components; rotational levels with even J correspond to I = 0, 2, 4 (*para*-state) and split into 15 components. For an odd (ungerade) electronic state such as the 0_u^+ state, the I, J combination rules are reversed [49]. For the states with $\Omega > 0$, one should take into account Ω-doubling.

The relationship between the total angular momentum J and the total nuclear spin momentum I follows from the (4.6.5, 4.6.6, 4.6.28):

$$(-1)^I = p\varepsilon \tag{4.6.28}$$

(p = + 1 and −1 for the g and u states, respectively [7], p. 39, [50]. Equation (4.6.28) means that:

$$\begin{cases} \text{For the } u \text{ negative}(\varepsilon = -1, p = -1)\text{and } g \text{ positive}(\varepsilon = +1, p = +1)\text{terms,} \\ \qquad \text{the total spin } I \text{ is even.} \\ \text{For the } u \text{ positive}(\varepsilon = -1, p = +1)\text{and } g \text{ negative}(\varepsilon = +1, p = -1)\text{terms,} \\ \qquad \text{the total spin } I \text{ is odd.} \end{cases} \tag{4.6.29}$$

(see 4.6.5, 4.6.6, 4.6.28 and Table 4.4).

Table 4.4 Correspondences between $^{127}I_2$ total nuclear spin momenta (I), symmetry indices (labels) (ε), and total angular momentum (J) for the 0_g^+, 0_g^-, 0_u^+, 0_u^-, 1_g and 1_u states[a] (see 4.6.1, 4.6.5, 4.6.6, 4.6.28, 4.6.29)

0_g^+	I	5	4	3	2	1	0
	ε	−	+	−	+	−	+
	J	odd	even	odd	even	odd	even
0_g^-	ε	−	+	−	+	−	+
	J	odd	even	odd	even	odd	even
0_u^+	ε	+	−	+	−	+	−
	J	odd	even	odd	even	odd	even
0_u^-	ε	+	−	+	−	+	−
	J	odd	even	odd	even	odd	even
1_g	ε	− (*f*)	+ (*e*)	− (*f*)	+ (*e*)	− (*f*)	+ (*e*)
	J	odd	even	odd	even	odd	even
1_u	ε	+ (*e*)	− (*f*)	+ (*e*)	− (*f*)	+ (*e*)	− (*f*)
	J	odd	even	odd	even	odd	even

[a]The same correspondences are valid for the Σ_g^+, Σ_u^-, Π_g and Π_u states

This leads to the fact that the summation over I in (4.6.27) does not include all possible I values and depends on the symmetry of the rovibronic state. The sign of the rotational components of the 0_g^+, 0_g^-, 0_u^+, 0_u^-, 1_g and 1_u and so on states, Hund's case (*c*)), ε, is determined by the number $(-1)^J$ (4.6.5) for the gerade (*g*) states, and that of the $(-1)^{J+1}$ (4.6.6) for the ungerade (*u*) states.

The assignment of quantum numbers, indices (labels) and parities to the molecular states in (4.6.27) for the 0_g^+, 0_g^-, 0_u^+, 0_u^-, 1_g and 1_u states are clarified in Table 4.4.

The hyperfine terms of both states mixed by hyperfine interaction must have the same parity: $+ \leftrightarrow +$, $- \leftrightarrow -$. Selection rules for hyperfine interaction matrix elements are:

$$|\Delta\Lambda| \leq 1(|\Delta\Omega| \leq 1), |\Delta J| \leq 1 \tag{4.6.30}$$

for the nuclear magnetic-dipole interaction with the electrons (*MD* interaction) and

$$|\Delta\Lambda| \leq 2(|\Delta\Omega| \leq 2), |\Delta J| \leq 2 \tag{4.6.31}$$

for the nuclear electric-quadrupole interaction with the electrons (*MQ* interaction) (see Table 4.1). The hyperfine Hamiltonian is of even parity, and therefore conserves sign (*p*) but not electronic parity ε. Since *p* is conserving and taking into account (4.6.28), one sees that:

$$\text{The } u \sim g \text{ hyperfine coupling implies an odd } \Delta I, \tag{4.6.32}$$

whereas

$$\text{The } u \sim u \text{ and } g \sim g \text{ hyperfine couplings imply an even } \Delta I. \tag{4.6.33}$$

One can use Table 4.4 to understand what hyperfine sublevels can interact. There is no precise selection rule for vibrational quantum number v. For a good approximation, propensity rule is defined by the Franck–Condon factors (*FCF*s) (see [7], p.60 and [39, 49, 50]).

4.6.1.5 Two-State Perturbation Model

It is convenient to use the two-level model to describe interactions of close-lying rovibrational states, hyperfine interaction (*HFI*), in particular. Let two close-lying rovibrational states $\Psi = |\Omega v J n\rangle$ and $\Psi' = |\Omega' v' J' n'\rangle$ are coupled by the interaction. According to the two-state perturbation model [7], p.61, [51, 52], the mixed $|1\rangle$ and $|2\rangle$ states are the following:

$$|1\rangle = cos|\Omega vJn\rangle + sin|\Omega' v' J' n'\rangle$$
$$|2\rangle = -sin|\Omega vJn\rangle + cos|\Omega' v' J' n'\rangle, \quad (4.6.34)$$

where the mixing angle θ can be obtained using (4.6.33)

$$tan\ 2\theta = \frac{2H}{\Delta E}, \quad (4.6.35)$$

$$H = \sqrt{FCF}\left\langle \Omega vJn \middle| \widehat{\boldsymbol{H}} \middle| \Omega' v' J' n' \right\rangle, \quad (4.6.36)$$

$\widehat{\boldsymbol{H}}$ is the operator of the interaction between the Ψ and Ψ' states, ΔE is an energy gap between unperturbed rovibrational states (Fig. 4.14).

Well-known local effects of perturbation, such as intensity borrowing or appearing of some extra lines, lifetime variations or anomalies of the rotational structure can be described using this model. Excitation of perturbed levels leads to the luminescence from both Ψ and Ψ' zero-order states. In the absence of saturation in excitation processes, the population ratio of the $|1\rangle$ and $|2\rangle$ states is equal to $cos^2\theta$ $/sin^2\theta$. Intensities of luminescence from the Ψ and Ψ' components are functions of the mixing angle θ, also. The intensity of the $\Psi \rightarrow i$ and $\Psi' \rightarrow k$ luminescence are:

$$I_{|1\rangle}(\Psi \rightarrow i) \sim \left(v^3_{|1\rangle i} cos^4\theta\right) \langle \Psi | \widehat{\boldsymbol{\mu}}_{\Psi \rightarrow i} | i \rangle^2 \quad (4.6.37)$$

$$I_{|2\rangle}(\Psi \rightarrow i) \sim \left(v^3_{|2\rangle i} sin^4\theta\right) \langle \Psi | \widehat{\mu}_{\Psi \rightarrow i} | i \rangle^2 \quad (4.6.38)$$

$$I_{|1\rangle}(\Psi' \rightarrow k) \sim \left(v^3_{|1\rangle k} cos^2\theta sin^2\theta\right) \langle \Psi' | \widehat{\boldsymbol{\mu}}_{\Psi' \rightarrow k} | k \rangle^2 \quad (4.6.39)$$

$$I_{|2\rangle}(\Psi' \rightarrow k) \sim \left(v^3_{|2\rangle k} \cos^2 \theta \sin^2\theta\right) \left\langle \Psi' \middle| \widehat{M}_{\Psi' \rightarrow j} \middle| k \right\rangle^2 \quad (4.6.40)$$

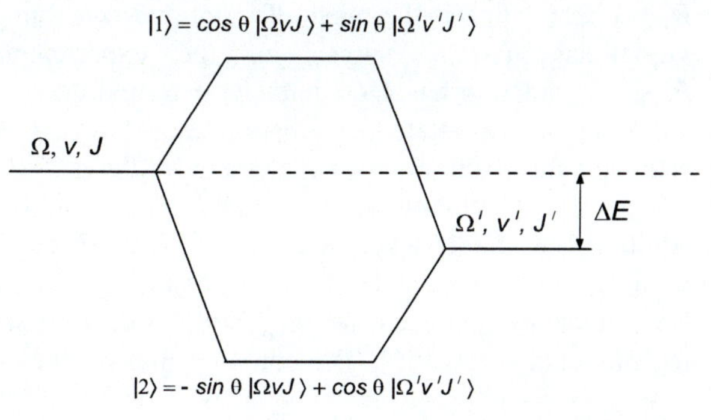

Fig. 4.14 The coupling scheme of two rovibrational states [7] p. 62

In (4.6.37)–(4.6.40), $\widehat{\mu}$ are dipole moment operators of the considered transitions, $\nu_{|1\rangle}$, $\nu_{|2\rangle}$ are transition frequencies.

Equations (4.6.37)–(4.6.40) are applicable for a weak interaction. If the transition to the Ψ state is allowed, an admixture of the Ψ' state results in extra lines corresponding to the $|2\rangle$ excitation, which borrows the intensity from the $|1\rangle$ transition.

4.6.1.6 Hyperfine Interactions Between Iodine Molecule Valence States

Among the ten valence states corresponding to the second $I_2(ab)$ dissociation limit, the only $B0_u^+$ state is deeply bound. Interactions between different electronic states can be observed in the near-dissociation region, where multiple accidental resonances occur. Low-lying rovibrational levels of the *B* state have a hyperfine structure, which has been perfectly described in the framework of the second-order PT. Higher lying rovibrational levels with $E > 20{,}000$ cm^{-1} energy (see Fig. 4.4) can interact directly with the rovibrational levels of other electronic states, which can be described within the first-order PT. Selection rules of the *HFI* define the possible coupling between the *B* state and 9 other ones with $\Omega = 0, \pm 1$ and ± 2 quantum numbers (see Sect. 4.2.3, [7], p. 67 and [50, 53]). Additionally, there is a superhyperfine structure at low-*J* rotational levels, also, where the energy gap between levels is lower than *HFI* matrix elements, and several rotational levels are coupled. These effects result in a complicated rovibrational structure of the *B* state.

Extensive analysis of the $B - X$ transition (more than 10,000 hyperfine lines) was performed in the range from dissociation limit down to 30 cm^{-1} below ($v_B = 71$–82) [50, 53]. It was shown that the perturbations caused by the $\Delta\Omega = 2$ selection rule (*MQ* interaction) are negligible.

The *HF* coupling is effective between near-lying rovibrational levels, which have sufficiently large *FCF*s. Due to differences in the potential curves of the deeply bound *B* and other shallow states, significant overlapping of their vibrational wavefunctions is localized at the *B* state outer branch. Calculation of the *FCF*s using accurate long-range ($R > 7$ Å) *PEC*s [54] shows strong $B0_u^+ \sim c1_g$ and moderate $B0_u^+ \sim (3)1_u$ (see Fig. 4.4) mixing that agrees with experiments. The $B \sim c$ interaction is the most effective and was found near at $v_B = 77$ and 78 [53] vibrational states. Analysis of the experimental data showed that the $B \sim c$ interaction has *MD* character essentially.

Using the two-state perturbation model described above, P. Jewsbury et al. estimated the electronic matrix element for the $B,59,J_B \sim c,14,J_c$ coupling of about 0.1 cm^{-1} [51], similar that obtained by Pique et al. (0.206 cm^{-1}) [50]. Later, several additional resonance levels with $v_B = 57$–59, 63, 69, 71, 74, and 76 were found and used as intermediate states for a population of the $\gamma 1_u$ and $H1_u$ IP states [54, 55]. The interaction between the $B0_u^+$ and (3) 0_u^- (*B'*) states was also found for some rovibrational levels [56]. The coupling matrix elements were not estimated.

Three weakly bound 0_g^+, 1_u and 0_u^- states correlating with the third (*bb*) dissociation limit have very similar potential curves and dense rovibrational structures (see Fig. 4.4). The bulk of accidental resonances between rovibrational levels makes this system an interesting model object for different perturbations studies. Indeed, as shown in [57, 59], this system can be considered a strongly coupled cluster of three electronic states.

Three-step three-color excitation scheme via the $B0_u^+$ and 0_g^+ (*bb*) states

$$I_2\left(\mathrm{IP} \overset{h\nu_2}{\leftarrow} 0_g^+, 0_u^-, 1_u(bb) \overset{h\nu_i^f}{\leftarrow} B, v_B J_B \overset{h\nu_1}{\leftarrow} X, v_X J_X\right) \quad (4.6.41)$$

was successfully realized for investigating the $I_2(bb)$ coupled states in the series of works [57, 58]. Despite the dense system of the rovibrational levels, a limited number of rovibrational levels of (*bb*) states could be populated in experiments if accidental resonances in the $0_g^+, 0_u^-, 1_u$ (*bb*) $\leftarrow B, v_B, J_B$ transitions occur.

The $0_g^+ \sim 1_u$ and $0_g^+ \sim 0_u^-$ are coupled by *HFI*, whereas the 0_u^- and 1_u states are coupled by heterogeneous interaction (see Sect. 4.6.1.3 and [42]). The second step of the multistep excitation scheme is available due to the $0_g^+ \leftarrow B0_u^+$ transition and results in population of the *ungerade* IP states. Two other interacting states, 1_u and 0_u^-, mixed with 0_g^+ one, and transitions to the *gerade* IP states become available, also. Figure 4.15 shows simultaneous excitation of the *F*, *G* and

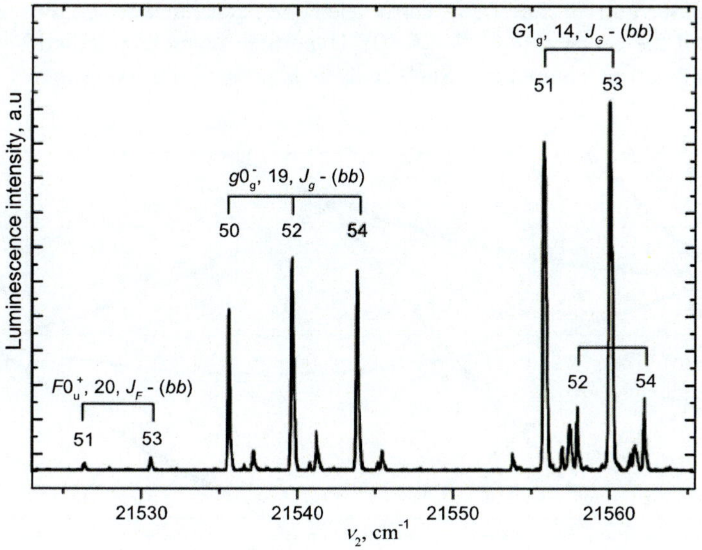

Fig. 4.15 Excitation *F*, *G* and *g* ← (*bb*) spectrum corresponding to the excitation pathways via the *B*, 21, 53 rovibrational state. Luminescence was collected at 285 nm. The weak unassigned lines are due to ‘satellite’ components of ν_2 [7], p. 70

g states from a single rovibrational level. Luminescence spectra of the $F \rightarrow X$, $G \rightarrow A,C$ and $g \rightarrow B'$ transitions (see Figs. 4.4 and 4.10), which arise from each perturbed (*bb*) state were measured to determine *HFI* and heterogeneous interaction matrix elements in the framework of the two-state model described above.

Hyperfine electronic matrix elements determined for the $0_g^+ \sim 1_u$ and $0_g^+ \sim 0_u^-$ couplings were estimated as 0.3(1) cm^{-1} and 0.5(2) cm^{-1}, respectively. The coupling of the 0_g^+ and 0_u^- (*bb*) states can occur by another, indirect mechanism via the 1_u intermediate state in the second-order perturbation approach. The 0_u^- state couples with the $1_u,v_1,J_1$ one by the heterogeneous interaction according to the $\Delta J = 1$ selection rule. The $1_u,v_1,J_1$ state is mixed with the $0_g^+,v_0,J_0$ one by *HFI* with the $|\Delta J| \leq 2$ selection rule. However, the observed 0_g^+,0,49 and 0_u^-,1,48/50 coupling states lie significantly lower than the 1_u, $v_1 = 0$ level ($\Delta E > 90$ cm^{-1}), so the 0_g^+ and 0_u^- states are coupled directly.

4.6.1.7 Hyperfine Interactions Between Iodine Molecule Ion-Pair States

All coupling effects induced by the *HFI* are weak and can be observed for a few rovibrational levels of some IP states despite the dense rovibrational structure of the ion-pair manifolds (see Fig. 4.10). So, the $E0_g^+ \sim \gamma 1_u$ [52], $2_g(^1D) \sim 1_u(^1D)$ [59], and $D0_u^+ \sim \beta 1_g$ [60] coupling were found for some rovibronic levels (see Fig. 4.16 as an example).

The procedure of the *HFI* matrix element determination is reduced to the solution of the coupled (4.6.37–4.6.40). Hyperfine interaction should be analyzed for each hyperfine component. Such analysis is possible for experiments with high

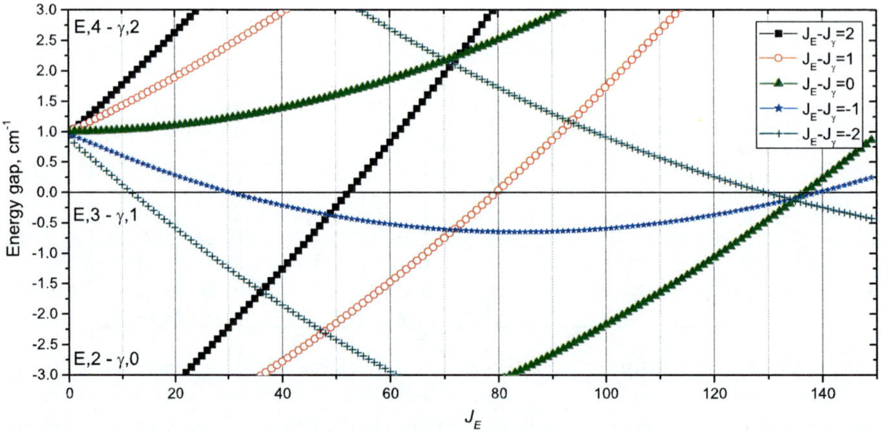

Fig. 4.16 The diagram of the energy gaps between E and γ rovibrational levels accessible for coupling by the hyperfine interaction [7], p. 79

resolution; however, it is often possible to estimate matrix elements averaged over hyperfine components. The coupling of the mentioned above IP states was studied using simpler methods.

The excitation of the coupled $E0_g^+ \sim \gamma 1_u$ states was performed by the standard optical–optical double resonance (OODR) $E \leftarrow B \leftarrow X$ scheme (see Sect. 2.12.1 in [7]). Two pairs of vibrational levels are close enough to find effective hyperfine coupling in the low-v range: $v_E = 3 \sim v_\gamma = 1$ and $v_E = 19 \sim v_\gamma = 18$. Selection rules $\Delta J = 0, \pm 1$, and ± 2 allow to couple different vibrational states at some rotational quantum numbers (Fig. 4.16).

The partly resolved spectrum of the E,3,41 $\sim$ γ,1,41 mixed state excitation is shown in Fig. 4.17 as an example. The $\gamma \rightarrow c,c'$ luminescence is observed at 474 nm due to admixture of the γ zero-order state to both $|1\rangle$ and $|2\rangle$ states (see 4.6.34).

In all experiments mentioned above, rovibronic levels are coupled according to the $\Delta J = 0, \pm 1$ selection rule (the *MD* interaction). No electric quadrupole hyperfine interaction ($\Delta J = \pm 2$) was found. The resolution of measured spectra is limited by laser bandwidth and was insufficient to resolve $|1\rangle$ and $|2\rangle$ perturbed states [52, 59]. In this case, transitions to both mixed states occur within the laser bandwidth, and the (4.6.37–4.6.40) should be considered jointly.

Often, the luminescence of the Ψ and Ψ' zero-order states can be easily separated spectroscopically. For example, the dominating transition from the E state is $E \rightarrow B$ transition in the 400–440 nm spectral range, while emission from the coupling partner $\gamma 1_u$ state occurs mainly at $\sim$350 nm ($\gamma \rightarrow a$ transition) and 450–480 nm ($\gamma \rightarrow c,c'$ transitions) (Fig. 4.18).

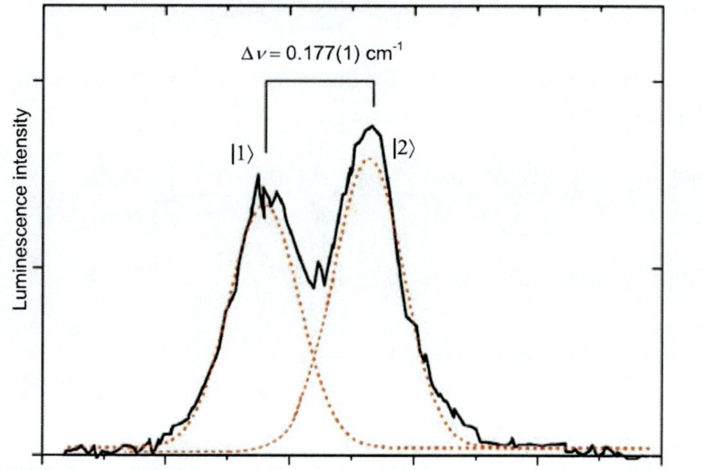

Fig. 4.17 Excitation spectra of the luminescence at 430 and 474 nm. Spectra recorded after population in the $E0_g^+$,3,40 $\leftarrow$ $B0_u^+$,19,21 $\leftarrow$ X,0,20 transition. Both components are approximated by Gauss profile [7], p. 79

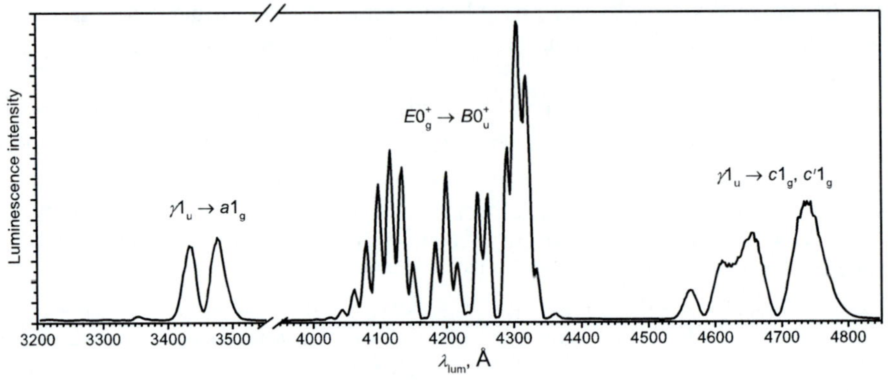

Fig. 4.18 Luminescence spectrum of the coupled $E0^{+}{}_{g}, v_E = 3, J_E = 32 \sim \gamma 1_u, v_\gamma = 1, J_\gamma = 33$ states [7], p. 80

Both $|1\rangle$ and $|2\rangle$ states have admixtures of the Ψ and Ψ' states and contribute to the luminescence (see Sect. 4.6.1.5). However, it is possible to extract admixture coefficients and determine interaction matrix elements using the ratio of the total luminescence of the Ψ and Ψ' components. In the framework of the two-state model (see above) the total luminescence intensity of the $\Psi \to i$ and $\Psi' \to k$ bands from both $|1\rangle$ and $|2\rangle$ are:

$$I_\Psi \sim \left(\cos^4 \theta + \sin^4 \theta\right)^3 \sum\nolimits_i \langle \Psi | \hat{\mu}_{\Psi \to i} | i \rangle^2, \tag{4.6.42}$$

$$I_{\Psi'} \sim 2 \cos^2 \theta \sin^2 \theta \, v^3 \sum\nolimits_j \langle \Psi' | \hat{\mu}_{\Psi' \to k} | k \rangle^2 \tag{4.6.43}$$

The total emission intensity ratio for the transitions, in the optical population of the Ψ' state case, is:

$$\frac{I_\Psi}{I_{\Psi'}} \equiv R_{\Psi/\Psi'} = \frac{2\cos^2 \ \sin^2}{\cos^4 \ + \sin^4} \frac{\int v^3 \mu^2_{\Psi \to i} d}{\int v^3 \mu^2_{\Psi' \to j} d} = \frac{2\tan^2}{1 + \tan^4} \frac{\sum_i A_{\Psi \to i}}{\sum_j A_{\Psi' \to j}}. \tag{4.6.44}$$

According to (4.6.33), it can be written as:

$$R_{\Psi/\Psi'}(J_\Psi) = \frac{1}{1 + \frac{\Delta E^2}{2H_{hf}^2}} \cdot \frac{\tau_{\Psi'}}{\tau_\Psi} \cdot K(J_\Psi) \tag{4.6.45}$$

Here, an additional factor K, if Ψ and Ψ' have different electronic parities ω, has to be introduced. Indeed, let us suppose that the u positive or g negative term of Ψ' is populated in the optical transition. For the u positive and g negative terms the I = 1, 3, 5 are odd (see 4.6.32). On the other hand, any $u \sim g$ hyperfine coupling implies an odd ΔI (see Sect. 4.6.1.4). Therefore Ψ' is coupled with the I = 0, 2, 4 nuclear spin states of Ψ. The total number of the nuclear spin states of Ψ' with odd I and Ψ with even I is $g_{\Psi'} = 21$ and $g_{\Psi} = 15$, respectively. One can construct 15 *bright* states from 21 spin states of Ψ' that interact with 15 spin states of Ψ, and 6 *dark* states which do not interact with Ψ. The population of the dark states of Ψ' does not lead to the optical transition from Ψ resulting in $K = \frac{g_{\Psi}}{g_{\Psi'}} = \frac{15}{21}$. In the opposite case, if the u negative or g positive term of Ψ' is populated in the optical transition, the number of nuclear spin states of Ψ' is less than the number of nuclear spin states of Ψ. In this case, each spin state of Ψ' interacts with Ψ, and therefore the population of every Ψ' hyperfine level leads to the optical transition from Ψ, which result in $K = 1$.

Normalized ratio of luminescence intensities

$$R'_{\Psi/\Psi'} = R_{\Psi/\Psi'}/K \tag{4.6.46}$$

can be approximated by a simple Lorentz function. If the Ψ' state is populated optically, the ratio of luminescence intensities $R_{\Psi'/\Psi}$ should be used:

$$R_{\Psi'/\Psi}(J_\Psi) = \frac{1}{1 + \frac{\Delta E^2}{2H_{hf}^2}} \cdot \frac{\tau_\Psi}{\tau_{\Psi'}} \cdot K(J_\Psi) \tag{4.6.47}$$

Such normalized ratio of luminescence intensities from near-resonant rovibrational levels of the D,12 and β,13 states as a function of energy gap ΔE is shown in Fig. 4.19. Fitting of the experimental data by (4.6.47) gives values of the electronic interaction matrix element $H_{hf}^{el} = 0.049(3)\ \text{cm}^{-1}$. The mixing of the close-lying rovibrational levels of the $\beta, v_\beta = 47$ and $D, v_D = 48$ states have been observed by the population of the β state [60]. The matrix element value is equal $H_{hf}^{el} = 0.063(7)$ cm^{-1} and increases slightly with the vibrational quantum number of the states.

The matrix element of interaction between the E,3,J and γ,1,J vibrational states has found to be similar value $H_{hf}^{el} = 0.084\ \text{cm}^{-1}$ in analogous experiments. Resolution of utilized lasers was insufficient to estimate the matrix element by (4.6.47, 4.6.48) for the E,19,J and γ,18,J coupled states [52]. However, it was shown that near-resonant rovibrational levels are $J_E = 81$ and $J_\gamma = 80$ and the electronic interaction matrix element can be estimated of about 0.02 cm^{-1}, significantly less than that of the $E,3 \sim \gamma,1$ interaction.

Beyond the two-state perturbation model, summation over all hyperfine components should be performed for the total emission intensities calculation. The ratio of luminescence intensities (4.6.45) for the $D \sim \beta$ coupling is given by

$$R_{D-X/\beta-A} = \frac{\sum_{i,j=1}^{3}\sum_{n=1}^{2}\sum\left|\left\langle X\middle|\hat{d}_i\middle|n\right\rangle\left\langle n\middle|\hat{d}_j\middle|B\right\rangle\right|^2}{\sum_{i,j=1}^{3}\sum_{n=1}^{2}\sum\left|\left\langle A\middle|\hat{d}_i\middle|n\right\rangle\left\langle n\middle|\hat{d}_j\middle|B\right\rangle\right|^2} \cdot \frac{\omega_{D\to X}^3}{\omega_{\beta\to A}^3}, \tag{4.6.48}$$

and, similarly, for $R_{\beta\text{-}A/D\text{-}X}$. Here $\sum$ means summation over all hyperfine components of initial, intermediate, and final states.

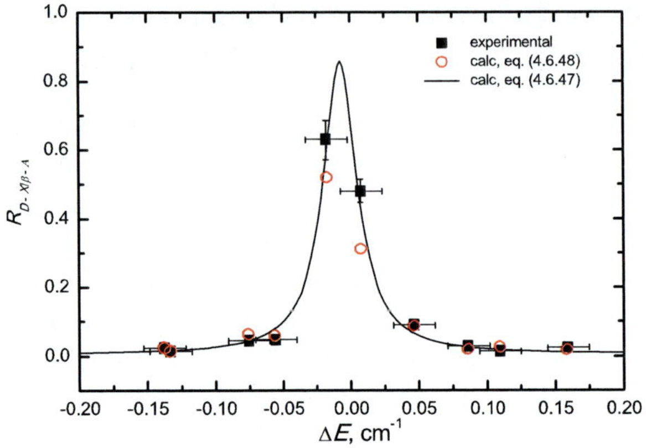

Fig. 4.19 The $R_{D\text{-}X/\beta\rightarrow A}$ normalized ratio determined at the coupled β,13,$J_\beta \sim$ D,12,J_D rovibrational states, plotted as a function of the energy gaps between zero-order states [60]. Squares are the experimental data, and circles are data calculated by (4.6.48). Curves are data obtained by (4.6.47) [7], p. 81 (Baturo, V.V., Cherepanov, I.N., Lukashov, S.S., Petrov, A.N., Poretsky, S.A., Pravilov, A.M.: Hyperfine coupling of the iodine $D0_u^+$ and β1g ion–pair states. J. Phys. B At. Mol. Opt. Phys. **51**, 095,101 (9 pp) (2018). https://doi.org/10.1088/1361-6455/aab6e3. © IOP Publishing. Reproduced with permission. All rights reserved)

In the case of the 2 $_g(^1D)$ $\sim 1_u(^1D)$ coupling, interaction matrix elements were determined by solving the equations in the two-state model. Laser bandwidth was also insufficient to resolve the splitting of the near-resonant rovibrational levels of the 2 $_g(^1D)$ and $1_u(^1D)$ IP states [59], and the authors employed another method. The line shapes were deconvoluted by two lines with the Lorentzian form to derive their transition frequencies for determining the energies of coupled rovibrational levels. The electronic interaction matrix element was determined as H_{hf}^{el} = 0.0482 (59) cm^{-1}, similar to that for $D \sim \beta$ and $E \sim \gamma$.

Due to the mixing of the zero-order states, which have different radiative lifetimes (τ and τ' for states Ψ and Ψ', respectively), the radiative lifetimes for the perturbed $|1\rangle$ and $|2\rangle$ states are determined as follows:

$$\tau_1^{-1} = \tau^{-1} cos^2\theta + \tau'^{-1} sin^2\theta \qquad (4.6.49)$$

$$\tau_2^{-1} = \tau^{-1} sin^2\theta + \tau'^{-1} cos^2\theta \qquad (4.6.50)$$

Direct measurements of the radiative lifetimes allow estimating the mixing angle, which is determined in (4.6.35) and hence, the matrix elements of the hyperfine interaction. Figure 4.20) demonstrates the reasonability of this method. Luminescence lifetimes observed for the $1_u(^1D)$ and 2 $g(^1D)$ mixing states strongly depends on the rotational quantum number near the avoiding crossing $J_{1u} = 25 \sim J_{2g} = 24$ range. The upper energy component has a lifetime longer than that of the lower one after avoiding crossing and vice versa.

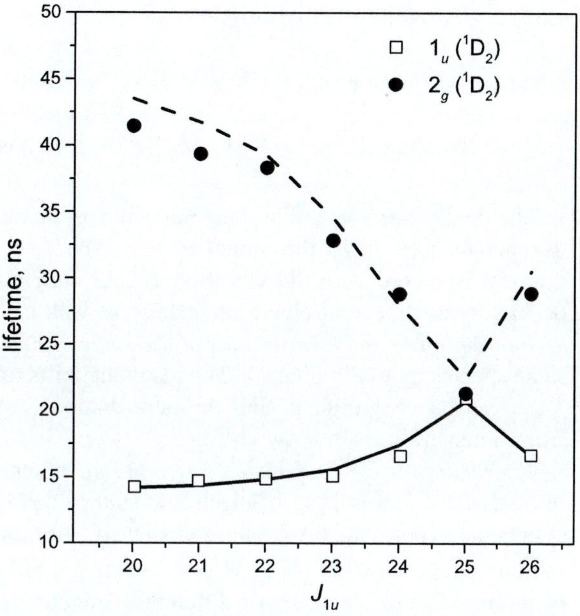

Fig. 4.20 Luminescence lifetimes of the $1_u(^1D)$ and $2_g(^1D)$ mixed states as a function of the J_{1u}. The solid and dashed lines show the calculated lifetimes with the assumption that the hyperfine coupling matrix element $H_{hf} = 0.0482(59)$ cm^{-1} [7], p. 83, [59] (Reproduced from Y. Nakano, H. Ukeguchi, T. Ishiwata, J. Chem. Phys. **121**, 1397–1404 (2004). https://dx.doi.org/10.1063/1.1756134 with the permission of AIP Publishing)

4.6.2 *Perturbations Between Bound States of Polyatomic Molecules. Anomalously Long Radiative Lifetime of Polyatomic Molecules*

As in diatomic molecules, electronic states of a polyatomic molecule lying fairly close may perturb each other. If the spin-orbit interaction between the states is negligible, both states have to possess the same orbital species and the same multiplicity, to be perturbed. If it is not the case, perturbations can occur between the states of different orbital species, but species of two electronic states must not differ not more than species of one of the normal vibrations [10], p. 66. For example, in the C_{2v} point group a $^1A_1$ state can be perturbed by a $^3A_2$ state, but one $^1A_1$ component of the $^3A_2$ state can mix with the former $((A_2 + B_1 + B_2) \times A_2 = A_1 + B_2 + B_1)$. If $^3A_1$ and $^2A_2$ states are perturbed, two, $B_2 + B_1$, components of the former are perturbed by the $B_2 + B_1$ components of the later (see Tables 56, 57 in [10]). Electronic states can also be perturbed due to vibronic and Coriolis interactions; the latter leads to coupling between rotation and vibration. Perturbations in polyatomic molecule are described in detail in [10, 61]. Therefore, the author will review them very briefly.

4.6.2.1 Vibronic Perturbations

A simple representation of vibronic wave function

$$\Psi_{ev}(r, Q) = \Phi(r, Q) \cdot \chi(Q), \tag{4.6.51}$$

similar to (4.2.4) for a diatomic molecule is invalid if two electronic states are degenerate, i.e., have the same energy. The two electronic eigenfunctions are thoroughly mixed with the vibrational wave functions. A partial mixing may also occur if two or several electronic states are well separated [10], p. 65.

Interaction of electronic states of the same symmetry. All electronic states of the same species mutually interact. The resultant shifts of rovibronic levels or change of *PES*s can be insignificant, and the perturbation leads to changes in over-all transition intensities.

Interaction of electronic states of different symmetry. As it has been mentioned in Sect. 4.2.2.1, a mixing of electronic states of polyatomic molecules of different symmetry types can be brought about by vibronic interaction. Identity of the vibronic wave functions Ψ^{i}_{ev}, Ψ^{j}_{ev} species is possible for suitable vibrational levels of the two electronic states of different symmetry types. The two electronic states may perturb each other if their species differ by than the species of one of the normal vibrations that is the direct product of a normal vibration species and that of one of the electronic states is equal to the species of the other electronic state (see Chap. III, and Table 57 in [10]).

4.6.2.2 Anomalously Long Radiative Lifetime of Polyatomic Molecules

Let us consider the behavior of a small polyatomic molecule excited state $\widetilde{B}$ populated in an optical transition from the ground state $\widetilde{X}$ and mixed with another state $\widetilde{A}$ isoenergetic with it (Fig. 4.21).

Let the $\widetilde{A} \leftarrow \widetilde{X}$ optical transition be forbidden. The mixing of the $\widetilde{A}$ and $\widetilde{B}$ states can be brought about by vibronic, rotational-electronic, rovibronic, and *S–O* interactions (see [10], p. 136–142 and Sects. 4.2.2, 4.2.3). The perturbation operator in the full Hamiltonian describing these states has the form corresponding to the perturbation. As noted in Sect. 3.6, the presence of any of these perturbations leads to the fact that the 'true' wave functions corresponding to the levels of the $\widetilde{A}$ and $\widetilde{B}$ states are 'no longer' a solution to the Schrödinger equation with $\widehat{H}_0$. Each of these states is a superposition of the states with the same total wave function species (see 3.6.3, 3.6.4, 4.2.15, 4.2.16). The intensities of transitions to the $\widetilde{B}$ state rovibronic levels 'become' lower than those calculated in the zero approximation, i.e., without the mixing. Accordingly, the intensities of transitions to the $\widetilde{A}$ state rovibronic levels, which are forbidden in the zero approximation, 'become' nonzero. The oscillator strength (see Sect. 4.3) 'has been pumped' from the $\widetilde{B} \leftarrow \widetilde{X}$ to the $\widetilde{A} \leftarrow$

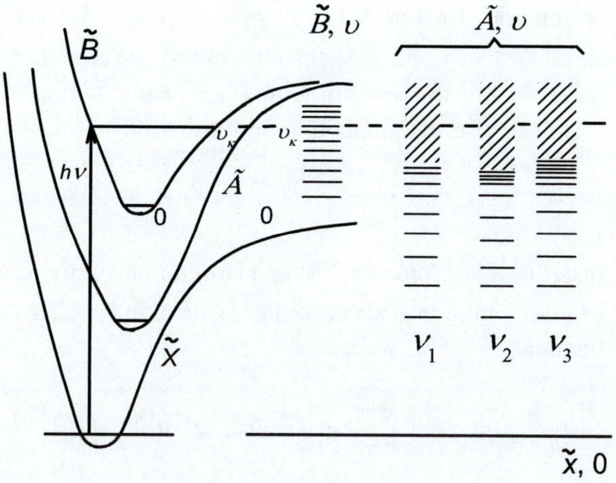

Fig. 4.21 To the explanation of an anomalously long radiative lifetime [3], p. 40

$\widetilde{X}$ one. In a fairly accurate approximation, the sum of the oscillator strength of the transitions to the $\widetilde{A}$ and $\widetilde{B}$ states remains unchanged and equal to the oscillator strength of the transition to the unperturbed $\widetilde{B}$ state.

The correctness of the last statement can be illustrated on the model problem under consideration, mixing of the $\widetilde{A}$ and $\widetilde{B}$ states. Let us calculate the probability of an optical transition from the ground state vibronic level v'' to all 'true' ones, i.e., mixed $\widetilde{A}$ and $\widetilde{B}$ state levels:

$$\left|\mu_{\widetilde{B},\widetilde{A},v'\leftarrow\widetilde{X},v''}\right|^2 = \sum\nolimits_{v'}\sum\nolimits_{n}|\langle\Psi_n^*|\widehat{\boldsymbol{\mu}}|\Psi_{v''}\rangle|^2, \tag{4.6.52}$$

Ψ_n are the mixed vibronic states.

Let us expand the wave functions of 'true' vibronic states Ψ_n in the set of wave functions similar to (3.6.3) that are solutions of the Schrödinger equation for Hamiltonian $\widehat{H}_0$, and normalize the expansion coefficients:

$$\Psi_n = \sum\nolimits_i a_{in}\Psi_i^0 \tag{4.6.53}$$

$$\sum\nolimits_i |a_{in}|^2 = \sum\nolimits_n |a_{in}|^2 = 1. \tag{4.6.54}$$

One has to take into account that the matrix element

$$\sum\nolimits_i a_{in}\langle\Psi_n^{0*}|\widehat{\boldsymbol{\mu}}|\Psi_{v''}\rangle \tag{4.6.55}$$

is non-zero for one $\Psi_i = \Psi_l$ state only (the matrix elements of the interaction of levels of the $\widetilde{A}$ and $\widetilde{B}$ states are not too large, and of all the levels corresponding to the $\widehat{H}_0$, only for one l-th admixture of the $\widetilde{B}$ state is large).

Let us use the adiabatic approximation

$$\Psi_l = \Phi(Q)\chi_l, \tag{4.6.56}$$

insert (4.6.56) into (4.6.55) and sum up the expression on the *rhs* of (4.6.52) from 0 to n, i.e., take into account the absorption on all $\widetilde{B}, v'$ vibronic levels. Considering this and (4.6.54), we get

$$\left|\mu_{\widetilde{B},\widetilde{A},v'\leftarrow\widetilde{X},v''}\right|^2 = \sum\nolimits_n |a_n|^2 \left|\left\langle\Phi_{nl}(Q)|\widehat{\boldsymbol{\mu}}|\Phi_{\widetilde{X}}(Q)\right\rangle\right|^2 |\langle\chi_{nl}|\chi_{v''}\rangle|^2 = 1\left|\mu_{\widetilde{B}\leftarrow\widetilde{X}}\right|^2 q(v', v'') \tag{4.6.57}$$

Let the unperturbed states $\widetilde{B}$ vibrational levels be so far apart from each other that the state (4.6.53) levels for each $\widetilde{B}, v'$ level do not overlap with each other. In other words, there is a set (4.6.53) for each $\widetilde{B}, v'$ level. Therefore, to obtain the intensity of the transition from the $\widetilde{X}, v''$ state to all levels of the $\widetilde{A}$ and $\widetilde{B}$ states, one has to sum up the (4.6.57). Taking into account (4.6.58)

$$\sum\nolimits_{v'} q(v', v'') = \sum\nolimits_{v''} q(v', v'') = 1, \tag{4.6.58}$$

(see 4.3.5a, 4.3.5b) we have:

$$\left|\mu_{\widetilde{A},\widetilde{B},v'\leftarrow\widetilde{X},v''}\right|^2 = |\mu_{B\leftarrow X}(Q)|^2 \sum\nolimits_{v'} q(v', v'') = |\mu_{B\leftarrow X}(Q)|^2. \tag{4.6.59}$$

Thus, the oscillator strength of the transition to the mixed state is equal to that of for transition to the unperturbed $\widetilde{B}$ state.

However, this does not in any way relate to spontaneous emission. In the case of two states mixing, one of which does not combine optically with the ground state, it is impossible to obtain the Einstein spontaneous emission coefficient for any v' level from the integral absorption coefficient, i.e., use Strickler-Berg formula [9], p. 40, [62]:

$$A(v') = 2.8 \cdot 10^{-8} \sum\nolimits_{v''} \widetilde{v}^3 (v'v) q(v'v) \int \frac{k(\widetilde{v})}{\widetilde{v}} d\widetilde{v}. \tag{4.6.60}$$

Indeed, the Einstein spontaneous emission coefficient for a transition from one v' level to any v'' level is

$$A_{\widetilde{B},v'\to\widetilde{X},v''} \sim |\mu_{v'v''}|^2 = a_n^2 \left|\mu_{\widetilde{B}\to\widetilde{X}}\right|^2 q_{v'v''} \tag{4.6.61}$$

i.e., $a_n^2/\sum_n a_n^2$ less than that of from unperturbed $\widetilde{B}, v'$ level (see 4.4.2 and 4.6.59). The $\widetilde{B}, v' \to \widetilde{X}, v''$ transition oscillator strength is 'spread' over n—l $\widetilde{A}, v'$ state levels. The authors of the [63, 64] have shown that anomalously long radiative lifetimes of some electronically excited states of the CS_2, SO_2, and NO_2 molecules are due to this factor, mainly (see a describing the similar effect in Sect. 4.8).

The radiative lifetime of the NO_2 molecule states populated at 4358 Å is ~44 μs, while that of calculated from the absorption coefficient in the 2600–7000 Å is 0.26 μs [65], ~170 times lower. The dissociation limit of the NO_2 ($\widetilde{X}^2$ A$_1$) state corresponds to 3997 Å [10], Table 64), and 4358 Å corresponds to transitions to rovibronic levels of the $^2B_1$ and $^2B_2$ states which lie lower than the dissociation limit. Therefore, rovibronic levels of the $^2B_1$ and $^2B_2$ states interact with those of the $\widetilde{X}^2$ A_1 state.

To describe the effect under discussion, one can apply a semi-classical approximation. Following the $\widetilde{B}, v' \leftarrow \widetilde{X}, v''$ transition, a nonradiative transition occurs from the $\widetilde{B}, v'$ state to the ensemble of the $\widetilde{A}, v$ levels, which cannot optically combine with the state $\widetilde{X}, v''$ and back. Let the number of levels in the $\widetilde{A}, v$ state be large. Therefore, most of the time, the molecule is in this state. Consequently, the radiative lifetime of the excited state 'must' increase.

One can note that the intermodal energy exchange in the $\widetilde{A}, v$ state can significantly increase the effect under discussion. Suppose that there are optical transitions to the vibrational levels $v' = 0 - k$ of the $\widetilde{B}$ state, $\widetilde{B}$,0-k, and, as noted above, to the vibrational levels of the $\widetilde{A}, v_1$ state, mixing with which is allowed by energy and symmetry considerations (active mode v_1 [66]) (see Fig. 4.21). Let mixing of the $\widetilde{B}, v'$ levels with the $\widetilde{A}, v_2, v_3$ levels be forbidden by the selection rules (v_2, v_3 are passive modes), and the number of active and passive levels is limited. If the k level is high enough and, therefore, the anharmonism is great, an effective intermodal energy exchange between the $\widetilde{A}, v_1$ and $\widetilde{A}, v_2, v_3$ levels can occur. This exchange leads to the fact that the excited state is no longer described by the superposition of the $\widetilde{B}$,k and $\widetilde{A}, v_1$ state wave functions. There is a 'drain' of excitation from these levels to levels that are not mixed with the $\widetilde{B}, v'$ state. This 'drain' leads to a decrease of the a_n coefficients in (4.6.57) and a greater increase in the radiation lifetime of the excited state. The described effect should be even stronger if the bottom of the potential well of the $\widetilde{A}$ state is lower than the level. This is the case in the NO_2 molecule.

In conclusion, the author notes the following. The above quantum–mechanical explanation of the anomalously long radiative lifetime was based on a rough adiabatic approximation, i.e., assuming that the optical transition moment is independent of the nuclear coordinates. However, in many cases, this may not be the case at all: the optical transition in absorption occurs in one nuclear configuration and spontaneous emission in another. For example, in CO_2 molecule, the absorption in the *A*-band is realized from the linear ground state to the linear one in the Franck–Condon zone, whereas the upper state is bent in the equilibrium configuration. Besides, radiation takes place from the bent excited state to the ground state bent configuration, though its equilibrium configuration is linear. Moreover, if the equilibrium configurations of the upper and lower states are different, then the transition in absorption can be carried out on one part of the *PES*, and the spontaneous emission occurs from the other. Part of the *PES*, from which the spontaneous radiative transition occurs, can be mixed with the metastable non-emitting state, and the part of the *PES*, to which the transition in absorption occurs, is not mixed. All this may lead to the fact that the matrix elements of transitions in absorption and radiation can be different.

4.7 Electronic Predissociation of Di- and Polyatomic Molecules. Types of Predissociation Processes

Electronic predissociation (otherwise, the first case of Herzberg predissociation [10], p. 458) is a *nonadiabatic* process (on the one hand) and a consequence of mutual perturbation of various electronic states (on the other). Here one of the interacting states belongs to the continuous spectrum in the considered energy region (Fig. 4.22, see Fig. 4.23, also).

Mixing of the bound $\Psi_{1,v,\,J}$ and repulsive (continuum) $\Psi_{2,v,\,J}$ states are governed by the interaction matrix element

$$H_{v,J;E,J} = \left\langle \Psi_{1,v,J} \middle| \widehat{\boldsymbol{H}} \middle| \Psi_{2,E,J} \right\rangle = \left\langle \Phi_1(r,R)\chi_{v,J} \middle| \widehat{\boldsymbol{H}} \middle| \Phi_2(r,R)\chi_{E,J}(R) \right\rangle \qquad (4.7.1)$$

The coupling operator for predissociation $\widehat{\boldsymbol{H}}$ in (4.7.1) has the same origin as for perturbations (see [31], p. 505, Sect. 4.6.1 and Table 4.1).

Vibrational and rotational predissociation (the second and third cases of predissociation, according to Herzberg) are *adiabatic* processes in the sense that they can be considered as proceeding along one *PES*.

We have examined the general laws for the mutual perturbation of electronic states in Sects. 3.6 and 4.6.1.1 and seen that two adiabatic states are mixed if:

1. The species of their electronic, vibronic, electron-rotational, or rovibronic wave functions are the same (see 3.6.5, 3.6.6a–3.6.6e);
2. In a particular region of the internal coordinates, the energy difference between their adiabatic *PEC*s (*PES*s) is relatively small, i.e., adiabatic *PEC*s (*PES*s) are close enough to each other (Fig. 4.22a), intersected (Fig. 4.22b, c)or quasi-intersected (Fig. 4.22 d);

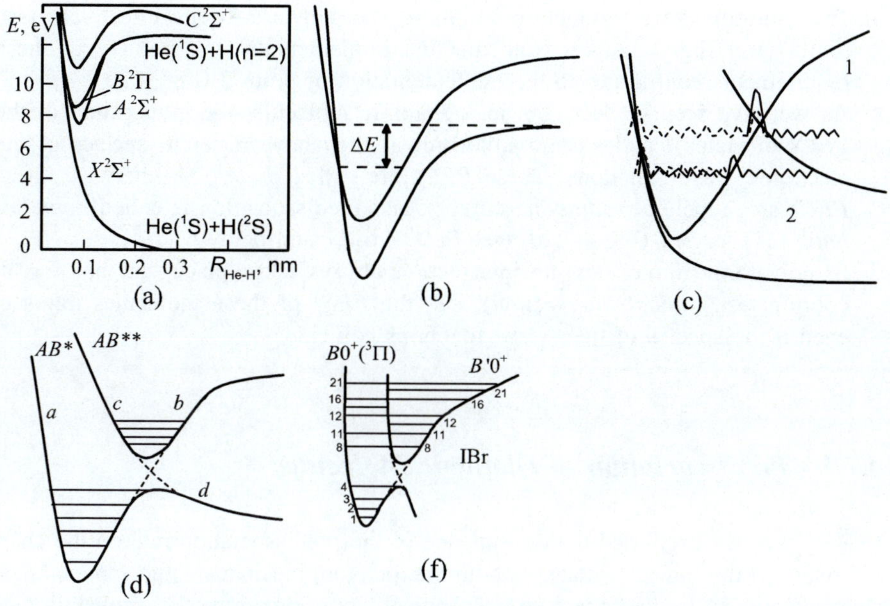

Fig. 4.22 Electronic predissociation of diatomic molecules: Vibrational wave functions of bound states are shown as dotted lines. Those of repulsive are shown as solid line [3], p. 45

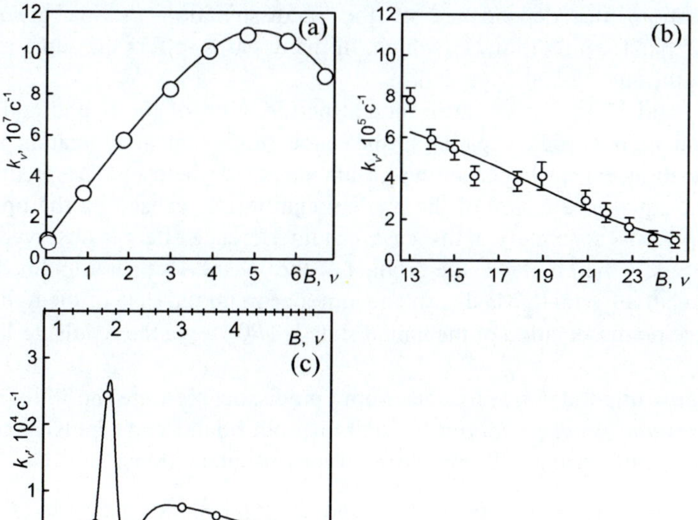

Fig. 4.23 Dependences of the gyroscopic predissociation rate constants on the vibrational quantum numbers of the HeH($B^2\Pi$) (**a**), and dihalogens $X_2(B0_u^+)$ states, X = Cl (**b**) and Br (**c**) [3], p. 50]

3. The vibrational wave functions of these states overlap quite effectively (see (3.6.8) and Fig. 4.22a–c). Note that the implementation of point 2 of these requirements contributes to the implementation of point 3 (Fig. 4.22).
As we have seen in Sect. 3.6, in a diatomic molecule, the intersection of the *PEC*s of states *i* and *j* is possible only if they have different species of the electronic wave functions. These *PEC*s are called *diabatic*. Otherwise, these *PEC*s are 'repelled' (*adiabatic* curves), and predissociation is called *nonadiabatic* [31], p. 519 (Fig. 4.22d) (see Sect. 4.6.1.1 and below).
In polyatomic molecules, an intersection always takes place, but not for all coordinates (*conical intersection*), i.e., the *PES*s of these molecules intersect even if the species of their wave functions coincide.

4.7.1 Predissociation of Diatomic Molecules

In Sect. 3.6, we examined the dependence of the predissociation rate on the characteristics of the molecule states in both the quantum mechanical and semiclassical approximations. In the quantum mechanical approximation, the probability of predissociation when passing the vicinity of the closest approximation of two *PEC*s depends on the matrix elements of the interaction of V_{ij}, i.e., in a certain approximation, on the values of the electronic interaction matrix elements A_{el} and the vibrational wave function overlap integral, see (3.6.8). In the semiclassical approximation, the dependence of the predissociation rate is given by the Landau-Zener formula (3.6.21), which, in many cases, offers the same result as in the quantum mechanical approach.

Let V_1 and V_2 operators cause mutual perturbation of the bound and repulsive states, and their complete parity is the same (these are all operators that cause predissociation except hyperfine and Stark ones (see below). The predissociation rate is not equal to the sum of the predissociation rate caused by the operators V_1 and V_2 operators separately in this case. An interference effect is observed, that is, a term appears, which includes the product of the V_1 and V_2 operators; most often its value is relatively small. Maybe, this feature occurs in the case of the hyperfine and gyroscopic predissociation of the bound state I_2 ($B0_u^+$) via the repulsive $1_u(^1\Pi)$ (see below).

The most important types of electronic predissociation are the following:

Electrostatic predissociation. It occurs if both bound and repulsive states have the same symmetry type. There are two types of interactions:

(a) Interactions between noncrossing (adiabatic) curves caused by the nuclear kinetic energy operator $\widehat{T}^N$. This type of predissociation is called *nonadiabatic*. The $N_2^+(C^2\ \Sigma_u^+ \sim B^2\ \Sigma_u^+)$ interaction is analysed in [31], p. 531) as an example of nonadiabatic predissociation.

(b) Interactions between crossing curves belonging the states of the same symmetry, $N_2(C^3\Pi_u \sim C'\,^3\Pi_u)$ [31], p. 514, as well as Rydberg and valence state if their configurations differ by two orbitals [31], p. 310), for example.

Spin–orbit predissociation. The spin-orbit interaction is the strongest among the fine-interaction components caused by the fine-structure operator $\widehat{\boldsymbol{H}}_{FS} = \widehat{\boldsymbol{H}}_{SO} + \widehat{\boldsymbol{H}}_{SS} + \widehat{\boldsymbol{H}}_{SR} + \widehat{\boldsymbol{H}}_{OR}$ (see Table 4.2). Accordingly, the predissociation rate is determined by the matrix elements of the *S–O* interaction $\widehat{\boldsymbol{H}}_{SO}$ (see 4.6.16). Three example of strong *S–O* predissociation are given in [31], Table 7.4, namely, OH $(A^2\Sigma^+ \sim {}^4\Pi)$, electronic matrix element, $H_{el} = 57$ cm^{-1}, $O_2(B^3\ \Sigma_u^- \sim {}^5\Pi_u)$, $H_{el} = 65$ cm^{-1}, Se_2 $(B0_u^+ \sim 0_u^+)$, $H_{el} = 373$ cm^{-1}.

Gyroscopic (Coriolis) predissociation. It is caused mainly by the interaction of the orbital and, to a lesser extent, the electron spin moment a with the momentum of the rotational motion of the molecule ($\widehat{\boldsymbol{L}}$*-uncoupling operator* (4.6.20), and leads to heterogeneous predissociation ($\Delta\Lambda = \pm 1$, $\Delta\Omega = \pm 1$). The rate of gyroscopic predissociation (before the discovery of hyperfine predissociation, see below, it was called simply heterogeneous)

$$1/\tau_{gyr} = k_v \cdot J \cdot (J+1) \tag{4.7.2}$$

where k_v is the gyroscopic predissociation rate constant, J is the rotational quantum number. For large values of J, k_v can vary

$$k_v = C_v^2[1 + p_v \cdot J(J+1) + \ldots] \tag{4.7.3}$$

$p_v \cdot << C_v^2$. Gyroscopic predissociation is manifested in iodine molecule $B0_u^+$ state. The B state high rovibrational levels are populated in the $B \leftarrow X$ transitions, and the rate of gyroscopic predissociation, $1/\tau_{gyr}$ could be large due to low rotational constants of the X state ($B_e^X = 0.037$ cm^{-1} see Table A1 in [7]). The $k_v \leq 250$ s^{-1}, (Fig. 4.24), and $1/\tau_{gyr} < 1.3 \cdot 10^5$ s^{-1} is much less than the radiative decay rate.

It was proved that $I_2(B)$ state gyroscopic predissociationis due to the $B \sim C1_u$ coupling [68] since the predissociation rate follows v-dependences of *B/C* Franck-Condon density (*FCD*) (Fig. 4.25).

Hyperfine predissociation. A coupling of the nuclear and electronic angular momenta (the hyperfine magnetic-dipole (*MD*) and electric-quadrupole (*MQ*) interaction, selection rules are $\Delta\Omega = \pm 1$ and ± 2, respectively) is the only non-Born–Oppenheimer term of molecular Hamiltonian that breaks *g/u* symmetry of electronic states of a homonuclear diatomic molecule and mixes near-degenerate rovibrational levels of the states of opposite electronic parity [50], see Sect. 4.6.1.4, also. Hyperfine predissociation (*HFP*) was discovered in 1976 [67] and has been fairly well studied using the example of the predissociation I_2 $(B0_u^+)$.

In principle, the $I_2(B)$ state *HFP* could be due to its perturbation by the all (*aa*) states except the 3_u (see Sect. 4.2.3). However, it was proved that it is due to the $B \sim C$ coupling since the predissociation rate follows v-dependences of *B/C* Franck-Condon density (*FCD*) (Fig. 4.25) just as those of B state gyroscopic

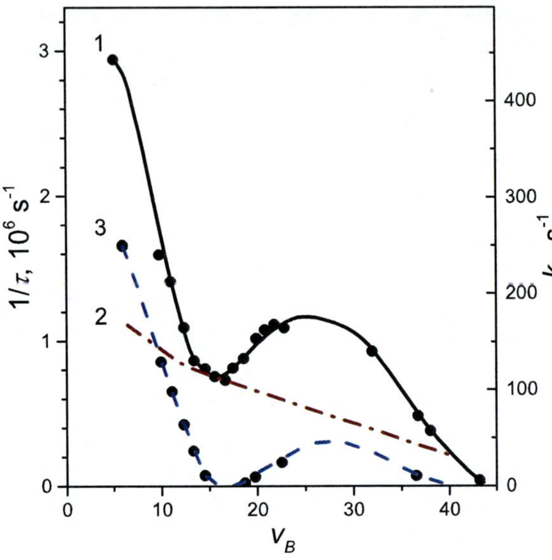

Fig. 4.24 The rate of spontaneous decay of the $I_2(B, v_B)$ vibronic levels, total for $J_B = 0$ (1), radiative (2) as well as gyroscopic predissociation constant k_v (3) [7], p. 63

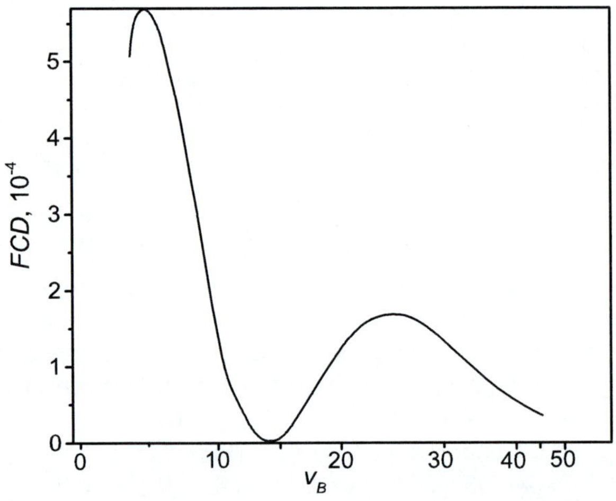

Fig. 4.25 The $I_2(B/C)$ Franck–Condon density as a function of v_B [7], p.64

predissociation. The *B/C FCD* has a strong maximum at $v_B = 5$ and another one at $v_B \approx 24$ and then decreases smoothly. The *B/a'* and *B/a FCD*s are highly oscillating functions of v_B [68].

The rate of $B \sim C$ gyroscopic and hyperfine predissociation is [7], p. 64, [69]:

$$\Gamma_{pred}(v, J, \varepsilon, F) = \sum\nolimits_I |\alpha(I\varepsilon JF)|^2 \Gamma_{IJF}. \tag{4.7.4}$$

Here.

ε is a label,

$$\Gamma_{IJF} = C_v^2 J(J+1) + \frac{\alpha_v^2}{3}\left(\mathrm{I}^2 + \frac{3(\mathbf{I.J})^2 + \frac{3}{2}\mathbf{I.J} - \mathbf{I}^2\mathbf{J}^2}{(2J-1)(2J+3)}\right) - \alpha_v C_v \sqrt{2}\mathbf{I.J}, \quad (4.7.5)$$

$$\mathbf{I.J} = (F(F+1) - J(J+1) - I(I+2)/2, \quad (4.7.6)$$

F is the total angular momentum [39]. For $I_2(B)$ state, $\Gamma_{HFP} \approx (0.4–1.7) \cdot 10^6\ c^{-1}$ in the v_B = 5–40 range, slightly different for the *ortho-* and *para*-states. The hyperfine (α_v^2) and gyroscopic (C_v^2) predissociation parameters follow the *B/C FCD*, and their ratio is independent of v_B [7], p.64.

According to Fermi's golden rule [31], p. 535, *HFP* rate is equal to

$$\Gamma_{HFP} = \frac{4\pi^2}{h}\left\langle \Psi_B \middle| \widehat{\boldsymbol{H}}_{HFI} \middle| |\Psi_C \right\rangle^2. \quad (4.7.7)$$

Therefore, the $I_2(B)$ *HFP* matrix element is

$$\left\langle \Psi_B \middle| \widehat{\boldsymbol{H}}_{HFI} \middle| \Psi_C \right\rangle = \sqrt{\frac{\Gamma_{HFP} \cdot h}{4\pi^2}} \approx (0.1\text{–}0.5) \cdot 10^{-2}\mathrm{cm}^{-1}. \quad (4.7.8)$$

Accidental (indirect) predissociation is caused by a perturbation of a rotational level J_1 of the bound (unperturbed by the repulsive) state by a predissociated level $J_2 = J_1$ that lies nearby (see details in [31], Sect. 7.13).

Several examples of mutual positions of the bound and repulsive states and their interactions are given in Fig. 4.22.

The HeH($A^2\Sigma^+$ and $C^2\Sigma^+$) bound states in Fig. 4.22a have the same symmetry type as the repulsive $X^2\Sigma^+$ one; they are repelled. The HeH($B^2\Pi$ and $X^2\Sigma^+$) do not cross, also, and the former undergoes the ($B^2\Pi \sim X^2\Sigma^+$) gyroscopic predissociation [70, 71] (see Fig. 4.23a).

Since the dependence of the *FCF*s on the vibrational quantum number of a bound state level are different for different cases of mutual positions of bound and repulsive states, the predissociation rate constants depend on the vibrational quantum number of the level in various ways. The most typical are those shown in Fig. 4.23a–c and obtained at the locations of the *PEC*s shown in Fig. 4.22a–c (*PEC*s 1, 2), respectively.

The intersection in Fig. 4.22b differs from the intersection in Fig. 4.22c (the left repulsive *PEC* wing); the former takes place at the energy less by ΔE than that necessary for predissociation. Therefore, the predissociation occurs higher than the intersection point. This is the case of the $Cl_2(B^3\Pi_u(0^+) \sim A^3\Pi_u(1))$ gyroscopic predissociation [27] (see Fig. 4.23b). Figures 4.22c and 4.23c corresponds to the

$Br_2(B^3\Pi_u(0^+) \sim C^1\Pi_u(1))$ gyroscopic predissociation. The dip and decrease of the $k_v = f(v)$ observed (Fig. 4.23c) is a consequence of the dephasing of the vibrational wave functions and (or) their different periods.

One must be aware that the separation of the concepts of adiabatic or diabatic *PEC*s is arbitrary; these coupling limits are applicable if the probability $P_{1,2}$ (Figs. 3.16, 3.17) is close to 1 or 0. Otherwise (interaction of an intermediate force), it is impossible to draw a boundary between these couplings.

Figure 4.22d shows the case of a 'strong' interaction of the adiabatic bound and repulsive states, In this case, the probability of a nonadiabatic transition from point *a* to point *b* is much less than from point *a* to point *d* and, accordingly, from point *c* to point *d* is much smaller than from point *c* to point *b*. The AB* states at energies greater than the corresponding barrier to the *PEC* of AB* practically belong to the continuous energy spectrum.

Often there is a case of interaction of states of the same type of symmetry of intermediate strength. Figure 4.22d shows the *PEC*s of the IBr($B0^+$ and $B'0^+$) states, which are formed as a result of the interaction of the bound and repulsive 0^+ states. The matrix element of the interaction of these states is not too large, and in the absorption spectrum at an energy larger than the corresponding intersection point, two progressions of the bands are observed, corresponding to transitions to the *B* and *B'* states, i.e., the *B* state in this energy region does not belong to the continuous energy spectrum. Accordingly, a transition to a part of the *PEC* that is above the dissociation threshold of *B* state with a probability greater than 0.62 leads to a nonadiabatic transition between the *B* and *B'* states. One can populate the *B* state rovibronic level, and observe luminescence from state *B'* and vice versa [27].

4.7.2 Predissociation of Polyatomic Molecules

The predissociation of polyatomic molecules is discussed in detail in Herzberg's monograph [10]. Therefore, in this book, the author will note only the following. The possibilities of predissociation in polyatomic molecules are much larger than in diatomic ones: this is a direct consequence of the multidimensionality of *PES*s, in particular, the increase in the lifetime of excited states due to the coupling of various vibrational modes and dimensions of the *PES* interaction regions. The motion of an image point in the predissociation is described as follows: when a *PES* quasi-intersection or intersection region passes, an image point with probability *P* jumps from one *PES* to another, so that when exiting this area, there will already be two trajectories, one on the original *PES*, the other on the 'new'. Further, the system is described by adiabatic motion until one of these trajectories comes up again to this 'interesting' area and again forks, etc.

4.8 Dissociation of Polyatomic Molecules. Long-Lived States of Polyatomic Molecules

Dissociation of polyatomic molecule excited states has a significantly different character than that of diatomic; the possibility of their consideration in one state framework is common only. It is true if there are no nonadiabatic processes during the decay of these states, see below). We have already discussed them in Sect. 3.5, and here we will dwell on anomalies of the dissociation processes.

Dissociation of a diatomic molecule, i.e., decay of a molecule in a state corresponding to a continuous spectrum (purely repulsive or bound states, but with an energy higher than the dissociation limit, taking into account potential barriers) proceeds in half of one vibration, i.e., for a time of the order of 10^{-13} s.

The dissociation of a polyatomic molecule, even in a purely repulsive state, as is seen in Sect. 3.5, cannot be considered as one vibration. Let us examine a dissociation of the bound state of a linear symmetric triatomic CO_2 molecule to O + CO (valley *c*) or OC + O (valley *d*) (Fig. 4.26). One can describe any vibration of a polyatomic molecule, including dissociation or vibrational predissociation as a motion of an image point on multidimensional *PES*. The vibration described by the motion of an image point along the bisector of the OCO angle (straight line *a-a* in Fig. 4.26) is symmetric valence: the OC and CO distances, in this case, are equal.

If an image point motion is carried out along the other half-axis of the ellipse (straight line *b-b*), then this is an antisymmetric stretching vibration (see Sect. 3.5). Let the absorption in CO_2 molecule from the minimum of the ground state corresponds to point *C* lying higher than the dissociation limits O–CO and OC–O. The CO_2 dissociation corresponds to the exit of an image point into the valleys *c* or *d*. Therefore, an image point motion when describing the dissociation cannot be described as straight lines *a-a* or *b-b*. It has to correspond to Lissajous figures, *C-D*

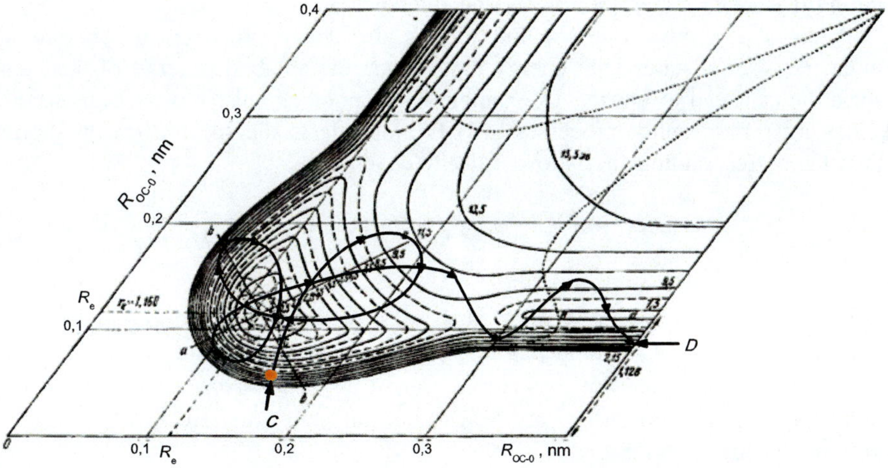

Fig. 4.26 A dissociation of a linear bound state of CO_2 molecule (the bending vibration is frozen)

line, and superposition of the symmetric and antisymmetric stretching vibrations. If the CO_2 *PES* is considered in other coordinates, then one can notice that in the general case, bending vibrations should also be observed. This feature leads to a relatively strong assertion that collinear decay of linear molecules is unlikely, i.e., it is a small fraction of noncollinear. Therefore, the decay of any triatomic molecule is described by the laws of the C_1 or C_s symmetry point group. Therefore, correlation diagrams between the electronic states of linear molecules and their dissociation products should be built not in the $C_{\infty v}$ and $D_{\infty h}$ symmetry groups but in C_1 or C_s groups, respectively. Due to the non-crossing rule, a decrease in symmetry during the decay of an excited state can lead to the appearance of potential barriers along the decay coordinate, which are absent during collinear decay of a molecule fragments. A similar effect is observed in CO_2, N_2O excited states, and we will discuss this feature below.

So, a polyatomic molecule can make quite a large (tens or more) oscillations before dissociation, during which stochastization of vibrational energy can occur, i.e., its pumping from one vibrational mode to another, followed by a statistical equilibrium (in this electronic state) distribution. If the coupling between the different modes in the AB_2 molecule is strong enough and excitation energy is close to the dissociation threshold, the stochastization time should be no more than 10^{-11} s (see [72, 73, 74] and references).

As is known, the absorption spectrum of a diatomic molecule, if it corresponds to a transition to the purely repulsive part of the bound state, is well approximated by a Gaussian or a Gaussian corrected by the frequency factor (see Sect. 4.3). In polyatomic molecules, this feature also takes place. Examples are *A* absorption bands of alkyl- and perfluoroalkyl iodides RI, R = CH_3, CF_3, etc. However, this is not always observed. According to calculations, for some properties of the upper repulsive state of linear symmetric triatomic molecules AB_2 (at least), a structure can be superimposed on the smooth dissociative absorption spectrum corresponding to the transition to this state due to vibrational excitation of the AB_2 fragment during dissociation (see [3, 75–77] and references).

A careless approach to the interpreting the decay processes of polyatomic molecule excited states with energy higher than the adiabatic energy of their dissociation can lead to errors. An example is the photochemistry of N_2O molecule. When interpreting photoprocesses in N_2O molecule in the spectral region shorter than the corresponding first allowed decay channel

$$N_2O(\tilde{X}^1\Sigma^+) + h\nu \rightarrow N_2\left(X^1\Sigma_g^+\right) + O(^1D) \tag{4.8.1}$$

several authors considered a priori that since luminescence is absent, then the quantum yield of N_2O photodecay is $\varphi_{N_2O}(\lambda) = 1$ at any pressure (see [4, 9] and references). As it turned out, this is not true. At $p_{N_2} > 1.5$ Torr, the $\varphi_{N_2O}(\lambda)$ value decreases with photon energy at $\lambda < 180$ nm, and close to zero in the $\lambda = 158$–162 nm spectral range Fig. 4.27.

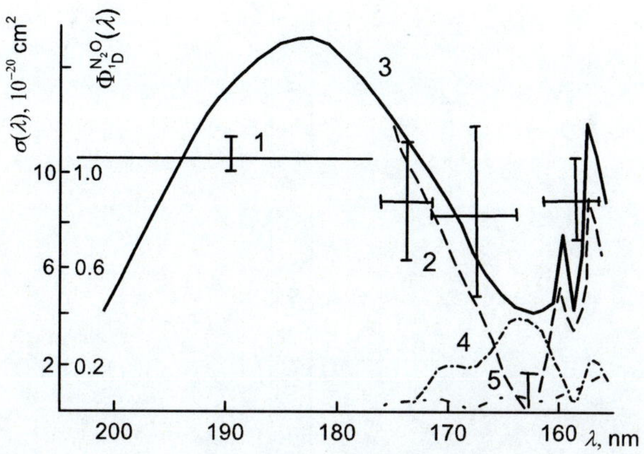

Fig. 4.27 Spectral functions of the convolution quantum yield (see 4.1.30) for N_2O photodecay into $N_2(X\ ^1\Sigma_g^+)$ + $O(^1D)$ (1), the partial cross-section of absorption associated with this process $\sigma^{N_2O}_{O(^1D)}(\lambda)$ (2), the total cross-section of absorption $\sigma_{N_2O}(\lambda)$ (3), the cross-section of absorption associated with the transfer into the bound state (4), $\sigma^{N_2O}_{O(^1S)}(\lambda)$ (5); T = 293 K (1), 273 K (2–5) [5], p. 91

A similar feature is observed for CO_2, NOCl, alkyl- and perfluoroalkyl iodide in the vacuum ultraviolet spectral range (see [3, 75–79] and references). The $\varphi_{CO_2}(\lambda)$ value depends on CO_2 pressure: $\varphi_{CO_2}(\lambda = 155 - 167nm) \rightarrow 1$ at $p_{CO_2} \rightarrow 0$, and less than 1 at $p_{CO_2} > 20$ Torr (see Table 4.5 and Fig. 4.28).

The mechanism of this effect may be different, but the dissociation of excited states should be impossible for energy reasons for any mechanism: the populated state does not correlate with the products of dissociation lying below the *PES* region populated in the optical transition (Fig. 4.29a). or there is a potential barrier on this state decay path, which, for example, is a consequence of the non-intersection rule (Fig. 4.29b). In this case, predissociation is either impossible for energy reasons (curves 1 in Fig. 4.29a, b) or proceed at a slow rate due to the existence of any prohibitions (curves 2 in Fig. 4.29a, b).

A situation similar to that shown in Fig. 4.24a (curve 1) is characteristic of optical transitions to Rydberg states, accompanied by the population of their lower vibrational levels (for example, the photodecay of alkyl- and perfluoroalkyl iodides in the VUV region of the spectrum. The situation, similar to that shown in Fig. 4.29b (curve 1) occurs at N_2O photodecay (Fig. 4.30).

A situation shown in Fig. 4.29b (curve 2) occurs in the *A*, λ = 175—139 nm, and *B*, λ = 139–122 nm, absorption bands of CO_2 (Figs. 4.28, 4.31).

In the *A* absorption band, transitions to the state $^1A_2$ ($^1\Sigma_u^-$) takes place, which dissociate to $CO(X^1\Sigma^+)$ + $O(^1D)$ even at p = 300 Torr, and long-lived $^1A_2$, $^1B_2$ ($^1\Delta_u$) states with a lifetime $\tau > 1.5 \cdot 10^{-7}$ s. At $p_{CO_2} \rightarrow 0$, quantum yields of the CO_2 decay is $\varphi_{CO_2} = 1$, due to $^1A_2$, $^1B_2$ ($^1\Delta_u$) predissociation

Table 4.5 Absolute (λ = 123.6 and 147 nm) and convolution (λ = 124–167 nm) quantum yields of the CO_2 decay [77]

λ, nm	167–160	160–165	155–150	150–143	143–124	147	123.6
φ_{CO_2}	0.62 ± 0.06	0.50 ± 0.10	0.42 ± 0.10	0.65 ± 0.10	0.47 ± 0.10	0.65 ± 0.05	0.70 ± 0.06

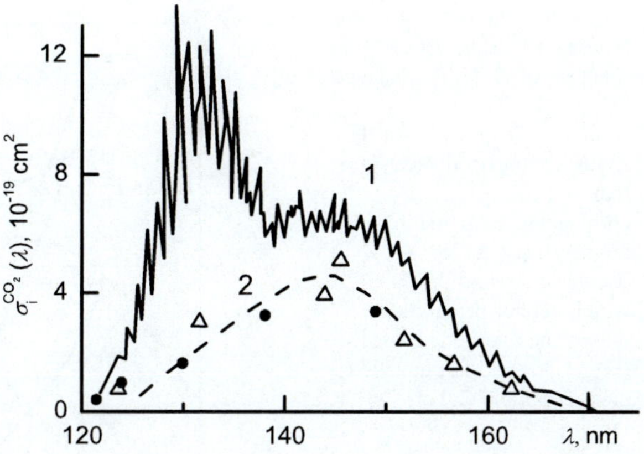

Fig. 4.28 Total (1) and partial, corresponding to transition to rapidly dissociating state (2), CO_2 absorprion cross-section [77]. Full circles are given from [80], open triangles are obtained in [77]

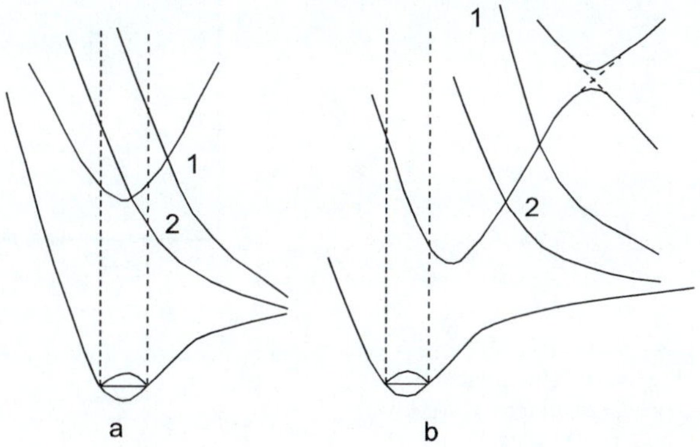

Fig. 4.29 To the existence of long-lived states of polyatomic molecules [5], p. 86

$$CO_2^* \rightarrow CO\left(X^1\Sigma_g^+\right) + O(^3P) \tag{4.8.2}$$

Luminescence of these states is not observed since the radiative transition rate is much lower than that of predissociation or (at higher pressures) collisional deactivation of states. In this case, the radiation lifetime can significantly exceed the value calculated from the absorption spectra due to state mixing (see Sect. 4.6.2.2). At $p_{CO_2} > 5$ Torr, the principal channel of the CO_2^* decay is collisional deactivation

$$CO_2^* + CO_2\left(\widetilde{X}^1\Sigma_g^+\right) \rightarrow \ldots \rightarrow 2CO_2\left(\widetilde{X}^1\Sigma_g^+\right), \tag{4.8.3}$$

and $\varphi_{CO_2} < 1$ (see Table 4.5). Optical transitions to the $^1A_2$ ($^1\Sigma_u^-$) and$^1A_2$, $^1B_2$ ($^1\Delta_u$) occur in the B absorption band, also (see Figs. 4.28, 4.31).

Let repulsive state is populated during the absorption of a photon, and it is known that fragments of the molecule is adiabatically correlated with it, then, in the general case, there is no reason to expect that this state will decay into these

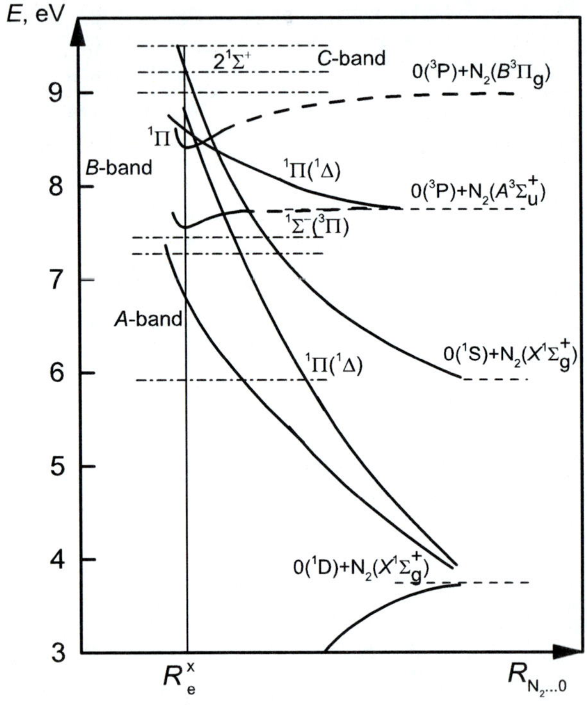

Fig. 4.30 The scheme of the N_2O excited state *PEC*s as a function of N–N--O distance [3], p. 249. The short-wavelength part of the *A* absorption band of N_2O (curve 4 in Fig. 4.27) corresponds to transition in a bound state ($^1\Sigma^-$ or $^3\Pi$, oscillator strength, $f \leq 10^{-4}$), which cannot predissociate due to the energetic reasons (Fig. 4.30) and is quenched even at p_{N_2O} = 2 Torr

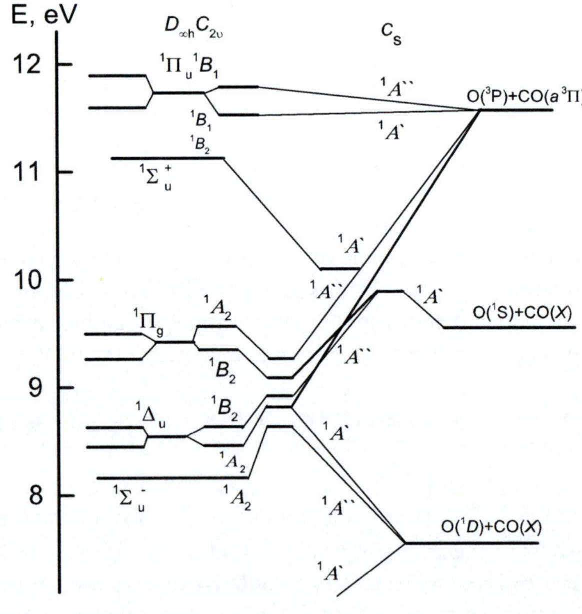

Fig. 4.31 The adiabatic correlation diagrams for lower singlet CO_2 states in the C_s symmetry point group [73]

fragments with a probability of 1. Nonadiabatic processes can occur during this state decay. A similar effect is observed in the *A* absorption band of alkyl- and perfluoroalkyl iodides.

Let in the Franck–Condon zone, the state correlating with fragments having lower excitation energy lies higher than the state correlated to products with higher excitation energy (Fig. 4.32a). In the process of decay of these states nonadiabatic processes are very likely, i.e., transitions from one *PES* to another occur.

The probabilities of these processes depend on interaction matrix elements and time of passage of an image point through the region of convergence of the states. Qualitative dependence of the probability of nonadiabatic processes on the photon energy, i.e., in fact, from the velocity of motion of the image point, obeys the Landau-Zener rule (see Sect. 3.6). If the 'intersection' of repulsive states occurs in the Franck–Condon zone, oscillations can be observed on partial absorption cross-sections corresponding to transitions to states dissociating through different channels. The curve of the total absorption cross-section remains smooth. The mechanism of this effect is illustrated in (Fig. 4.32b).

Suppose that the oscillator strengths of optical transitions to the interacting states A and B converging diabatically to $A_2 + B_2$ and $A_1 + B_1$, respectively, are approximately the same, and the *PES*s of these states in the vertical transition region approach and repel along the decay coordinate, i.e., the probabilities of transition from point *a* to point *b* and from point *c* to point *d* are small. Then the spectral dependencies of the partial absorption cross-sections corresponding to transitions to states dissociating on $A_2 + B_2$ (1) and $A_1 + B_1$ (2) have an oscillating character. Indeed, the optical transition to the $\tilde{B}$ state *PES* corresponds to a wider absorption spectrum with a lower maximum value $\sigma_i(\lambda)$ (the left side of curve 1 and the right side of curve 2). Since the partial absorption cross-section corresponding to the decay into $A_2 + B_2$ is due to the transition of the state $\tilde{B}$ to the intersection zone, and state $\tilde{A}$ after the intersection zone on the *PES*s, on curve 1 at the point corresponding to the intersection zone, a sharp change in the absorption cross-section

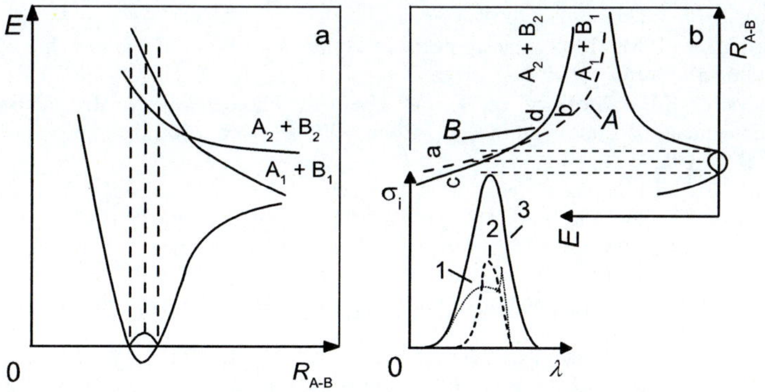

Fig. 4.32 Interaction of repulsive states in the process of their dissociation

should be observed, as a result of the change in the *PES* steepness. Curve 2 should behave similarly. For oscillation appearance, a small velocity of the image points is necessary, which is ensured by the 'intersection' of the *PES* in the Franck – Condon zone and the small angle at which they intersect. In this case, the total absorption cross Sect. 3 is a smooth curve without features.

The presence of this effect accounts for the oscillations of the partial absorption cross-sections corresponding to the photodecay of CH_3I into CH_3 + $I(^2P_{1/2})$ and CH_3 + $I(^2P_{3/2})$ [81].

References

1. Franck, J., Rabinovitsch, E.: Some remarks about free radicals and the photochemistry of solutions. Trans. Far. Soc. **30**, 120–130 (1934)
2. Pravilov, A.M.: A procedure for the measurements of quantum yields of photoprocesses. High Energy Chemistry (USSR, Eng. transl.).**15**, 261-264 (1981)
3. Pravilov, A.M.: Fotoprozessy v molekulyarnykh gasakh (Photoprocesses in molecular gases). Energoatomizdat, Moscow (1992).(in Russian)
4. Calvert, J.G., Pitts, J.N., Jr.: Photochemistry, J. Willey & Sons, New York, London, Sydney (1966)
5. Pravilov, A.M.: Radiometry in Modern Scietific Experiments. SpringerWienNew York (2011). https://doi.org/10.1007/978-3-7091-0104-9
6. Tscherbul, T.V., Buchachenko, A.A., Akopyan, M.E., Poretsky, S.A., Pravilov, A.M., Stephenson, T.A.: Collision-induced non-adiabatic transitions between the ion-pair states of molecular iodine: A challenge for experiment and theory (Invited Article). Phys. Chem. Chem. Phys. **6**, 3201–3214 (2004). https://doi.org/10.1039/b402655a
7. Lukashov, S., Petrov, A., Pravilov, A.: The Iodine Molecule: Insights into Intra- and Intermolecular Perturbation in Diatomic Molecules. Springer (2018). https://doi.org/10.1007/978-3-319-70072
8. Noyes, W.A., Leighton, P.A.: The photochemistry of gases. Dover Publication, New York (1966)
9. Okabe, H.: Photochemistry of small molecules. A Willey Interscience Publication, New York, Chichester, Brisbane, Toronto (1978)
10. Herzberg, G.: Molecular Spectra and Molecular Structure III. Electronic Spectra and Electronic Structure of Polyatomic Molecules. Krieger Publishing Company, Second Edition (1966)
11. Hochstrasser, R.M.: Molecular Aspects of Symmetry. W.A. Benjamin. Inc., New York, Amsterdam (1966)
12. Braslavsky, S.E.: Glossary of Terms Used in Photochemistry, 3rd edition (IUPAC Recommendations 2006). Pure Appl. Chem. **79**, 293–465 (2007). https://doi.org/10.1351/pac200779030293
13. Gardner, P.J., Kasha, M.: Electronic consequences of vibrational deficiency in polyatomic molecules. J. Chem. Phys. **50**, 1543–1552 (1968). https://doi.org/10.1063/1.1671240
14. McGlynn, S.P., Azumi, T., Kinoshita, M.: Molecular Spectroscopy of the triplet State. Prentice-Hall (1969)
15. Mulliken, R.S.: The interpretation of bound spectra. Part IIc. Empirical band types. Rev. Mod. Phys. **3**, 89-155 (1931)
16. Mulliken, R.S.: The halogen moleciles and their spectra. *J-J*-like coupling. Molecular ionization potentials. Phys. Rev. **46**, 549–571 (1934). https://doi.org/10.1103/PhysRev.46.549

17. Mulliken, R.S.: Structure, ionization and ultraviolet spectra of nethyl iodide and other molecules. Phys. Rev. **47**, 413–415 (1935). https://doi.org/10.1103/PhysRev.47.413
18. Mulliken, R.S.: Halogen molecule spectra. II. Interval relations and relative intensities in the long wave-length spectra. Phys. Rev. **57**, 500–515 (1940). https://doi.org/10.1103/PhysRev.57.500
19. Mulliken, R.S.: Iodine revisited. J. Chem. Phys. **55**, 288–309 (1971). https://doi.org/10.1063/1.1675521
20. Van Vleck, J.H.: The theory of the magnetic quenching of iodine fluorescence and of Λ-doubling in $^3\Pi_0$ state. Phys. Rev. **40**, 544–568 (1932). https://doi.org/10.1063/1.1749853
21. Barrow, R.F., Long, D.A., Millen, D.J., Coxon, J.A.: The low-lying electronic states of diatomic halogen molecules. Molec. Spectrosc. **1**, 177–228 (1973). https://doi.org/10.1039/9781847556684-00177
22. Pravilov, A.M.: Spectroscopy and primary processes in the photolysis of the iodides for iodine photodissociation lasers (Review). Soviet J. Quant. Electronics (English translation) **11**, 847–862 (1981)
23. Herzberg, G.: Molecular Spectra and Molecular Structure I. Spectra of Diatomic Molecules. D. van Nostrand Co., Inc, Second Edition (1950)
24. Nadler, I., Mahgerefteh, D., Reisler, H., Wittig, C.: The 266 nm photolysis of ICN: Recoil velocity anisotropies and nascent E, V, R, T excitations for the CN + I(2P3/2) and CN + I (2P1/2) channels. J. Chem. Phys. **82**, 3885–3893 (1985). https://doi.org/10.1063/1.448879
25. Ashfold, M.N.R., Macpherson, M.F., Simons, J.P.: Photodissociation spectroscopy and predissociation pathways in HCN and DCN. J. Photochem. **9**, 134–136 (1978). https://doi.org/10.1016/0047-2670(78)80085-9
26. Clear, R.D., Riley, S.J., Wilson, K.R.: Energy partitioning and assignment of excited states in the ultraviolet photolysis of HI and DI. J. Chem. Phys. **63**, 1340–1347 (1975). https://doi.org/10.1063/1.431520
27. Heaven, M.C.: Fluorescence decay dynamics of the halogens and interhalogens. Chem. Soc. Rev. **15**, 405–448 (1986). https://doi.org/10.1039/cs9861500405
28. Tellinghuisen, J.: Least-squares analysis of overlapped bound-free absorption spectra and predissociation data in diatomics: The C($^1\Pi_u$) state of I_2. J. Chem. Phys. **135**, 054301 (2011). https://doi.org/10.1063/1.3616039
29. Li, J., Balasubramanian, K.: Spectroscopic properties and potential energy curves of I_2 and I_2^+. J. Mol. Spectrosc. **138**, 162–180 (1989). https://doi.org/10.1016/0022-2852(89)90108-2
30. de Jong, W.A., Visscher, L., Nieuwpoort, W.C.: Relativistic and correlated calculations on the ground, excited, and ionized states of iodine. J. Chem. Phys. **107**, 9046–9058 (1997). https://doi.org/10.1063/1.475194
31. Lefebvre-Brion, H., Field, R.W.: The Spectra and Dynamics of Diatomic Molecules, Elsevier (2004)
32. Fraser, P.A.: A method of determining the electronic transition moment for diatomic molecules. Can. J. Phys. **32**, 515–521 (1954). https://doi.org/10.1139/p54-054
33. Tellinghuisen, J.: The Franck-Condon Principle in Bound-Free Transitions. In: Photodissociation and Photoionization, ed. K.P. Lawley, Adv. in Chem. Phys. **60**, 299–369 (1985)
34. Lendi, R.: On quantum beats in polyatomic molecules. Chem Phys. **46**, 176–190 (1980)
35. Mulliken, R.S.: Role of kinetic energy in the Franck-Condon principle. J. Chem. Phys. **55**, 309–314 (1971). https://doi.org/10.1063/1.1675522
36. Akopyan, M.E., Baturo, V.V., Lukashov, S.S., Poretsky, S.A., Pravilov, A.M.: Dipole moment functions of the iodine $D'2_g - A'2_u$, $D0_u^+ - a'0_g^+$, $D0_u^+ - X0_g^+$ and $E0_g^+ - B0_u^+$ transitions. J. Phys. B: At. Mol. Opt. Phys. **44**, 20510 (9 pp) (2011). https://doi.org/10.1088/0953-4075/44/20/205101
37. Tellinghuisen, J.: The $E \to B$ structured continuum in I_2. Phys. Rev. Letts. **34**, 1137–1140 (1975). https://doi.org/10.1103/PhysRevLett.34.1137

38. Lefebvre-Brion, H., Field, R.W.: Perturbations in the Spectra of Diatomic Molecules, Academic Press, Inc. (1986)
39. Vigué, J., Broyer, M., Lehmann, J.C.: Natural hyperfine and magnetic predissociation of the I_2 B state. I. Theory. J. Physique **42**, 937–947 (1981). https://doi.org/10.1051/jphys:01981004207093700
40. Brown, J.M., Hougen, J.T., Huber, K.-P., Jones, J.W.C., Kopp, I., Lefebvre-Brion, H., Merer, A.J., Ramsay, D.A., Rostas, J., Zare, R.N.: The labeling of parity doublet levels in linear molecules. J. Molec. Spectrosc. **55**, 500–503 (1975). https://doi.org/10.1016/0022-2852(75)90291-X
41. Alexander, M.A., Corey, G.C.: Collision induce transitions between $^2\Pi$ and $^2\Sigma$ states of diatomic molecules: Quantum theory and collisional propensity rules. J. Chem. Phys. **84**, 100–113 (1986). https://doi.org/10.1063/1.450831
42. Akopyan, M.E., Baturo, V. V, Lukashov, S.S., Mikheev, L.D., Poretsky, S.A., Pravilov, A. M., Vasyutinskii, O.S.: Hyperfine interaction in molecular iodine between the 0_g^+, 1_u and 0_u^- states correlating with the $I(^2P_{1/2}) + I(^2P_{1/2})$ dissociation limit. J. Phys. B At. Mol. Opt. Phys. **48**, 25102 (14pp) (2015). https://doi.org/10.1088/0953-4075/48/2/025102
43. Perrot, J.P.P., Femelat, B., Broyer, M., Chevaleyre, J.: Ω doubling in 1g state and electronic transition moments for $B0_u^+$ towards 0_g^+ and 1_g ($^3P_2$) states of iodine. Mol. Phys. **61**, 97–108 (1987). https://doi.org/10.1080/00268978700101011
44. Ishiwata, T., Motohiro, S., Kagi, E., Fujiwara, H., Fukushima, M.: Optical-optical double-resonance spectroscopy of the $1_u(^3P_2)$ and $2_u(^3P_2)$ states of I_2 through the $A^3\Pi(1_u)$ State. Bull. Chem. Soc. Jpn. **73**, 2255–2261 (2000). https://doi.org/10.1246/bcsj.73.2255
45. Kagi, E., Yamamoto, N., Fujiwara, H., Fukushima, M., Ishiwata, T.: Optical–optical double resonance spectroscopy of the $1_g(^3P_1)$-$A^3\Pi(1_u)$-$X^1\Sigma_g^+$ transition of I_2. J. Mol. Spectrosc. **216**, 48–51 (2002). https://doi.org/10.1006/jmsp.2002.8676
46. Motohiro, S., Ishiwata, T.: Optical–optical double resonance spectroscopy of the $H1_u(^3P_1)$ – $A^3\Pi1_u$ – $X^1\Sigma_g^+$ transition of I_2. J. Mol. Spectrosc. **204**, 286–290 (2000). https://doi.org/10.1006/jmsp.2000.8229
47. Motohiro, S., Umakoshi, A., Ishiwata, T.: Perturbation-facilitated optical–optical double-resonance spectroscopy of the $1_u(^1D)$ and $2_u(^1D)$ ion-pair states of I_2 through the parity mixing intermediate state. J. Mol. Spectrosc. **208**, 213–218 (2001). https://doi.org/10.1006/jmsp.2001.8379
48. Baturo, V.V., Lukashov, S.S., Poretsky, S.A., Pravilov, A.M., Vasyutinskii, O.S.: Mixing of the D 0_u^+ and $\delta 2_u$ ion-pair states of iodine molecule. Chem. Phys. Lett. **638**, 244–248 (2015). https://doi.org/10.1016/j.cplett.2015.08.047
49. Levenson, M.D., Shawlou, A.L.: Hyperfine interaction in molecular iodine. Phys. Rev. **A6**, 10-20 (1972). https://doi.org/10.1103/PhysRevA.6.10
50. Pique, J.P., Hartmann, F., Churassy, S., Bacis, R.: Hyperfine interaction in homonuclear diatomic molecules and u-g perturbation. I. Theory J. Physique **47**, 1909–1916 (1986). https://doi.org/10.1051/jphys:0198600470110190900
51. Jewsbury, P.J., Ridley, T., Lawley, K.P., Donovan, R.J.: Parity mixing in the valence states of I_2 probed by optical-optical double-resonance excitation of ion-pair. J. Mol. Spectrosc. **157**, 33–49 (1993). https://doi.org/10.1006/jmsp.1993.1003
52. Akopyan, M.E., Lukashov, S.S., Maslennikova, Y.D., Poretsky, S.A., Pravilov, A.M.: Hyperfine coupling of the iodine $E0_g^+$,v_E = 19 and $\gamma 1_u$,v_γ = 18 ion-pair states. J. Phys. B At. Mol. Opt. Phys. **40**, 1173-1181 (2007). https://doi.org/10.1016/j.cplett.2015.08.047
53. Saute, M., Aubert-Frécon, M.: Calculated long-range potential-energy curves for the 23 molecular states of I_2. J. Chem. Phys. **77**, 5639–5646 (1982). https://doi.org/10.1063/1.443770
54. Ishiwata, T., Yotsumoto, T., Motohiro, S.: Optical–optical double resonance spectroscopy of I_2 through the parity mixed valence states. Bull. Chem. Soc. Jpn. **74**, 1605–1610 (2001). https://doi.org/10.1246/bcsj.74.1605

55. Motohiro, S., Nakajima, S., Ishiwata, T.: Perturbation-facilitated optical-optical double resonance spectroscopy of the h 0_u^- ($^3P_1$) and $H1_u(^3P_1)$ ion-pair states of I_2. J. Chem. Phys. **117**, 187–196 (2002). https://doi.org/10.1063/1.1481391
56. Motohiro, S., Nakajima, S., Aoyama, K., Kagi, E., Fujiwara, H., Fukushima, M., Ishiwata, T.: Analysis of the 0_g^- ($^3P_1$)–$B'^3\Pi(0_u^-)$ system of I_2 by perturbation-facilitated optical–optical double resonance. J. Chem. Phys. **117**, 9777–9784 (2002). https://doi.org/10.1063/1.1516790
57. Baturo, V.V., Cherepanov, I.N., Lukashov, S.S., Poretsky, S.A., Pravilov, A.M., Zhironkin, A.I.: Heterogeneous and hyperfine interactions between valence states of molecular iodine correlating with the $I(^2P_{1/2}) + I(^2P_{1/2})$ dissociation limit. J. Chem. Phys. **144**, 184310–184318 (2016). https://doi.org/10.1063/1.4948630
58. Baturo, V.V., Cherepanov, I.N., Lukashov, S.S., Poretsky, S.A., Pravilov, A.M.: Spectroscopic constants and potential energy curves of some iodine valence ungerade weakly bound states. J. Phys. B At. Mol. Opt. Phys. **48**, 055101 (8pp) (2015). https://doi.org/10.1088/0953-4075/48/5/055101
59. Nakano, Y., Ukeguchi, H., Ishiwata, T.: Observation and analysis of the $2_g(^1D)$ ion-pair state of I_2: The g/u mixing between the $1_u(^1D)$ and $2_g(^1D)$ states. J. Chem. Phys. **121**, 1397–1404 (2004). https://doi.org/10.1063/1.1756134
60. Baturo, V.V., Cherepanov, I.N., Lukashov, S.S., Petrov, A.N., Poretsky, S.A., Pravilov, A. M.: Hyperfine coupling of the iodine $D0_u^+$ and $\beta 1g$ ion–pair states. J. Phys. B At. Mol. Opt. Phys. **51**, 095101 (9 pp) (2018). https://doi.org/10.1088/1361-6455/aab6e3
61. Herzberg, G.: Molecular Spectra and Molecular Structure. II. Infrared and Raman Spectra in Polyatomic Molecules. D. van Nostrand Co., Inc. (1945)
62. Strickler, S.J., Berg, R.A.: Relationship between absorption intensity and fluorescence lifetime of molecules. J. Chem. Phys. **37**, 814–822 (1962). https://doi.org/10.1063/1.1733166
63. Douglas, A.E.: Anomalously long radiative lifetimes of molecular excited states. J. Chem. Phys. **45**, 1007–1015 (1966). https://doi.org/10.1063/1.1727650
64. Bixon, M., Jortner, J.: Long radiative lifetimes of small molecules. J. Chem. Phys. **50**, 3284–3290 (1969). https://doi.org/10.1063/1.1671552
65. Neuberger, D., Duncan, A.B.F.: Fluorescence of nitrogen dioxide. J. Chem. Phys. **22**, 1693–1696 (1954). https://doi.org/10.1063/1.1739877
66. Medvedev, E.S., Osherov, V.I.: Radiationless Transitions in Polyatomic Molecules. Springer (1995). https://doi.org/10.1002/bbpc.19961000329
67. Broyer, M., Vigué, J., Lehmann, J.C.: Hyperfine predissociation of molecular iodine. J. Chem. Phys. **64**, 4793–4794 (1976). https://doi.org/10.1063/1.432067
68. Tellinghuisen, J.: Potentials for weakly bound states in I_2 from diffuse spectra and predissociation data. J. Chem. Phys. **82**, 4012–4016 (1985). https://doi.org/10.1063/1.448841
69. Vigué, J., Broyer, M., Lehmann, J.C.: Natural hyperfine and magnetic predissociation of the I_2 B state II. – Experiments on natural and hyperfine predissociation. J. Phys. Fr. **42**, 949–959 (1981). https://doi.org/10.1051/jphys:01981004207094900
70. Alekseev, A.B., Vilesov, A.F., Ivanov, V.S., Pravilov, Sidorov, I.I., A.M., Shestakov, O.D.: The mechanism and kinetics $HeH(B^2\Pi)$, $HeD(B^2\Pi)$ molecules formation and decay during recombination $H(^2P)$, $D(^2P)$ and He. Soviet J. Quant. Electronics (Eng.transl.) **17**, 1236–1241 (1990)
71. Alekseev, A.B., Ivanov, V.S., Pravilov, A.M., Shestakov, O.D.: Theoretical study of the HeH ($B^2\Pi$) fluorescence and predissociation. Chem. Phys. **155**, 173–178 (1991)
72. Farantos, S.C., Murrel, J.N.: On the transition from quasiperiodiv to stochastic classical motion on real polyatomic energy surfaces. Chem. Phys. **55**, 205–214 (1981). https://doi.org/10.1016/0301-0104(81)85021-5
73. Pravilov, A.M., Shulpyakov, I.O.: Mechanism of CO_2 photolysis in the 167–155 nm spectral range. Sov. J. Chem. Phys. **3**, 2432–2451 (1986)
74. Pack, R.T.: Simple theory of diffuse vibrational structure in continuous uv spectra of polyatomic molecules. I. Collinear photodissociation of symmetric triatomics. J. Chem. Phys. **65**, 4765–4770 (1976). https://doi.org/10.1063/1.432931

75. Heller, E.J.: Photofragmentation of symmetric triatomic molecules: Time dependent picture. J. Chem. Phys. **68**, 3891–3896 (1978). https://doi.org/10.1063/1.436197
76. Schinke, R., Engel, V.: Reactive resonances in the photodissociation of symmetric triatomic molecules: an interpretation of collinear CO_2 in terms of polar coordinates. Chem. Phys. Letts. **124**, 504–508 (1986). https://doi.org/10.1016/0009-2614(86)85064-3
77. Pravilov, A.M., Shulpyakov, I.O.: Photodecomposition of O_2, CO_2 and N_2O in the wavelength region $\lambda > 155$ nm. High Energy Chemistry (USSR, Eng.transl.) **19**, 410–414 (1985)
78. Pravilov, A.M.: Gas phase actinometers for UV and VUV spectral region (a review). High Energy Chemistry (USSR, Eng.transl.) **21**, 243–255 (1987)
79. Elokhin, V.A., Ivanov, V.S., Pravilov, A.M., Ryabov S.E.: Photodissociation of some perfluoroalkyliodides in the vacuum ultraviolet part of spectrum. Soviet J. Quant. Electronics (Eng.transl.) **15**, 1306–1307 (1985)
80. Slanger, T.G., Sharpless, R.L., Black, G., Filseth, S.V.: Photodissociation quantum yields of CO_2 between 1200 and 1500 . J. Chem. Phys. **61**, 5022–5027 (1974). https://doi.org/10.1063/1.1681842
81. Alekseev, A.B., Pravilov, A.M.: Spectral dependences of absolute quantum yields for formation of $I(^2P_{1/2})$ and $I(^2P_{3/2})$ atoms during the photolysis of CH_3I. High Energy Chemistry (USSR, Eng.transl.) **18**, 75–78 (1984)

Chapter 5
Energy Transfer in Collisions

Abstract Inelastic collision processes, namely, processes of energy exchange of translational motion (*T-T* exchange), a transformation of translational, and rotational energy (*R-T* exchange), a transformation of vibrational energy (*V-T, V-R and V-V* exchange) are discussed briefly. Then perturbation-facilitated and perturbation-irrelevant collision-induced non-adiabatic transitions (CINATs), and other processes in which CINATs occur, are considered. CINATs between halogen ion-pair states are examined in detail.

We are starting to discuss the mechanism of a reasonably wide range of processes of transfer of excitation energy (translational, rotational, vibrational, vibronic (electronic-vibrational) of molecules and radicals, and electronic of atoms) in collisions of atoms, as well as diatomic and polyatomic species with each other and atoms. Processes of translational and rotational excitations transfer have been studied quite well by now. As we will see soon, they are very fast, and very often we deal with local thermodynamic equilibrium for these degrees of freedom. Therefore, we will pay the most attention to the processes of transfer of vibrational and, especially, vibronic excitation energy.

5.1 Translational-Translational Energy Transfer ($T \leftrightarrow T$ Exchange)

We already touched on these processes when we discussed elastic atomic collisions. From the course of mechanics, we know that even when material points or hard balls collide, the kinetic energy of the motion of the particle as a whole or, in other words, translational energy is exchanged. If a material point A having kinetic energy E_A collides with a fixed material point B then the first loses, and the second acquires energy:

A. Pravilov, *Gas-Phase Photoprocesses*, Springer Series in Chemical Physics 123,
https://doi.org/10.1007/978-3-030-65570-9_5

$$\Delta E_{\mathrm{B}} = -\Delta E_{\mathrm{A}} = \frac{4m_{\mathrm{A}}m_{\mathrm{B}}}{(m_{\mathrm{A}} + m_{\mathrm{B}})^2} E_{\mathrm{A}} \tag{5.1.1}$$

The elastic collisions of atoms and molecules are described very similarly. Of course, they are not rigid balls, since attractive forces can act between them, and in any case, the repulsive forces do not increase abruptly (see Sect. 3.4, and Fig. 3.6). However, in this case, the model of rigid balls with a diameter equal to the gas kinetic one works quite well, and the rate constant of elastic collisions of species A and B is.

$$k_{gk} = \overline{V}_{\mathrm{AB}}\pi d_{\mathrm{AB}}^2, \tag{5.1.2}$$

where $\overline{V}_{\mathrm{AB}} = \sqrt{\frac{8R\mathrm{T}}{\pi\mu_{\mathrm{AB}}}}$ is the average relative velocity of motion of A and B, μ_{AB} is the reduced mass of A and B, d_{AB} is the gas-kinetic diameters of A and B species, which we can take from monograph tables (see Sect. 2.3).

If we will change material points to species whose interaction potential is described by a certain effective potential U_{eff} (see Sect. 3.5 and Figs. 3.6, 3.8) and take into account the scattering of species A by species B in the angle $\vartheta = 0 - \pi$, then instead of (5.1.1) we should consider the energy loss by the (5.1.3) [1], p. 70:

$$\Delta E_{\mathrm{B}}(\vartheta) = -\Delta E_{\mathrm{A}}(\vartheta) = \frac{2m_{\mathrm{B}}}{(m_{\mathrm{A}} + m_{\mathrm{B}})^2} E_{\mathrm{A}}(1 - cos\vartheta) \tag{5.1.3}$$

One sees that energy exchange is maximal when the scattering angle is $\vartheta = \pi$, i.e., species A is back-scattered with respect to the initial direction of motion (*head-on impact*). In this case, species A loses $\frac{4m_{\mathrm{A}}m_{\mathrm{B}}}{(m_{\mathrm{A}} + m_{\mathrm{B}})^2}$ part of its initial energy. For collisions with a small scattering angle, the amount of transferred energy is also small.

If we go over to a more realistic system and take into account that the B energy is also nonzero, then the transfer of the kinetic energy of species motion as a whole can be characterized by the mean square of the energy transferred in a single collision $\langle(\Delta E_{\mathrm{A}})^2\rangle$ (averaging is carried out over all relative orientations of the direction of motion of A and B species). For the case $m_{\mathrm{A}} << m_{\mathrm{B}}$ this is:

$$\left\langle(\Delta E_{\mathrm{A}})^2\right\rangle = \frac{8m_{\mathrm{A}}}{m_{\mathrm{B}}} E_{\mathrm{A}}E_{\mathrm{B}}, \tag{5.1.4}$$

and the reciprocal relaxation time included in the dependence $\Delta E_{\mathrm{A}} = \Delta E_{\mathrm{A}}^0 \exp(-t/\tau)$ is

$$1/\tau = k_{gk}[\mathrm{B}]m_{\mathrm{A}}/m_{\mathrm{E}} \qquad (5.1.5)$$

We see again that the relaxation time is the shorter, the closer m_A is to m_B. Although this equation was obtained for $m_A < < m_B$, it gives the correct order of magnitude for $m_A \approx m_B$, as well. It follows from this equation, in particular, that the 'rate constant' of the relaxation of translational energy is close to gas-kinetic. If p_B = 1 Torr, i.e., [B] = $3.3 \cdot 10^{16}$ cm^{-3}, then for $m_A = m_B$, $\tau = (310^{-10}\ 3.3 \cdot 10^{16})^{-1} = 10^{-7}$ s.

5.2 Rotational-Translational Energy Transfer ($R \leftrightarrow T$ Exchange)

The simplest model describing the *R-T* exchange is the collision of a ball with a dumbbell, and from it, on an intuitive level we can understand that these processes should be very fast when the mass of the ball and dumbbell is comparable, and a noncollinear collision takes place (Fig. 5.1).

Everything is very similar to the *T-T* exchange. This is understandable because the processes are quite close to mechanics. In addition to the model described above, there are several others: the model of spherical cylinders (two hemispheres connected by a cylinder), loaded spheres (the molecule is modeled by a sphere whose center of gravity does not coincide with the geometric center), rough spheres (surfaces at the time of impact do not slip relative to each other). From a consideration of the model of two rough spheres (which is suitable for describing the rotational relaxation of symmetric polyatomic molecules) with the diameter *R*, one of which models an atom and the other a molecule, we can obtain that the average square of the energy transferred from the translational degrees of freedom to rotational at the initial local equilibrium distribution is:

$$\left\langle (\Delta E_{\mathrm{A}})^2 \right\rangle = (k\mathrm{T})^2 \frac{8}{3} \frac{I \cdot I'}{(I + I')^2} \qquad (5.2.1)$$

where I is the moment of inertia of the molecule (*radical*) BC, and I' is the moment of inertia of the complex formed during the collision. We see that this formula is somewhat reminiscent of (5.1.1), which describes the fraction of energy transferred

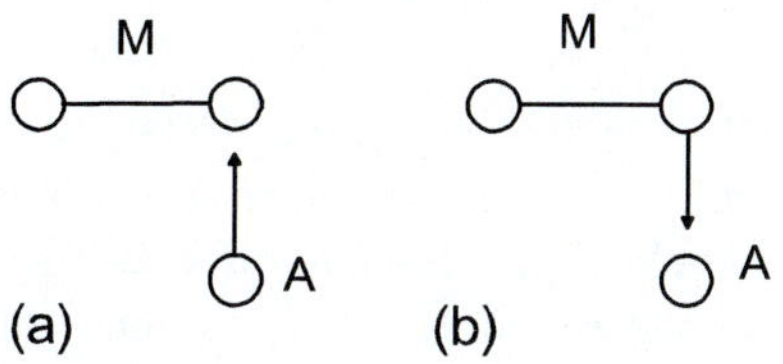

Fig. 5.1 Model of energy exchange of translational and rotational motions: **a**—rotational excitation, **b**—rotational relaxation (see [2], p. 160)

in the collision of hard balls or material points. Again, $\langle(\Delta E_A)^2\rangle$ is maximal when I and I' are close, and this may be when the reduced masses BC and BC-A are close. The (5.2.1) describes the $T \rightarrow R$ transfer; but we must remember that if we are able to describe this process, we can also use the principle of detailed equilibrium to describe the opposite, i.e., the $R \rightarrow T$ process.

We will not go further into the jungle of semiclassical calculations of the quantities $\langle(\Delta E_A)^2\rangle$ during rotational relaxation of various classes of molecules, because these processes are very fast, and very often we need to know only the order of magnitude of the time of rotational relaxation, i.e., time required to establish local thermodynamic equilibrium for these degrees of freedom. Below, only the order of magnitude of the number of collisions required for complete rotational relaxation, information on the order of magnitude of the energy transferred in the collision, and on the temperature dependence of the relaxation efficiency, are presented.

H_2, D_2. These molecules stand apart because their rotational quanta are large ($B_e^{H_2} = 61\ \text{cm}^{-1}$), and their dipole moment is equal to zero. The relatively large magnitude of the quanta 'makes relaxation quantized', and has little resemblance to a relaxation of non-quantized translational energy. Correspondingly, the probability of relaxation is relatively small: to establish local thermodynamic equilibrium, approximately $Z_{rot} = 200$–300 collisions are required at $T = 300$–1000 K [1], p. 71.

In the process of rotational relaxation of any homonuclear diatomic molecules, if it is described in the framework of quantum mechanics, rather than in the semiclassical approximation, the selection rule is strictly followed, which we will often recall later: $s \leftrightarrow s$, $a \leftrightarrow a$, i.e., symmetric rotational levels commute in collision with symmetric one, antisymmetric with antisymmetric one; the process $s \leftrightarrow a$ is forbidden. This rule is a consequence of the weakness of the hyperfine interaction, i.e., interactions of nuclear spins with other types of intramolecular motion (see Sect. 4.6.1). This rule is quite simply proved in the monograph of G. Herzberg [3], p. 131: the probability of such a transition is determined by the integral $\langle\Psi_s|\widehat{V}|\Psi_a\rangle$, and it is clear that the operator $\widehat{V}$ is symmetric with respect to the exchange of nuclei; 'it does not distinguish between them.' Since Ψ_a changes its sign when replacing the nuclei, this operation should also lead to a change of the integral sign. Since the collision partner does not matter which of the same atoms is closer to it, i.e., the result of the collision does not depend on this, it must be assumed that this integral is generally 0, i.e., the process $s \leftrightarrow a$ is forbidden. For the $H_2\left(X^1\Sigma_g^+\right)$ state, in which even levels are symmetric and odd are antisymmetric, the selection rule for rotational relaxation (excitation) $\Delta J = 0, \pm 2$ follows from the selection rule discussed above.

Di-, triatomic hydrides. The values of the rotational quanta of HCl, OH, and H_2O molecules are also relatively large ($B_e^{HCl} = 11\ \text{cm}^{-1}$). However, these molecules possess nonzero dipole moments, and in collisions with a similar molecule a dipole–dipole interaction takes place. This circumstance leads to a high probability of the *R-T* process: the number of collisions required for complete relaxation of

HCl, H_2O, and OH is Z_{rot} = 7, 3, and 10, respectively, at T = 300 K. Here, the rule $\Delta J = \pm 1$ is usually fulfilled [3], p. 71.

The remaining molecules. For them, as well as for hydrides, the number of collisions required for relaxation is less than ten, Z_{rot} = 4.8 and 3.6 for $N_2(X)$ and $O_2(X)$, respectively.

All these data are related to the molecule ground state, low rotational quantum numbers, and collisions with the same molecules. The Z_{rot} values increase if the collision of molecules with high and low J values occurs due to the difference of rotational quanta [1], p. 71, [4]. These molecules are characterized by small values of rotational quanta $\left(B_e^{O_2(X)} = 1.5\ \text{cm}^{-1} - B_e^{I_2(X)} = 0.04\ \text{cm}^{-1}\right)$, therefore, in the process of rotational relaxation, the 'little ΔJ' propensity rule is violated: during rotational relaxation of $I_2\left(B0_u^+, v_A, J_A\right), \Delta J \leq 40$ were recorded [5].

The relaxation efficiency decreases with increasing temperature; different works give a dependence on $1/T$ to $(1/T)^{1/2}$; some works indicate the absence of this dependence.

5.3 Vibrational Energy Transfer

5.3.1 $V \leftrightarrow T$, $V \leftrightarrow R$, T Processes

Qualitatively, it can be understood that the probability *T-V*, (and, consequently, *V-T* processes, recall the detailed balance principle, Sect. 3.1) have to be much lower than $T \leftrightarrow T$ or $T \leftrightarrow R$, because even in a collinear collision a 'hot' (having large translational energy) species A with a 'cold' BC molecule is much more likely transfer to the energy of A into the motion of the BC molecule as a whole, i.e., $T \leftrightarrow T$ exchange occurs (Fig. 5.2a).

One can also easily understand, considering the non-collinear collision of an atom and a diatomic particle, that there are no pure $V \leftrightarrow T$ processes; part of the translational or vibrational energy always transfers to rotational excitation

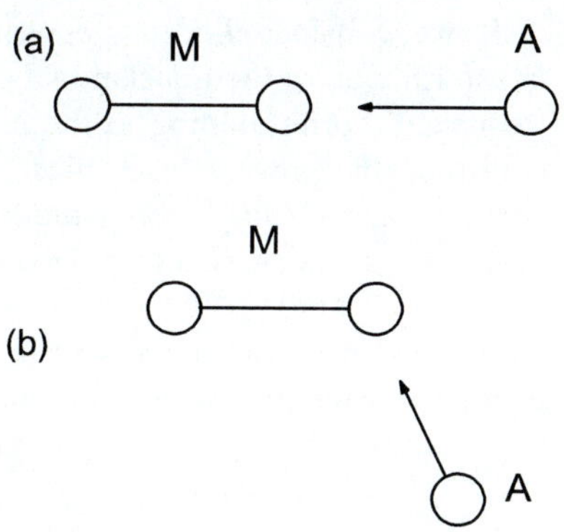

Fig. 5.2 Model of vibrational excitation: **a**—collinear collision, **b**—non-collinear collision

(Fig. 5.2b). Nevertheless, the interaction of translational and rotational degrees of freedom can often be neglected. These models give the correct dependencies of the rate of $V \leftrightarrow T$ processes on molecular constants and collision conditions. To some extent, this is because that all theories of $V \leftrightarrow T$ processes are semi-empirical since the exact interaction potentials of even an atom and a diatomic molecule are unknown. And by fitting the parameters to achieve the coincidence of the calculation results with experimental data, it is often possible to eliminate these errors implicitly. Usually, but not always, and then you must take into account other types of interaction, namely, $V \leftrightarrow R$, $V \leftrightarrow V$, if molecules collide, or nonadiabatic processes occur during collisions (see below).

Qualitatively, we already considered the problem of the $V \leftrightarrow T$ process when discussing the semiclassical description of the collinear collision of an atom and a diatomic particle by moving an image point on the *PES* (see Sect. 3.5 and Fig. 3.11). The solution of this problem in this way is quite difficult since it is necessary to know the *PES*; besides, it can only be solved numerically. There are simpler approaches.

The vibrational and rotational energies are quantized, and it is not always possible to use the semiclassical or adiabatic approximations when considering their relaxation (excitation). Here, as in some other cases, the so-called *adiabatic principle* can be applied, according to which the process is adiabatic, i.e., a process without changing the quantum states of particles in a collision process, if the rate of change of the perturbing action of the collision is negligible compared to the rate of periodic quantized motion corresponding to internal degrees of freedom, i.e.,

$$1/\tau \ll \omega \tag{5.3.1a}$$

or the collision time leading to a change of periodic motion is much longer than the period of the latter. Here τ is the time of the 'productive' collision, ω is the angular frequency of the periodic motion. Otherwise, this principle can be expressed as follows:

$$\Delta E \gg \hbar\, \tau, \tag{5.3.1b}$$

since $\Delta E = \hbar\omega$.

If we consider *T*-*V* processes, then (5.3.1b) means that the probability of transformation of the translational energy $1/\tau$ (s^{-1}) is much less than the 'transition frequency' corresponding to the internal energy $\Delta E/\hbar\omega$ (s^{-1}), and far from every collision, this process takes place. If the quantized energy is high, for example, when we are dealing with vibrational excitation, then the classical approach for describing a change in a quantized state is not applicable, but the probability of this change in a single collision is small, (5.3.1a, 5.3.1b) are valid; the adiabatic principle is also true with them. This principle is also applicable when considering other energy transfer processes. The above considerations relate to the transfer of the energy of translational motion (non-quantized) to quantized degrees of freedom; however, following the detailed balance principle, this approach is fully suitable for considering the processes of energy relaxation of quantized degrees of freedom. You see that the semiclassical approximation and the adiabatic principle are, in a sense, antipodes. In the future, we will recall the adiabatic principle from time to time.

It should be said that the approaches to solving the problem of the probability of the $V \leftrightarrow T$ processes for low and high vibrational levels close to the dissociation limit are very different since the values of vibrational quanta and vibration amplitudes differ. At the lower vibrational levels, the vibration amplitude is small compared to the characteristic sizes of the intermolecular interaction, comparable with $1/\alpha$ (α is the parameter of the Morse potential, see Sect. 3.5):

$$\mathrm{U}(r) = \varepsilon\{1 - \exp\{-\alpha(r - r_e)]\}^2 - \varepsilon; \tag{5.3.2}$$

Then, assuming that the interaction potential of an atom A with a diatomic species BC is spherically symmetric (which of course is an approximation), expanding it in a series the interaction potential A and BC − U(R, r) (R is the distance A—BC, r is the BC internuclear distance) in powers of $x = r - r_e$, where r_e is the equilibrium r, and assuming that a parabola describes the potential of the BC molecule, and the collision is almost adiabatic for vibrations (inequality (5.3.1b) holds), it can be obtained that the probability of vibrational excitation of a BC molecule from level v to level $v + 1$ in a collision with atom A is:

$$\langle P_{v,v+1} \rangle = (v+1)C(\mathrm{T},\omega)exp\left[-3\left(\frac{\omega l_0\sqrt{\mu}}{\sqrt{2k\mathrm{T}}}\right)^{2/3}\right] \tag{5.3.3}$$

$$P_{v,v+1} = 0 \text{ for } \Delta v > 1 \tag{5.3.4}$$

Here, l_0 is a specific characteristic radius of interaction, which is of the order of magnitude or larger than $1/\alpha$, $C(\mathrm{T},\omega)$ is a particular temperature function that changes with its change weaker than the exponent (ω is the angular frequency of vibration). This model is called the model of 'breathing spheres', and (5.3.3) is the well-known Landau-Teller formula (see [1], p. 72 and references). It is applicable if the exponential term is small.

Let us analyze (5.3.3, 5.3.4).

The $\langle P_{v,v+1} \rangle$ value:

- is proportional to (v + 1) (due to an increase in the amplitude and period of oscillation);
- depends exponentially on -τ/T, that is, on the ratio of the collision time to the oscillation period (the $\frac{l_0\sqrt{\mu}}{\sqrt{2k\mathrm{T}}}$ is proportional to the collision time);
- depends on temperature as $ln\ \langle P_{v,v+1} \rangle \sim \mathrm{T}^{-1/3}$;

– transitions with $\Delta v > 1$ are forbidden.

What additional conclusions can be drawn?

– the higher the vibration frequency, i.e., the larger the vibrational quantum, the lower the probability of the *T-V* process;
– the lighter the diatomic species BC or relaxer, the higher the probability of the process.

The rate constant of vibrational excitation is

$$k_{v,v+1} = k_{gk} < P_{v,v+1} > \tag{5.3.5}$$

The rate constant of vibrational relaxation, *V-T*, as follows from (3.1.17), is equal to the rate constant of excitation divided by the exponent $exp(-\Delta E_{v+1,v}/k\mathrm{T})$.

The problem of vibrational excitation or relaxation has been repeatedly considered quantum mechanically, corrections for anharmonicity, etc. have been introduced. These theories give interesting results; however, if only *V-T* takes place in a collision, the conclusions from them are qualitatively the same as those given above. A significant difference is the fact that when anharmonicity is taken into account, transitions with $\Delta v > 1$ are allowed. The numerical values of the transition probabilities, as it has been noted above, depend on the parameters introduced very significantly, i.e., the probability of relaxation is impossible to calculate ab initio.

What are the probabilities of vibrational relaxation of diatomic molecules for cases that are well described as *V-T* processes (T = 300 K). They are 9 10^{-9} (N_2) (approximately the same for CO), 6 10^{-8} (O_2), 2 10^{-5} (Cl_2) [6, 7]. The ω_e values for these molecules are 2358, 1580, and 557 cm^{-1}, respectively [8]. The probability of vibrational relaxation of CO(X,v) in collisions with helium atoms is approximately 2 orders of magnitude greater than for CO [9].

V-T processes in polyatomic molecules. They have a set of vibrational frequencies, and, at first glance, very complex relaxation kinetics should be expected. However, due to the presence of repeatedly mentioned intermode exchange stochastization of vibrational energy occurs. In the relaxation process, low-frequency vibrational modes, which are 'fed' from higher-frequency ones, should most actively participate. The stochastization time is much shorter than the relaxation time, and the relaxation probability of the polyatomic molecule $< P_{\mathrm{M}} >$ is related to the probability of relaxation of the low-frequency vibrational mode $< P_v >$ by the relation:

$$\langle P_{\mathrm{M}} \rangle = \langle P_v \rangle \frac{C_1^v}{C_{\mathrm{M}}^v} \tag{5.3.6}$$

where C_{M}^v and C_1^v are the total vibrational heat capacity and low-frequency heat capacity, respectively [10].

It is empirically established that the probabilities of vibrational relaxation of molecules (on themselves) containing a hydrogen atom and not containing, differ

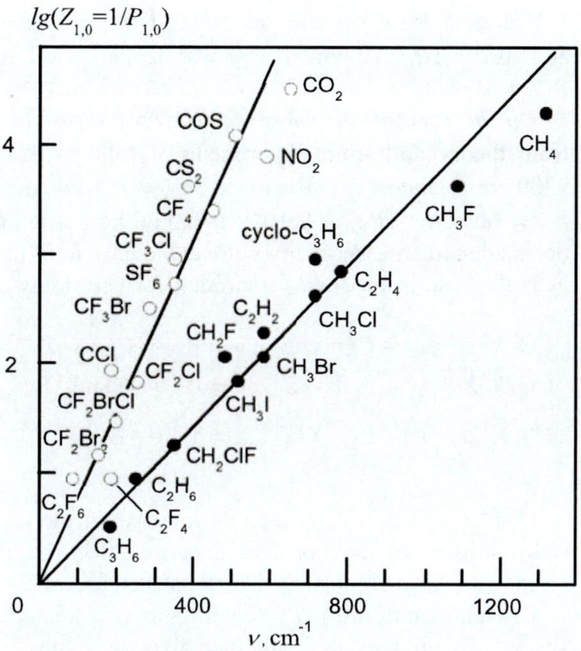

Fig. 5.3 Dependence of $lgZ_{1,0} = 1/P_{1,0}$ on the lowest vibrational frequency of a polyatomic molecule (see [10])

quite strongly. In this case, the linear dependence of $lg(1/ < P_{1,0} >)$ is observed on the minimum vibration frequency, which we discussed above (Fig. 5.3),

as well as on the number of hydrogen atoms in the molecule: at $\tilde{\nu}$ = 550 cm^{-1} for COS molecules, CH_2F_2 and CH_3I, the probabilities of vibrational relaxation are approximately 5 10^{-5}, 6 10^{-4} and 1.5 10^{-2}, respectively. The temperature dependence of $P_{v+1,v}$, as a rule, is the same as for diatomic molecules, i.e., $lgP \sim T^{-1/3}$.

To conclude this section, one should point out that the theories described above are entirely satisfactory, especially for diatomic particles, are consistent with the experiment. If this is not the case, then it is necessary to take into account the processes, which we will discuss below.

5.3.2 *V-V Exchange*

First, let us discuss diatomic molecules. *V-V* exchange probabilities are much higher than those of *V-T* process even if there is no exact resonance, $\tilde{\nu}_2 - \tilde{\nu}_1 = \Delta\tilde{\nu} << \tilde{\nu}_2, \tilde{\nu}_1$, between the vibrational frequencies of the colliding species. The first conclusion that we must draw from a consideration of the experimental data is that the probability of the *V-V* exchange is far from 1, even with a very exact resonance because the probability of collinear collision required for energy transfer is 1/3·1/3 = 0.1 in this case. The second conclusion: the probability

V-V process decreases with increasing v; ceteris paribus $P \sim 1/\tilde{\nu}$. For $\tilde{\nu} = 300\ \text{cm}^{-1}$ and $\Delta\tilde{\nu} = 10 - 20\ \text{cm}^{-1}$, $P_{1,0} = 0.02$; at $\tilde{\nu} = 2300\ \text{cm}^{-1}$ and $\Delta\tilde{\nu} = 10\ \text{cm}^{-1}$ $P_{1,0} = 10^{-3}$.

On the effect of an energy defect on the probability of the V-V process. Both from theory and from experiments it follows that $lg(1/P) \sim \Delta\tilde{\nu}$. So, if at $\tilde{\nu} = 2300\ \text{cm}^{-1}$ and $\Delta\tilde{\nu} = 10\ \text{cm}^{-1}$, $P_{1,0} = 1\ 10^{-3}$, then at the same frequency and $\Delta\tilde{\nu} = 600\ \text{cm}^{-1}$ $P_{1,0} = 2\ 10^{-7}$. It should be borne in mind that if an energy defect occurs due to anharmonicity with an increase in of the vibrational quantum number, as is the case in processes important for acronomy

$$\begin{aligned}
\text{CO}(X^1\Sigma^+, v) + \text{CO}(X^1\Sigma^+, v = 0) &\to \text{CO}(X^1\Sigma^+, v-1) + \text{CO}(X^1\Sigma^+, v = 1)\\
\text{CO}(X^1\Sigma^+, v) + \text{N}_2\left(X^1\Sigma_g^+, v = 0\right) &\to \text{CO}(X^1\Sigma^+, v-1) + \text{N}_2\left(X^1\Sigma_g^+, v = 1\right)\\
\text{N}_2\left(X^1\Sigma_g^+, v\right) + \text{N}_2\left(X^1\Sigma_g^+, v = 0\right) &\to \text{N}_2\left(X^1\Sigma_g^+, v-1\right) + \text{N}_2\left(X^1\Sigma_g^+, v = 1\right)
\end{aligned}$$

(Tables 5.1, 5.2, 5.3), then an increase in the energy defect does not lead to a sharp drop in the rate constant since the effect described above is compensated by an increase in the constant due to an increase of the anharmonicity (see below).

Comparison of the V-T rate constants (see Sect. 5.1) in which initially unexcited species remain unexcited with data given in Tables 5.1, 5.2, 5.3 shows that the latter are several orders higher even there is no exact resonances in an exchange.

V-V processes in polyatomic molecules. Here we must add collision-induced intramolecular energy transfer processes. We know that there are spontaneous processes of stochastization of vibrational energy, but no energy loss occurs. For quasi-resonant processes, there is no significant difference between these processes in diatomic and polyatomic molecules (see Sect. 5.3 in [16]).

Table 5.1 The rate constant V-V, R of the process $\text{CO}(X^1\Sigma^+, v) + \text{CO}(X^1\Sigma^+, v = 0) \to \text{CO}(X^1\Sigma^+, v-1) + \text{CO}(X^1\Sigma^+, v = 1)$, T = 300 K (see [9, 11–14])

v	ΔE^a, cm^{-1}	k, 10^{-12} cm^3/s			
		[9]	[11]	[12]	[13]
2	26.6		1.8	1.9	3.3 ± 0.5
3	53.2		2.2		
4	79.7	2.0	2.1		
5	106.3	1.3	1.5		
6	132.9	0.63	0.85		
7	159.5	0.34	0.4		

[a]Energy deficit

Table 5.2 The rate constant *V-V*, *R* of the process $CO(X^1\Sigma^+, v) + N_2\left(X^1\Sigma_g^+, v = 0\right) \to CO(X^1\Sigma^+, v - 1) + N_2\left(X^1\Sigma_g^+, v = 1\right)$, T = 300 K (see [9, 13, 14])

v	ΔE^a, cm^{-1}	k, 10^{-15} cm^3/s		
		[9]	[13]	[14]
1	186.5			
2	213.3			
3	239.9			
4	266.4	2.4		
5	293.0	1.7		
6	319.6	1.0		
7	346.2	0.85		

[a]Energy deficit

Table 5.3 The rate constant *V-V*, *R* of the process $N_2\left(X^1\Sigma_g^+, v\right) + N_2\left(X^1\Sigma_g^+, v = 0\right) \to N_2\left(X^1\Sigma_g^+, v - 1\right) + N_2\left(X^1\Sigma_g^+, v = 1\right)$, T = 200, 300 K (see [14, 15])

v	ΔE^a, cm^{-1}	k, 10^{-15} cm^3/s	
		200 K	300 K
1	0	4.2	4.2
2	28.6	7.0	7.5
3	57.3	7.3	8.4
4	85.9	7.5	9.3
5	114.6	6.1	8.1
6	147.3	6.4	7.6
7	171.9	4.1	6.2

[a]Energy deficit

5.3.3 V-R Exchange

That such processes take place is beyond doubt. You can understand this if you imagine a noncollinear collision of two molecules, one of which oscillates. The presence of such processes is evidenced, for example, by a different relaxation efficiency of $CO(X, v = 1)$ in a collision with p-H_2 and o-H_2 having different equilibrium rotational energies: $P_{1,0}^{p-H_2}/P_{1,0}^{o-H_2} = 2$. A higher probability of relaxation in a collision with p-H_2 is explained, apparently, due to resonance of the p-$H_2(J = 2 \to J = 6)$ and $CO(v = 1 \to v = 0)$ transitions (see [17, 18] and references).

It is complicated to study such processes, since the probability of rotational relaxation, as a rule, is orders of magnitude higher than the probability of vibrational relaxation. These studies can be carried out directly if the conditions of single collisions are satisfied, i.e., for example, in molecular beams.

5.3.4 V-T Processes at High Vibrational Levels

The vast majority of data on the probabilities of vibrational relaxation refers to the $v = 1$ vibrational levels. As we have seen more than once, in the problems that we solve, very often there is no local thermodynamic equilibrium in terms of vibrational degrees of freedom; therefore, information on the value of transition probabilities for high vibrational levels is of importance. It must be remembered that if the vibrational quantum is $h\nu < < k\mathrm{T}$, then, according to the semiclassical approximation, the vibrational motion is very similar to the translational one with all the ensuing consequences: when such a species collides with M, energy of the order of $k\mathrm{T}$ is transferred with a probability close to 1. In this case, the selection rule $\Delta v = 1$ is violated.

It follows from the Landau-Zener formula (5.3.3) that with vibrational relaxation of a harmonic oscillator $\Delta \mathrm{v} = \mathrm{const}$), the transition probability $P_{v+1,v} \sim (v + 1)$, i.e., in some approximation

$$k_{v,v-1} = v \cdot k_{1,0}. \tag{5.3.7}$$

Equation (5.3.7) is the simplest version of the so-called ‘scaling relation’, which describes the dependence of the vibrational relaxation rate constant on the vibrational quantum number. If we take into account that the energy separation between successive vibrational levels is [3], p. 95:

$$\Delta G_{v,v+1} = \omega_0(1 - 2\omega_e x_e - 2\omega_e x_e v),$$

then the exponential term in the Landau-Teller formula should increase rapidly, and the above expression does not work (remember, however, that (5.3.3) is valid only for small exponential terms).

In general, $P_{v+1,v}$ at the upper levels must be large, and can reach 1, at the lower levels it is small following the Landau-Zener formula, for example, and at intermediate levels, its value can be estimated using some semi-empirical scaling relation, for example [19]:

$$\langle P_{v+1,v}\rangle = \begin{cases} a\{\frac{kT}{\Delta E} + exp[-\frac{b\Delta E}{kT}]\} \\ a + exp[-\frac{b\Delta E}{kT}] \\ akT\Delta E \end{cases} \tag{5.3.8}$$

or others. Parameters a, b are selected as follows: $< P_{v+1,v} > = 1$ for $v = v_{\max}$; for $v = 1$, the $< P_{1,0} >$ value is taken from experiment. It should be noted that besides

V-T processes, there can be others, in particular, those discussed below. So, the data obtained in this way are nothing more than a subject for the start of a serious study. It should be said that there is currently no theory with which you could calculate the vibrational relaxation constants at intermediate levels. You can get them only in the experiment by populating a single level and examining the dependence of the population of this and other levels on the M pressure. But in the general case, this also does not guarantee your success, since in addition to vibrational relaxation, a bunch of other collisional and spontaneous processes can take place. Numerical modeling can be useful for establishing their mechanism and kinetics. But even in this case, with success, one must say: 'We were able to describe the dependence observed in the experiment consistently.'

The mechanism of vibrational relaxation is interesting if the vibrational excitation energy *E*, which transforms into kinetic energy in a collision, is close to *k*T. This mechanism is realized at the vibrational relaxation of the $I_2(B0_u^+)$ molecule [20]. In this case, the highest probability of relaxation is observed if the collision time t_{coll} is equal to the oscillation period T_v. The vibrational relaxation constant can be equal to k_{gk} in this case.

In general, the following pattern can be traced. Rate of *V-T* processes:

- - increases with increasing of collision partner reduced mass μ if the energy taken away in a collision $\Delta E << k$T;
- falls if $\Delta E >> k$T;
- if $\Delta E = k$T, then it is maximum when $t_{coll} = T_v$ for a given μ.

5.4 The Influence of Nonadiabatic Effects on the Vibrational Relaxation Rate

Consider the collision of NO molecule in the ground state with He atom, for example. It is known that the ground state of $NO(X^2\Pi_{1/2,3/2})$ can be considered as degenerate. Whether it is degenerate or not, it should be spoken, bearing in mind the energy scale. This state is non-totally symmetric both in the orbital function and in the spin: $\Lambda = 1$, $S = 1/2$. Therefore, there are two Ω components of this state: $\Omega = 1/2, 3/2$. The splitting between them is very small, 0.0148 eV $\sim$ ½ *k*T, and it can be neglected in many cases, i.e., consider the ground state as degenerate. However, this cannot be done a priori, considering vibrational relaxation within the NO ground state. Here we must take into account that in fact, we deal with two states NO – $X^2\Pi_{1/2}$ and $X^2\Pi_{3/2}$, whose *E* values differ by 0.0148 eV, and two *PEC*s correlated in the zero approximation to the of $N(^4S) + O(^3P)$ limit. In the zeroth approximation since there is a fine structure of the term $O(^3P_J)$, $J = 2, 1, 0$, splitting $\Delta E_{2\text{-}0} \approx 0.03$ eV (see [8] and references). If we take this effect into account, the picture will be even more complicated, so let us stop for now.

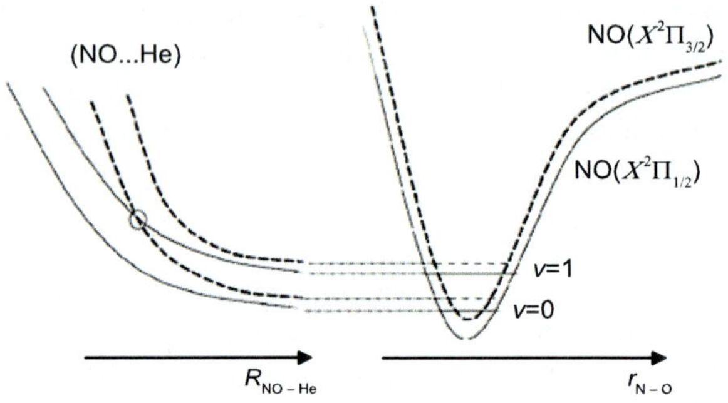

Fig. 5.4 Vibrational relaxation of the NO($X^2\Pi_{1/2,3/2}$,v) molecule

We already know that strictly speaking, describing the collision of NO($X^2\Pi_{1/2,3/2}$,$v > 0$) molecule with helium atom, we must describe this process in the framework of the NO-He complex *PES*. It is clear that this, or rather, these *PES*s are repulsive, but this does not change the matter. What is very important, we have the right to a collision complex with each vibronic level of NO($X^2\Pi_{1/2,3/2}$,v) to 'give' its *PES*. The picture will be interesting (Fig. 5.4):

For $R \to \infty$ we have a set of asymptotes (dissociation limits) corresponding to a set of NO levels ($X^2\Pi_{1/2,3/2}$,v) (note, we neglect rotational excitation, if we take it into account, then ...!). With decreasing *R*, these asymptotes 'bend' upwards, but (the most interesting begins here) they are not necessarily equidistant, since they correlate with NO in different electronic states. And since they are not equidistant, then with such a meager splitting of the *PES*s, which takes place at the asymptotic behavior, at some distance *R* they can intersect or quasi-intersect. And now recall that we discuss various types of nonadiabatic processes, on which their rate depends. We discussed mainly the mixing of bound and repulsive states accompanied by the predissociation of the former. But the selection rules for mixing is $\Delta J = 0$, which in the case of predissociation is performed automatically, are the same for both bound and repulsive states. The 'degree of mixing' in both cases is higher, the higher the overlap of the vibrational wave functions, and the latter, ceteris paribus, increases when the *PEC*s (*PES*s) intersect, or they come closer, more than with their equidistant passage. We conclude that in this case, the probability of vibrational relaxation can be much higher than in a collision of the same molecule, but not in a degenerate state. If we consider only two vibrational levels, the picture will be clearer: the image point moves, for example, by the *PES* of the NO($X^2\Pi_{3/2}$,v) -He complex *PES*, in the zone of intersection of the *PES* NO($X^2\Pi_{3/2}$,$v = 1$) -He and NO ($X^2\Pi_{1/2}$,$v = 0$) -He *PES*s a spontaneous non-adiabatic transition occurs from the upper *PES* to the lower one, and then the NO($X^2\Pi_{1/2}$,$v = 0$) and He atoms expand. Kinetic energy would be enough to get to this zone. Vibrational relaxation has occurred. Everything is determined by the difference of the *PES* curvature, the matrix elements of mixing of the complex states, and the kinetic energy of the collision.

Here, the author is forced to philosophize again. We must remember all the time: all the *PES*s that we draw which we are being discussed are nothing more than an approximation to the real, accurate picture of the process; we are not given to know the latter. Hereinafter, we will discuss various relaxation models in which nonadiabatic processes take place, and the author will convince you: '... and then a spontaneous process occurs in a molecule' or '... and then a spontaneous process occurs in a complex.' But we must remember all the time: the adiabatic process, the non-adiabatic process are all approximations with which we create models that help us get closer to understanding the essence of the process and somehow describe it in a consistent (at this stage of our development). Of course, a more accurate model is the one in which we consider the transition in a complex, rather than in an isolated molecule. But the statement is also obvious that it is much simpler to describe processes in a diatomic molecule than in a triatomic complex. And if we are able to describe the process of energy transfer, assuming that the partner that collides with the molecule produces a perturbation, and the process occurs in a free (isolated) molecule, and not in a complex, i.e., if the conclusions from our theory describe the experimental data well, then this is what should be done. Naturally, all models of vibrational relaxation, which we have discussed earlier, and which fit into the framework of the adiabatic principle, are also approximate.

And now we turn to the consideration of various models that describe the processes of energy transfer and do not fit into the framework of adiabatic approximations.

5.5 Collision-Induced Nonadiabatic Transitions

The first systematic research in this area of chemical physics was carried out more than 40 years ago. Almost simultaneously, the first attempts were made to create theoretical models describing the data obtained in experiments. Several models of collision-induced nonadiabatic transitions (CINATs) have been created, but even now, only a few of them have at least some predictive power. Most of the models, at best, explain the results of experiments, for which they were created.

The CINATs in different molecules obey laws that can be roughly divided into several groups. In some cases, non-adiabatic transitions occur between rovibronic states which mixing is also allowed in a free molecule, in others, the transitions between the states combining in the CINAT is strictly forbidden in a free molecule; an example is CINATs in some diatomic homonuclear molecules between electronic states of different *g*, *u* parities. Sometimes, it is assumed that large *FCF*s of combining levels (of course, in a free molecule) facilitate the transition in collisions, sometimes this correlation is absent.

CINAT model, in the author's opinion, can be considered truly successful if it has *predictive power*. It should make it possible to calculate the CINAT cross-sections or to show in which cases these cross-sections can be zero or very small, i.e., establish the CINAT *selection* or *propensity* rules. The term 'propensity rule' is undoubtedly more suitable for many collisional transitions because the mechanisms of these processes are not well studied, and the rules themselves often have a phenomenological character.

How is the problem of CINAT cross-sections in a collision AB* + M solved correctly? Ideally, it is necessary to calculate all the potential energy surfaces of the AB*···M complex correlating with the initial (input channel) and final (output channel) states, calculate the dependences of the matrix elements of the interaction of these states on the complex internal coordinates. Then one can solve the dynamic problem of the image point motion along these surfaces, taking into account the presence of non-adiabatic transitions between them. This ideal is unattainable even to describe the collision of a light diatomic molecule with helium atom; in any case, as we will see below, even the best of the existing models are unable to explain some of the effects observed in the experiment.

You already know that a non-adiabatic transition between adiabatic *PES*s only occurs when the matrix elements of the interaction of states corresponding to these *PES*s are nonzero. Therefore, the CINATs selection (propensity) rules are determined by the symmetry of these states and the interaction operator. We met with these interactions in Sect. 3.3. At small distances, these are exchange interactions, then direct electrostatic (multipole-multipole) and polarization (calculated in the second order of perturbation theory) interactions are 'added' to them (become comparable in magnitude).

On the classification of CINATs. They can be divided into two large groups:

- CINATs between rovibronic levels of a molecule, the mixing of which is allowed to one degree or another. These are the so-called *perturbation-facilitated* [21], p. 313, [22], p. 445 (*perturbation-assisted* [23]) processes. The role of a species colliding with a molecule between which states the CINAT is realized can be quite primitive: collisions lead to the fact that one or more (many) of the mutually perturbed levels is populated as a result of $R \leftrightarrow T$ or $V, R \leftrightarrow T$ processes, and then the non-adiabatic transition occurs.
- CINAT between rovibronic levels of a molecule, the mixing of which is either weak, so that it does not affect the CINAT rate, or is rigorously forbidden. These are *perturbation-irrelevant* [21], p. 313, [22], p. 447 (*perturbation-independent* [23]) processes indifferent to perturbation in a free molecule. Here the partner's participation in the CINAT is fundamental: the transition cannot occur without it. Just in the similar CINATs analysis, one cannot do anything without solving the problem of an image point moving along the *PES* of the complex, which was discussed above.

The cross-section dependences of these two types of processes on the level characteristics between which the transitions occur and the colliding particle properties are different. We will consider them separately. Let us start with the perturbation-facilitated processes.

5.5.1 *Perturbation-Facilitated Processes*

Gateway (doorway) model (see [21], p. 313, [22], p. 445, [23] and references). This model is the simplest to understand, but not to confirm in an experiment. Gateway (doorway) is the way through the gate (door). The term describes the essence of the model very accurately. What does it consist?

Suppose that one deals with a specific diatomic (for simplicity) molecule AB, in which there are at least two electronic states 1 and 2, which correspond to two *PEC*s in a particular energy range (Fig. 5.5a), and from one of these states (for example, 1) the optical transition to the lower states is forbidden. Let some high rovibronic levels of this metastable state be populated. Within the framework of our agreements, the only way to dissipate the energy of its excitation is a collision-induced loss. Let the mixing of these two electronic states be rigorously forbidden, for example, because of their different multiplicity and the 'lightness' of the atoms of which this molecule consists. Therefore, the wave function of state 1, for example, has a 'noticeable impurity' of state 2,

$$\Psi_1 = \Psi_1^0 + \frac{V_{1,2}}{E_1^0 - E_2^0}\Psi_2^0 \tag{5.5.1}$$

(see 3.6.3); here Ψ_1^0 and Ψ_2^0 are the wave functions of states 1 and 2 obtained in the zeroth approximation, $V_{1,2}$ is the matrix element of the perturbation operator of these states which in this case is very small (see 3.6.4). It is necessary for a mixing that there be a 'very-very accurate' resonance between levels with the same total moment J (see Fig. 5.5 b). Indeed, for the coefficient before the wave function of the 'mixing' states for a small matrix element of the perturbation operator

$$V_{1,2} = \langle\Psi_1^0|\widehat{V}|\Psi_2^0\rangle \tag{5.5.2}$$

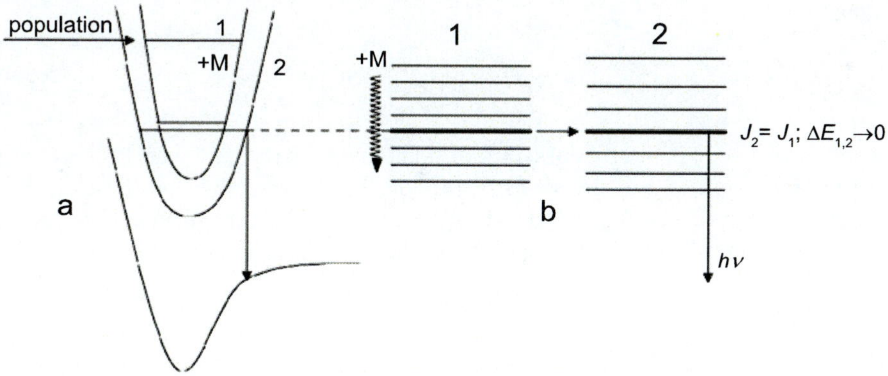

Fig. 5.5 To the gateway model

was large enough, we must have a very small denominator in (5.5.1). We must also remember that often $V_{1,2}$ can be expressed as

$$V_{1,2} = A^{el}_{1,2}\langle\chi^0_1|\chi^0_2\rangle \tag{5.5.3}$$

where $A^{el}_{1,2}$ is the electronic matrix element of the interaction of the states 1, 2, $\langle\chi^0_1|\chi^0_2\rangle$ is the overlap integral of the vibrational wave functions of these states, equal to the square root of the Frank-Condon factor for these states (see (3.6.8)). As we have agreed, $V_{1,2}$ is small, and for almost all rovibronic levels, $\Psi_1 = \Psi^0_1$.

Let this diatomic molecule collides with the He atom, for example, and the result of these collisions is vibrational–rotational relaxation in the state 1. And let some rovibronic levels of states 1 and 2 having the same angular momentum $J_1 = J_2 \equiv J$, are in a very accurate resonance, such that the coefficient before Ψ^0_2 in (5.5.1) $\frac{V_{1,2}}{E^0_1 - E^0_2}$ is nonzero, and the 'addition' $\frac{V_{1,2}}{E^0_1 - E^0_2}\Psi^0_2$ is comparable to Ψ^0_1. States 1 and 2, as we have agreed, are bound. Therefore, for levels with $J_1 \neq J_2$, the matrix element $V_{1,2}$ is equal to 0 (this selection rule is absolute; we remember that it is automatic for unbound states). A rigorous framework is $J_1 = J_2$, $(E^0_1 - E^0_2) \to 0$, and the latter is stricter, the smaller is the value of $V_{1,2}$.

What can this feature lead? If an optical transition to some other state is possible from state 2, it can occur (Fig. 5.5b), and its intensity will be determined by the ratio of the probability of an optical transition from this mixed level to the collision rate of its settlement directly proportional to the He concentration. If state 2 can predissociate, then this predissociation can happen. In any case, state 1 may 'disappear' with some probability.

Why do we not describe this feature as non-adiabatic transitions in the AB···He complex? And why do we consider spontaneous non-adiabatic transitions in an isolated AB molecule and not in a complex? There are many such questions. The reason is a simple one: this model is elementary, and when using this model, it is relatively easier to describe the process we are considering: Its cross-section is equal to

$$\sigma_{1,2}(J) = \sigma(J, J-1)C^2_{1,2} \tag{5.5.4}$$

where $\sigma(J, J-1)$ is the cross-section of rotational relaxation in state 1, and $C_{1,2}$ is the mixing coefficient of the states $\Psi_1(J)$ and $\Psi_2(J)$, depending on the electronic matrix element of mixing of states 1 and 2, the overlap integral vibrational wave functions and resonance (see 5.5.1–5.5.3). The1 $\to$ 2 transition rate constant at this level is

$$k_{1,2} = k_{gk}C^2_{1,2} \tag{5.5.5}$$

That is, at $C_{1.2} \to 1$, it is close to k_{gk}.

So, the model is straightforward. It is only necessary to establish when it can be applied. This work was carried out by its creators in the 70-s of the last century. It was shown that for its applicability, the following requirements must be met:

1. The matrix element of the interaction of the considered levels $V_{1,2}$ should be so small that

$$2\pi\tau V_{1,2} \leq \hbar \tag{5.5.6}$$

   (here τ is the time of the AB and M collision). Since the collision time is of the order of 10^{-12} s, $\hbar$ = 5.3 10^{-12} cm^{-1} s, then $V_{1.2} \leq 1$ cm^{-1} should be. If this is not so, then we must take into account the evolution of the complex during its existence, i.e., consider the motion of an image point along the *PES*s of the complex. The task immediately becomes difficult, and the transition cross-section cannot be calculated using (5.5.4). It is said that the collision should be '*sudden*', i.e., during the collision, the molecule should not rotate (this is not a very strong condition since the rotation period is approximately two orders of magnitude longer than the oscillation period).
2. Collision should be '*weak*'; in other words, the *PES*s corresponding to states 1 and 2 have to be equidistant, the gap between them should not depend on the distance AB···M. Otherwise, at least the coefficient $C_{1,2}$ depends on *R*, the $C_{1,2}$ value calculated for a *free* molecule cannot be used. In short, in the event of a '*strong*' collision, we must consider the transition in the AB···M complex.

Let us discuss the gateway model as it has been studied for the NO ($a^4\Pi \to B^2\Pi$, $b^4\Sigma$) transitions in [24, 25]. The scheme of the experimental set-up utilized is shown in Fig. 5.6. In general terms, it consists of a device forming a supersonic molecular beam (a pulse valve, skimmer, diaphragms) and a chamber with a gas target placed in a large chamber equipped with pumping systems, pumping speed is 3400 l/s.

Excited molecules are generated by means of dc discharge between a nozzle and filament. Charged species can be deflected by a field (electric or magnetic) so that, if necessary, they do not enter into the chamber. The important thing is that one can obtain a supersonic molecular beam of excited metastable (having a sufficiently long lifetime for radiation decay or predissociation) molecules, study the processes that occur when an excited molecule collides *once* (this is very important) with the target species, excites or 'takes away' it by laser radiation, etc. Of course, there are several excited states in a beam, but sometimes this fact does not interfere.

So, what will happen if one forms in this way a NO molecule molecular beam? The NO($X^2\Pi, v_X$) and NO($a^4\Pi, v_a$) *metastable (long-lived)* states get in the camera with target species (Fig. 5.7); the $A^2\Sigma^+$, $B^2\Pi$, $C^2\Pi$, and many other states decay due to rather fast radiative transitions.

What will happen if molecules excited in this way collide with a target species? In principle, in this case, a collision-induced transition to the $A^2\Sigma^+$, $B^2\Pi$, $C^2\Pi$ states

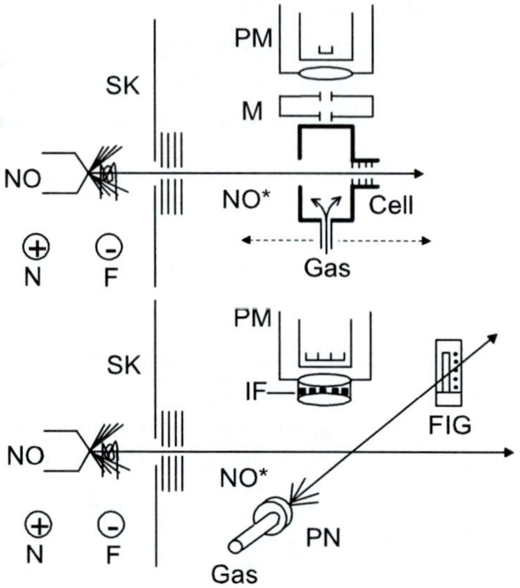

Fig. 5.6 Schematic representation of an experimental setup for studying collision-induced non-adiabatic transitions. Above, the collision partner is located in the collision chamber; below, a collision partner is formed in a molecular beam (see [26]). N, nozzle, F, filament, SK, skimmer, M, monochromator, PM, photomultiplier tube; IF, interference filter; PN—pulse nozzle forming a colliding partner molecular beam; FIG—pulse ionization manometer (Reproduced from R. Bachmann, Ch. Ottinger, A.F. Vilesov, J. Chem. Phys. **96**, 5151–5164 (1992). https://doi.org/10.1063/1.462756 with the permission of AIP Publishing)

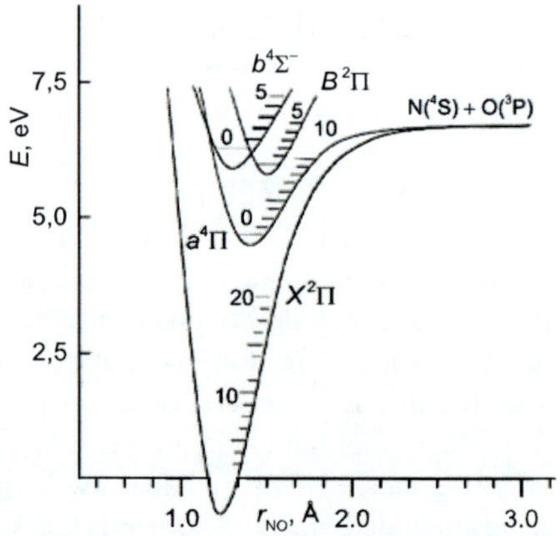

Fig. 5.7 Potential energy curves of the NO lower states [24] (Reproduced from Ch. Ottinger, A.F. Vilesov, J. Chem. Phys. **100**, 1805–1814 (1994) https://doi.org/10.1063/1.466533 with the permission of AIP Publishing)

decayed by this moment can occur and, therefore, optical transitions from them can be observed. Where can settlement come from? From metastable states. What did the authors of [24, 25] observe? It was found that for target species, M = He, Ar, N_2, only two rotational levels of only two vibrational levels of two states of the NO multiplet turned out to be populated, namely, $B^2\Pi_{3/2},v_B = 0,J_B = 10.5$, and $B^2\Pi_{3/2}$, $v_B = 3,J_B = 17.5$, and a broad set of rotational levels of NO($b^4\Sigma^-,v_b = 3$–5).

Let us now consider the B state. Everything is very similar to the case described in the framework of the gateway model. The analysis of the obtained and published data, as well as the study of the lifetimes of the NO($B^2\Pi_{3/2},v_B = 0,J_B = 10.5$, and $B^2\Pi_{3/2},v_B = 3,J_B = 17.5$) clarified the following picture:

Unperturbed rovibronic states NO($B^2\Pi_{3/2},v_B = 0,J_B = 10.5$ and $a^4\Pi_{5/2},v_a = 8$, $J_a = 10.5$, as well as NO($B^2\Pi_{1/2},v_B = 3,J_B = 17.5$) and $a^4\Pi_{5/2},v_a = 12,J_a = 17.5$) are randomly in the exact resonance, $\Delta E \approx 10^{-4}$ cm^{-1}. The electronic matrix element of the interaction of these states A_{el} is small since the spin momenta of these states are different and the nuclei in the molecule are relatively light: $A_{el} \approx 50$ cm^{-1}. Besides, the overlap integrals of the vibrational wave functions of the B,0 $\sim$ a,8 and B,3 $\sim$ a,12 states are of the order of 5 10^{-5} (Frank–Condon factors of the order of 2 10^{-9}). Consequently, the matrix element of their interaction is negligible, of the order of $n{\cdot}10^{-3}$, and resonance of the order of 10^{-3} cm^{-1} is required for mixing these levels of the B, and a states. A stream of NO molecules flies onto the target, in which a huge number of the NO(a) metastable state rovibronic levels are, including those in the vicinity of $v_a = 8,J_a = 10.5$ and $v_a = 12,J_a = 17.5$. The latter are *just absent*, since they are mixed with the $v_B = 0,J_B = 10.5$ and $v_B = 3,J_B = 17.5$ levels and have decayed in radiation transitions along the way to the observation zone. What will happen if molecules excited in this way collide with a target species? Collisions lead to T-R (from bottom to top) and R-T (from top to bottom) to populate the $v_a = 8,J_a = 10.5$ and $v_a = 12,J_a = 17.5$ levels. Then spontaneous transitions to the NO($B^2\Pi_{3/2},v_B = 0,J_B = 10.5$ and NO($B^2\Pi_{1/2},v_B = 3,J_B = 17.5$) states occur, followed by optical transitions, which we discussed above. Since the exact ($< 10^{-3}$ cm^{-1}) resonance and the fulfillment of the selection rule $J_1 = J_2$ takes place only for the levels indicated above, only they find themselves in the radiation.

It must be borne in mind, of course, that a complex is formed during a collision, and as a result, ΔE, and the matrix element of the interaction of the complex states correlating with the rovibronic levels depend on $R_{NO*...M}$. But still, it remains very small, since on the asymptote, it is negligible ($< 10^{-3}$ cm^{-1}). The gateway model is applicable if $V_{1,2} < 1$ cm^{-1}. Here this inequality holds.

Other perturbation facilitated processes. What will happen if $V_{1,2} > 1$ cm^{-1}? In the case of the NO(B,v_B $\sim$ a,v_a), mixing this can happen if NO collides with a heavy atom, like Xe, or with a paramagnetic molecule such as NO or O_2, for example. In the general case, this can be, if either the electronic part of the matrix elements of the interaction of states in a free molecule, or/and the overlap integrals are sufficiently large. The gateway model is not applicable in this case, and one must consider the processes already within the complex. Such effects were observed in [24].

In particular, when M = Xe, O_2, and especially NO, the NO(a,v_a)$\cdots$M complex is formed, the matrix elements of the interaction in which are so large that inequality (5.5.6) is certainly not satisfied. The interaction of a heavy or paramagnetic species with NO leads to the fact that the spin momentum conservation rule is violated (Xe), or the total spin of the colliding partners is retained, due to the nonzero M spin momentum, while that of NO changes. One way or another, the requirements for resonance are reduced due to the large value of $V_{1,2}$, many rotational levels are populated, not only $B, v_B = 0, 3$, but also other vibrational levels. The same feature is observed in the NO($a^4\Pi \rightarrow b^4\Sigma^-$, $v_b = 3$–5) CINATs between quartet states, since the forbiddance of this transition is much weaker than in the processes considered above. But the fee for this is 'heavy': the gateway model is not applicable in this case. This type of process needs additional researches. In the author's opinion, a decent model that adequately describes perturbation-facilitated processes in the case of large quantities of interaction matrix elements has not yet been created.

5.5.2 *Perturbation-Irrelevant Transitions*

Now we will consider some experimental data concerning processes indifferent to perturbations and models that explain the observed features. We will move from simple to complex.

5.5.2.1 Collision-Induced Predissociation of the $I_2(B0_u^+)$ State

Let us start with the most, perhaps, the oldest and most simple model of these processes. It can describe CINATs only qualitatively. A collision leads to the deformation of the electron shells, an appearance of electric and magnetic fields and, therefore, perturbations induced by them (Stark or Zeeman effects) [22], p. 418. These perturbations may cause CINATs. The collision-induced predissociation (CIP) of the $I_2(B0_u^+)$ state

$$I_2(B, v_B, J_B) \xrightarrow{M} I_2(RS) \rightarrow I(^2P_{3/2}) + I(^2P_{3/2}) \tag{5.5.7}$$

(RS means repulsive state (states)) is an example of such processes. The CIP perturbation operator can be obtained in the same way as for the van der Waals (vdW) interactions (see Sect. 6.3); its cross-section is [27]:

$$\sigma_{CIP} \sim \mu^{1/2} I_M \alpha_M / R_C^3;$$

here μ is the reduced mass, I_M and α_M are the M ionization potential, and polarizability, $R_c = (1/2)\cdot(\sigma_M + \sigma_{I_2(B)})$ is a half of the M and $I_2(B)$ gas-kinetic diameter sum. The form of the perturbation operator determines the selection rules. If the transition is caused by an induced electric field, the perturbation operator has the same symmetry type as that of transition dipole moment. Accordingly, the selection rules for the CINAT become the same as for the optical ones, in particular, the strict selection rule for intramolecular perturbations of homonuclear molecules:

$$g \leftrightarrow g, u \leftrightarrow u \tag{5.5.8}$$

reverses

$$g \leftrightarrow u, \tag{5.5.9}$$

exactly the same as for optical transitions. It was suggested that the $I_2(B0_u^+)$ state CIP (5.5.7) causes by the $a'0_g^+$ and $a1_g$ states, i.e., by the states of the opposite parity (see [28], p. 95 and (Fig. 4.4).

5.5.2.2 The $CN(A^2\Pi \rightarrow^{Rg} X^2\Sigma^+)$, $N_2^+\left(A^2\Pi_u \rightarrow^{Rg} X^2\Sigma_g^+\right)$ CINATs

These processes were studied with a high spectral resolution, allowing one to obtain information on the populations of individual rovibronic levels and even components of their fine structures. To study CINATs in these species, one used optical–optical double resonance, OODR, (CN, N_2^+) and resonance-enhanced multiphoton ionization spectroscopy, REMPI, (N_2). Briefly about the technique of only one a series of experiments performed by Dagdigian et al. [23, 29–32].

Cyanide free radicals, $CN(X^2\Sigma^+)$, were prepared by the addition of cyanogens, C_2N_2, to a flow containing metastable He, Ar atoms, which were produced in a low-current d.c. discharge. C_2N_2 molecules dissociate, and radicals $CN(X, v_X, J_X)$ are formed. A pump laser radiation is transferred $CN(X, v_X, J_X)$ to the $CN(A^2\Pi, v_A = 3$ or $7, 8; J_A)$ rovibronic levels, for example (Fig. 5.8),

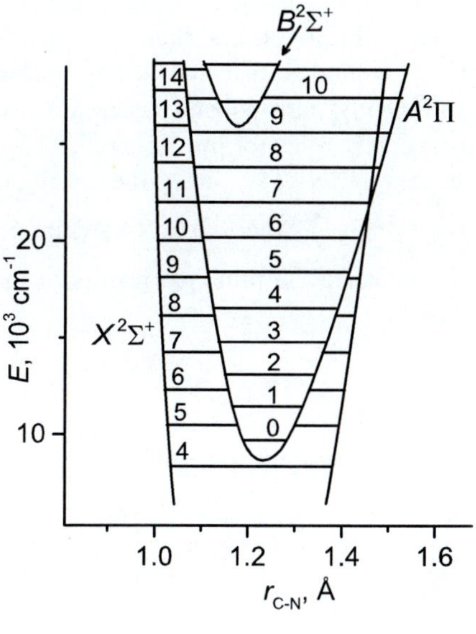

Fig. 5.8 Potential energy curves of the lower CN states (see [23] and references)

$$\mathrm{CN}(A, 7, J_A) \overset{h\nu_p}{\longleftarrow} \mathrm{CN}(X, v_X, J_X), \tag{5.5.10a}$$

$$\mathrm{CN}(A, 3, J_A) \overset{h\nu_p}{\leftarrow} \mathrm{CN}(X, v_X, J_X), \tag{5.5.10b}$$

and after a delay time, using another (probe) laser, the population of A and X state quasi-resonance levels is checked, for example, after a population of the CN(A, 7, J_A):

$$\mathrm{CN}(A, 7, J_A) \overset{h\nu_{pr}}{\longrightarrow} \mathrm{CN}\left(B^2\Sigma^+, v_B, J_B\right), \tag{5.5.11}$$

$$\mathrm{CN}(B, v_B, J_B) \rightarrow \mathrm{CN}(X, v_X, J_X) + h\nu_{\mathrm{lum}}, \tag{5.5.12}$$

(checking the population of the initial level);

$$\mathrm{CN}\left(B, v_B', J_B'\right) \overset{h\nu_{pr}}{\leftarrow} \mathrm{CN}(X, 11, J_X), \tag{5.5.13}$$

$$\mathrm{CN}\left(B, v_B', J_B'\right) \rightarrow \mathrm{CN}(X, v_X, J_X) + h\nu_{\mathrm{lum}}, \tag{5.5.14}$$

(checking the population of the level forming in the

$$\mathrm{CN}\left(A, 7, J_A \overset{Rg}{\longrightarrow} X, 11, J_X\right), \tag{5.5.15}$$

CINAT).

If, during the delay time, a collision-induced transition (5.5.15) to the probed level X, 11, J_X occurs, then the transition (5.5.13) can occur, and the luminescence (5.5.14) can be detected at appropriate $h\nu_{pr}$ energy for the transition (5.5.13). If sufficiently narrow-band lasers are used for pumping and probing, one can study propensity rules for processes (5.5.15). By reducing the delay time, it is possible to study CINATs in the single-collision conditions. Similarly, the $N_2^+\left(A^2\Pi_u \overset{Rg}{\longrightarrow} X^2\Sigma_g^+\right)$ CINAT (Fig. 5.9) was studied (see [33, 34] and references).

What are the principal results of the

$$\mathrm{CN}\left(A, v_A, J_A \overset{\mathrm{Rg}}{\longrightarrow} X, v_X, J_X\right) \tag{5.5.16}$$

CINAT studies [29–31]?

Fig. 5.9 Potential energy curves of the lower N_2^+ states (see [23] and references)

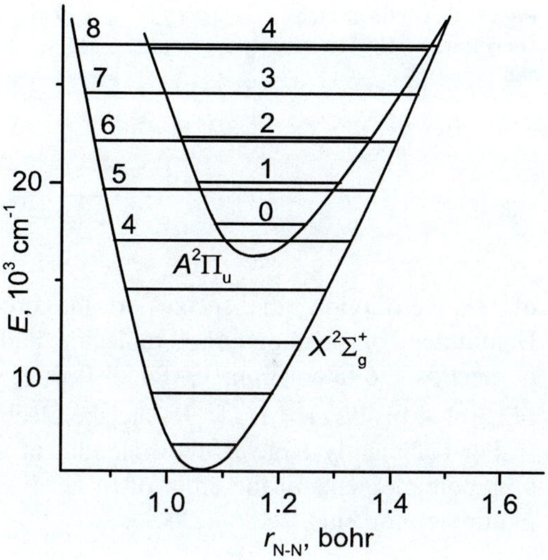

1. The cross-sections of the quasi-resonant CINATs (5.5.16) for $A, 7 \xrightarrow{\text{Rg}} X, 11$ and $A, 8 \xrightarrow{\text{Rg}} X, 12$ are close to that of rotational relaxation, i.e., are large, and for $A, 3 \xrightarrow{\text{Rg}} X, 7$ (the energy gap is $\sim 600\ \text{cm}^{-1}$) is $\sim$2–5 times lower.
2. The propensity rules $\Delta J \approx 0$ is valid, and larger cross-sections correspond to transitions accompanied by a change of the symmetry index of the level $\varepsilon_X = \varepsilon_A \pm 1$ (see Sect. 4.6.1 and Table 4.3). For the CINATs in a homonuclear molecule, $N_2\left(a^1\Pi_g \xrightarrow{\text{Rg}} a'^1\Sigma_u^-\right)$ [35], and homonuclear ion, N_2^+ $\left(A^2\Pi_{ui} \xrightarrow{\text{Rg}} X^2\Sigma_g^+\right)$ [32, 33], almost everything is very similar, but the propensity rule $\varepsilon^{\text{out}} = \varepsilon^{\text{in}} \pm 1$ becomes strict (selection rule).

These CINATs were investigated theoretically by M.H. Alexander and H.-J. Werner with colleagues [32, 36, 37].

The Hamiltonian of a system of three atoms in a laboratory coordinate system (LCS) which describes the collision of an atom with a diatomic molecule can be written as:

$$\hat{\boldsymbol{H}}(\boldsymbol{r},\boldsymbol{R}) = -\frac{\hbar^2}{2\mu R^2}\frac{d}{dR}\left(R^2\frac{d}{dR}\right) + \frac{L^2}{2\mu R^2} + \hat{\boldsymbol{H}}_{mol}(\boldsymbol{r}) + \hat{\boldsymbol{V}}_{el}(\boldsymbol{R},\boldsymbol{r}) \qquad (5.5.17)$$

Here: R is the distance from the atom to the center of mass of the molecule, r is the internuclear distance in the molecule (Fig. 5.10), μ is the reduced mass of the entire system, $\boldsymbol{L} = \mu \boldsymbol{R} \boldsymbol{V}_{\text{rel}}$ is the orbital angular momentum, $\boldsymbol{V}_{\text{rel}}$ is the relative velocity vector. The first term of the Hamiltonian corresponds to the kinetic energy

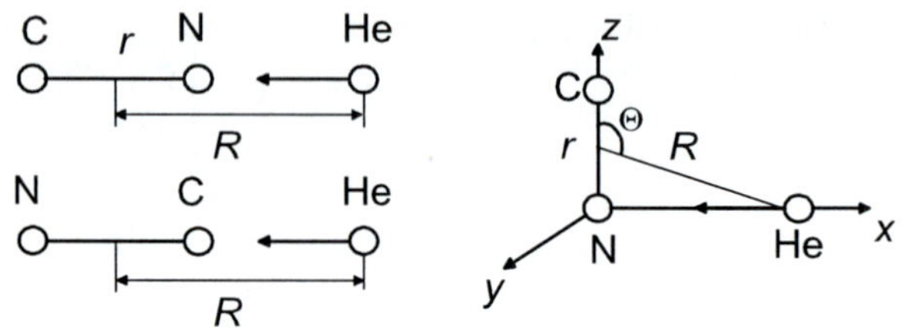

Fig. 5.10 Collinear and noncollinear collision of CN and He

of relative motion, the second to the orbital energy. The third term is the Hamiltonian of the unperturbed molecule, and the fourth one is the potential energy of electrostatic interaction; V_{el}(R) $\to 0$ as $R \to \infty$. The last two terms of (5.5.17) $\widehat{\boldsymbol{H}}_{mol}(\boldsymbol{r}) + \widehat{\boldsymbol{V}}_{el}(\boldsymbol{R},\boldsymbol{r}) = \widehat{\boldsymbol{H}}_{el}(\boldsymbol{r},\boldsymbol{R})$ are the electronic Hamiltonian of the system.

For sufficiently large R, the potential of two collision partners (i = 1, 2) with multipole moments of the order of l_1 and l_2 can be represented as an expansion in multipole moments (see Sect. 3.3):

$$V_{el}(r,R) = \sum_{l_1,l_2} \frac{C_{l_1,l_2}}{R^{l+1}} \sum_{m,m,m_2} \langle l_1 m_1 l_2 m_2 | l_1 l_2 l m \rangle Q_{l_1 m_1} Q_{l_2 m_2} Y_{lm}(\theta, \varphi) \tag{5.5.18}$$

Here: C_{l_1,l_2} are the numerical coefficients determined by the type and magnitude of the electrostatic interaction (for dispersion interaction $\frac{C_{l_1,l_2}}{R^{l+1}} = \frac{C_6}{R^6}$), $\langle l_1 m_1 l_2 m_2 | l_1 l_2 l m \rangle$ are the Clebsch-Gordan coefficients, m_i are the projections of l_1, l_2, l = l_1, l_2 on the selected axis in the LCS, $Q_{l_i m_i}$ are the multipole moments of the collision partners, $Y_{lm}(\theta, \varphi)$ are the spherical functions of angles that determine the position of $\boldsymbol{R}$ in the LCS, θ is the angle between $\boldsymbol{R}$ and $\boldsymbol{r}$.

CINAT corresponds to transitions between the states of the continuous spectrum of the wave equation with the Hamiltonian (5.5.17), and the wave function $\psi(\boldsymbol{r}, \boldsymbol{R})$ is linear combination of the basic functions, for which the wave functions of the *diabatic* representation corresponding to $V_{el}(r,R)$= 0 are used. For an input channel at $R \to \infty$, this function is equal to.

$$\psi_j^d(\boldsymbol{r},\boldsymbol{R}) = \varphi_i(\boldsymbol{r}) \left[\exp(ik_j R) + \frac{f_{ij}}{R} \exp(ik_j R) \right]. \tag{5.5.19}$$

The first term (plane wave) describes the incident stream of species moving to $R \to 0$ and having a wave vector $k_i = \frac{2\mu(E-\varepsilon_i)}{\hbar^2}$ (ε_i is the internal energy of a free molecule with the wave function $\varphi_i(\boldsymbol{r})$. The second term (spherical wave) describes elastic scattering. The asymptotic form of the wave function of the output channel contains only the spherical wave.

$$\psi_j^d(\boldsymbol{r},\boldsymbol{R}) = \varphi_i(\boldsymbol{r}) \frac{f_{ij}}{R} \exp(ik_j R) \tag{5.5.20}$$

The functions $f_{ij}(R, \Theta, r)$ are the amplitudes of the probability of scattering from channel i to channel j. The transition probability and the CINAT cross-section are proportional to f_{ij}^2.

The wave functions of the diabatic representation are not eigenfunctions of $\widehat{\boldsymbol{H}}_{el}(\boldsymbol{r}, \boldsymbol{R})$. The electronic states in the *diabatic* representation are connected by off-diagonal matrix elements $\langle \psi_i^d | \widehat{\boldsymbol{V}}_{el} | \psi_j^d \rangle$ (potential binding), i.e., a non-adiabatic transition between them occurs as it does in the case of weak interaction (Fig. 3.16).

The set of the $\widehat{\boldsymbol{H}}_{el}(\boldsymbol{r}, \boldsymbol{R})$ eigenfunctions corresponds to the *adiabatic* representation. In this case, there is no potential binding, and a *nonadiabatic* transition between these states occurs as it does in the case of the strong interaction (see Sect. 3.6, Fig. 3.17 and Sect. 4.6.1.1). The wave functions of the *adiabatic* and *diabatic* representations are related by a linear transformation mixing the *diabatic* functions of the same symmetry (see below).

Now, let us consider the results obtained in [36] on the calculation of processes in the CN + He system (see [37], also). What happens when a He atom collides with the CN radical (Fig. 5.10)? If the collinear collisions, He–CN, or CN–He, occur, then the symmetry type does not change compared to the CN symmetry type, i.e., remains $C_{\infty v}$, if non-collinear one, it is lifted to C_s point group. In this symmetry group, the state Σ^+ corresponds to A' state and the degenerate Π state splits into two, $A' + A''$ states. The mixing of the Σ^+, Π states in the $C_{\infty v}$ point group, and A', A'' in C_s one is allowed only taking into account the electron-rotational or vibronic interactions.

Let R is the distance from the atom to the center of mass of the molecule, r is the internuclear distance of the molecule, and the z axis coincides with CN internuclear axis (Fig. 5.10). Then the totally-symmetric component of the state Π, A', corresponds to Π_x, and non-totally symmetric A'' to Π_y. The resulting *adiabatic* wave functions of symmetry A' are linear combinations of the *diabatic* wave functions of the Σ^+, and Π states (see above):

$$\psi_1(R, r, \theta) = \cos\chi|\Sigma^+\rangle - \sin\chi|\Pi_x\rangle \tag{5.5.21}$$

$$\psi_2(R, r, \theta) = \sin\chi|\Sigma^+ - \cos\chi|\Pi_x\rangle \tag{5.5.22}$$

where χ is the *mixing angle*, a coefficient depending on the geometry of the complex, i.e., bond length C-N, distance R, and angle of attack (see below). Here $|\Sigma^+ >$, $|\Pi_x >$ are *diabatic* wave functions obeying the relation

$$\langle \Sigma^+ | \frac{\partial}{\partial q} | \Pi_x \rangle = 0 \tag{5.5.23}$$

for all three internal coordinates of the complex (see Sect. 4.6.1.1).

The matrix elements of the electronic Hamiltonian $\widehat{\boldsymbol{H}}_{el}$, by definition being potential surfaces, can be represented as:

$$\langle\Sigma^+|\widehat{\boldsymbol{H}}_{el}|\Sigma^+\rangle \equiv T_\Sigma(r) + V_\Sigma(R,r,\theta) = T_\Sigma(r) + \sum\nolimits_{l=0}^{\infty} V_\Sigma^l(r,R)d_{00}^l(\theta) \quad (5.5.24)$$

$$\Pi_{x,y}|\widehat{\boldsymbol{H}}_{el}|\Pi_{x,y} \equiv T_\Pi(r) + V_\Pi(R,r,\theta) = T_\Pi(r) + V_\Pi(R,r,\theta) \mp V_2(R,r,\theta) = T_\Pi(r) + \sum_{\infty}^{l=0} V_\Pi^l(r,R)d_{00}^l(\theta) \mp \sum_{\infty}^{l=2} V_2^l(r,R)d_{20}^l(\theta) \quad (5.5.25)$$

$$\langle\Sigma^+|\widehat{\boldsymbol{H}}_{el}|\Pi_x\rangle \equiv \langle\Pi_x|\widehat{\boldsymbol{H}}_{el}|\Sigma^+\rangle = V_1(R,r,\theta) - \sqrt{2}\sum\nolimits_{l=1}^{\infty} V_1^l(r,R)d_{10}^l(\theta) \quad (5.5.26)$$

The terms $T_\Sigma(r)$, $T_\Pi(r)$ in (5.5.24, 5.5.25) are the potential energy of isolated molecules in the corresponding states, and the *diabatic* interaction potentials $V_\Sigma(R, r, \Theta)$, $V_\Pi(R, r, \Theta)$, $V_1(R, r, \Theta)$, $V_2(R, r, \Theta)$ describe the change in the electronic Hamiltonian as species approach each other. The d_{mn}^l is reduced rotation matrix elements being functions of the angle Θ, which determines the relative orientation of the vectors R and r (Fig. 5.10); the minus sign belongs to the V_{Π_x} potential. The second term in (5.5.24, 5.5.25) describes the perturbation resulting from the approaching of the Σ^+ and Π potentials, and the third term in (5.5.25) describes the splitting of the degenerate Π state. We see that (5.5.24, 5.5.25) are responsible for relaxation within the same state and transitions between levels of Λ doubling in the Π state. The off-diagonal potential $V_1(R, r, \Theta)$ (5.5.26) is responsible for the potential binding of the Σ^+ and Π_x states of a free molecule, i.e., for CINATs. These matrix elements are primarily interested for us. In the first order of perturbation theory, for a fixed $l = l_1 + l_2$, the CINATs are realized between rotational levels with $\Delta J \leq l$. The matrix elements V_Σ, V_Π, V_1, and V_2 obtained using the energies of the adiabatic states Ψ_1, Ψ_2 and Ψ_3 (the latter is the A" component of the Π state) are as follows:

$$V_\Sigma = E_1 cos^2\chi + E_2 sin^2\chi \quad (5.5.27)$$

$$V_\Pi = 1/2 \cdot \left(E_1 sin^2\chi + E_2 cos^2\chi + E_3\right) \quad (5.5.28)$$

$$V_1 = \langle\Sigma^+|\hat{\boldsymbol{H}}_{el}|\Pi_x\rangle = 1/2(E_2 - E_1)\sin 2\chi \quad (5.5.29)$$

$$V_2 = 1/2\left(-E_1\sin^2\chi - E_2\cos^2\chi + E_3\right). \quad (5.5.30)$$

The mixing angle χ is

$$\chi(\Theta) = \int_0^\Theta \left\langle \psi_1 \left| \frac{\partial}{\partial\Theta'} \right| \psi_2 \right\rangle d\Theta' \quad (5.5.31)$$

for fixed values of R and r, and the matrix elements of *nonadiabatic* mixing of the *adiabatic* states $\Psi_1(A')$ and $\Psi_2(A')$ are

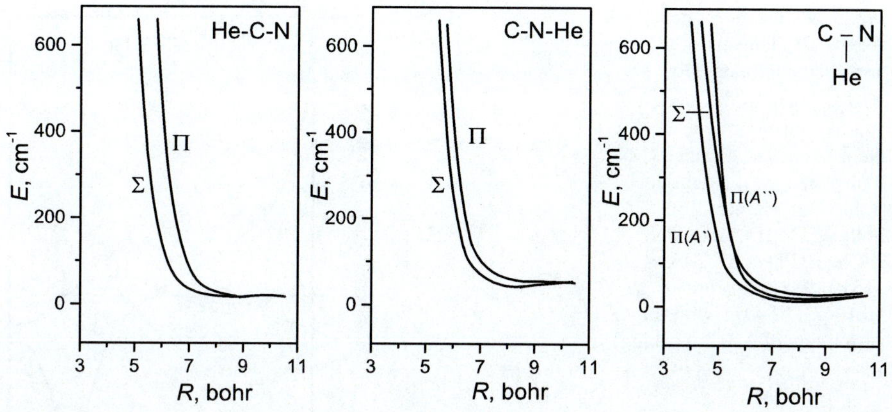

Fig. 5.11 Adiabatic CN-He interaction potentials as a function of distance R for collinear and noncollinear $\Theta = 90°$ collisions of CN and He; $r = 2.396$ bohr (see [36]) (Reproduced from H.-J. Werner, B. Follmeg, M.H. Alexander, J. Chem. Phys. **89**, 3139–3151 (1988) https://doi.org/10.1063/1.454971 with the permission of AIP Publishing)

$$\left\langle \psi_1 \left| \frac{\partial}{\partial q} \right| \psi_2 \right\rangle = -\left\langle \psi_2 \left| \frac{\partial}{\partial q} \right| \psi_1 \right\rangle = \frac{\partial \chi}{\partial q}, \mathrm{q} \equiv r, R, \Theta \tag{5.5.32}$$

Now about the principal results of *ab-initio* calculations carried out in [36].

The cross-sections of the *adiabatic PES*s corresponding to the Σ^+, Π_x, and Π_y states at fixed angles and internuclear distances r are given in Fig. 5.11.

We can note that the distance R thermally accessible at $T = 300$ K ($\sim$200 cm^{-1}) is greater than 5.5 bohr = 2.9 Å, and this distance is shorter in a noncollinear collision. Matrix elements of nonadiabatic interaction of *adiabatic* states A', A' (5.5.32) characterizing the probability of a non-adiabatic transition between *adiabatic PES*s (Fig. 5.12) are especially large in collinear collisions ($\theta = 0$, C-N — He) and $r = 2.8$ bohr (1.48 Å), i.e., at CN internuclear distance r_e at which the *PEC*s of the Σ^+ and Π states intersect in the isolated CN molecule (Fig. 5.8). Therefore, near the collinear geometry of the CN-He complex, when the He atom approaches the N atom, and at the C-N internuclear distance corresponding to the intersection of the *PEC*s of the Σ^+ and Π states, there is a conical intersection of the $^2A'$ ($^2\Sigma^+$, $^2\Pi$) states at the C_s point group. The largest non-adiabatic coupling occurs along the seam with $r_c(R, \theta)$ at which the two CN$\cdots$He ($^2A'$) *PES*s cross.

Now about the matrix elements of the interaction, characterizing the probability of a *nonadiabatic* transition between *adiabatic PES*s. It can occur if the coupling matrix element is not less than a certain *preset* value comparable with the energy gaps between the *adiabatic PES*s of the $A'(\sum^+)$ and $A'(\Pi_x)$ complex (see Figs. 5.11, 5.13, 5.14). For thermal energy CN-He collisions, the He atom can approach to CN radical within $R = 5$–5.5 bohr. At these distances, the vibrationally averaged V_1 diabatic potentials at r close to $r_c = 2.8$ bohr and $\theta \approx 30°$ and 150°,

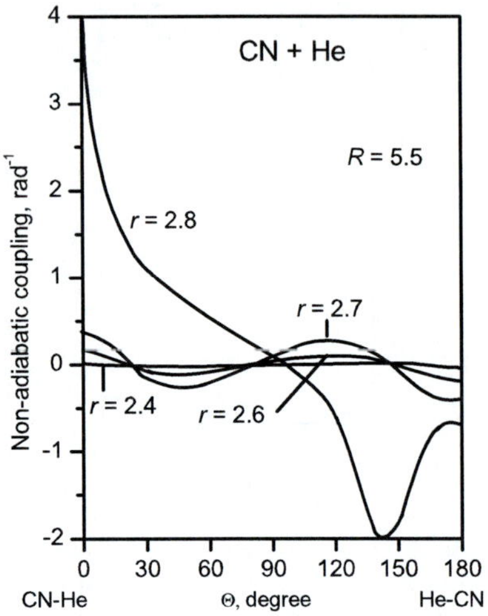

Fig. 5.12 The dependence of the matrix elements of non-adiabatic interaction, $\langle \psi_1 | \frac{\partial}{\partial \Theta} | \psi_2 \rangle$, on the angle of attack θ for various values of the internuclear distance C-N (bohr) (see [36]) (Reproduced from H.-J. Werner, B. Follmeg, M.H. Alexander, J. Chem. Phys. **89**, 3139–3151 (1988) https://doi.org/10.1063/1.454971 with the permission of AIP Publishing)

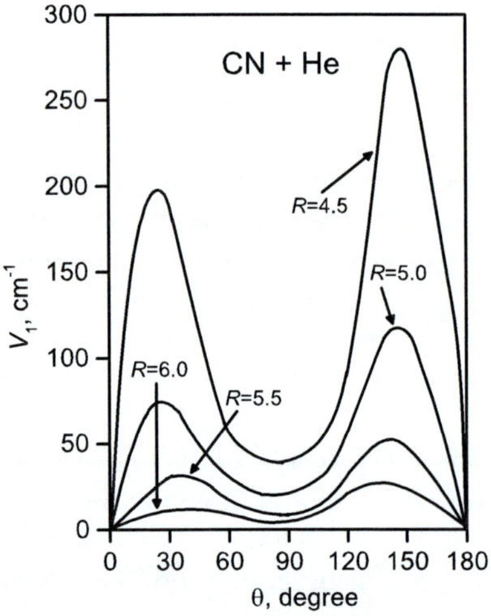

Fig. 5.13 The dependence of the vibrationally-averaged matrix elements of the diabatic interaction V_1 on the angle of attack Θ for different values of R (bohr) (see [36]) (Reproduced from H.-J. Werner, B. Follmeg, M. H. Alexander, J. Chem. Phys. **89**, 3139–3151 (1988) https://doi.org/10.1063/1.454971 with the permission of AIP Publishing)

$V_1(R, \theta) \approx 60$ and $100\ \text{cm}^{-1}$, (see Table VI in [36]) is comparable with the energy gaps between these *PES*s, and the CINAT cross-section is high [29].

A similar problem was solved for the $N_2^+ \left(A^2\Pi_u \xrightarrow{\text{He}} X^2\Sigma_g^+ \right)$ CINATs. This case differs from the previous one in that collisional transitions with even ΔJ are strictly forbidden. The symmetry of the perturbation operator determines the completely symmetrical behavior of the dependence $V_1 = f(\theta)$ to $\theta = 90°$ (Fig. 5.15).

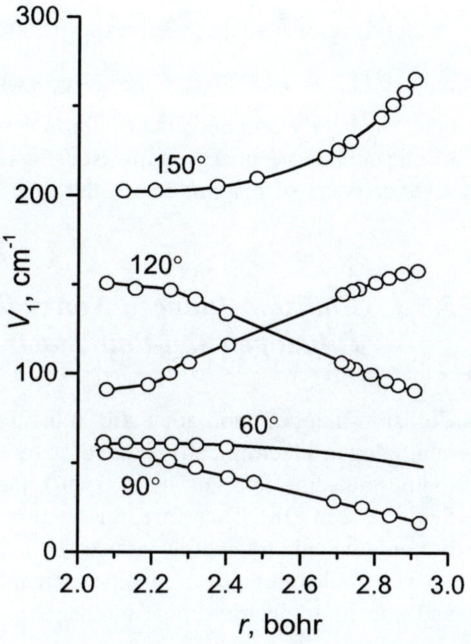

Fig. 5.14 The dependence of the matrix elements of the CN-He diabatic potential $V_1(R,r,\theta)$ on r for several θ values and $R = 5.5$ bohr (see [36]) (Reproduced from H.-J. Werner, B. Follmeg, M. H. Alexander, J. Chem. Phys. **89**, 3139–3151 (1988) https://doi.org/10.1063/1.454971 with the permission of AIP Publishing)

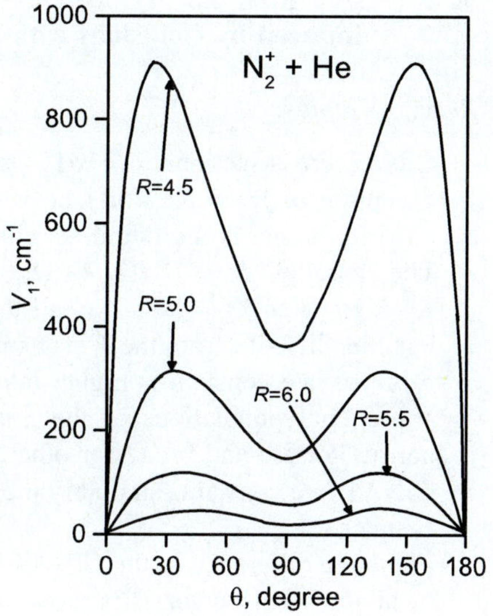

Fig. 5.15 The dependence of the matrix elements of the diabatic interaction V_1 on the angle of attack at different values of R (bohr) and $r_{N-N^+} = 2.11$ bohr (see [37]) (Reproduced from A. Berning, H.-J. Werner, J. Chem. Phys. **100**, 1953–1967 (1994) https://doi.org/10.1063/1.466548 with the permission of AIP Publishing)

One sees that the angular dependences V_1 for CN and N_2^+ are similar, which is determined by the isoelectronicity of these molecules: CN is very similar to a homonuclear molecule.

Good agreement between experimental and calculated $CN\left(A^2\Pi \xrightarrow{He} X^2\Sigma^+\right)$, $N_2^+\left(A^2\Pi_u \xrightarrow{He} X^2\Sigma_g^+\right)$ CINAT cross-sections were obtained for transitions between nearly isoenergetic rovibronic states. However, for transitions between levels with large energy gaps (see Figs. 5.8, 5.9), calculated cross-sections are several orders of magnitude smaller than experimentally observed (see [29–37]).

5.5.3 *Collision-Induced Non-Adiabatic Transitions Between Dihalogen Ion-Pair States*

Collision-induced non-adiabatic transitions (CINATs) between dihalogen and interhalogen electronically-excited states studied most well using the example of iodine molecule ion-pair states, I_2(IP), (see Fig. 4.10). They discussed in detail in Sect. 5.3.2 in [28]. Therefore, this section provides a summary of the main patterns of non-adiabatic transitions induced by collisions I_2(IP) with rare gas atoms, as well as with molecules possessing permanent electric quadrupole or dipole moments, and transition electric dipole moments.

5.5.3.1 Non-Adiabatic Transitions Between Iodine Ion-Pair States Induced by Collisions with Different Partners

Rare gas atoms.

- CINAT cross-sections, $\sigma = k/V$, are *appr.* independent on Rg nature, with the exception of Rg = Xe, for which reactive quenching leading to XeI(A, X) + I (^2P_J) formation is the dominant channel in the $I_2(E)$ + Xe collisions.
- The ‘parallel’ $E \leftrightarrow D$ ($\Omega_g \leftrightarrow \Omega_u$) transitions with $\Delta\Omega = 0$ are dominant for Rg = He. For Rg = Ar – Xe, all the electronic states of the first tier are populated in CINATs from the E states in a statistical way, and total, for β, D', γ and δ states, rate constant is higher than that of corresponding to the D state.
- Vibrational populations of the final states depend on Rg. It is resonant and narrow for He and broad for other Rg atoms.
- CINAT rate constants smoothly increase with initial vibrational excitation of the E state;
- The large difference of the CINAT branching ratios is explained by the different behavior of the potential curves. The dipole polarizability of the He atom is $0.2 \cdot 10^{-24}\,cm^3$, whereas for Ar–Xe they are about one order of magnitude larger, $(1.6 - 4.0) \cdot 10^{-24}\,cm^3$. Since the dipole polarizability determines the magnitude of the long-range interaction, the depth of the He–I_2 (E) *PES* is much smaller than the mean thermal energy at 300 K, while for the Ar–I_2 (E) it is

410 cm^{-1}. In addition, the mean collision velocity is smaller for heavier Rg atoms. Both factors increase the mean collision time and favor statistical population of the CINAT channels for Rg = Ar − Xe.

Molecules possessing permanent electric quadrupole moments.

The long-range, ~ R^{-4}, interaction between the $I_2\left(E0_g^+ - D0_u^+\right)$ or $I_2\left(f0_g^+ - F0_u^+\right)$ transition electric dipole moment and the permanent electric quadrupole moments of M is strong in these cases (R is the distance between centers of mass of the colliding partners). In pure iodine vapor, the $I_2\left(0_g^+ \overset{I_2(X)}{\leftrightarrow} 0_u^+\right)$ are dominant due to large $I_2(X)$ permanent electric quadrupole moment. The $u \leftrightarrow g$ and $\Delta\Omega = 0$ propensity rules are valid in these cases;

- Total, for all D state vibronic levels rate constants exhibit sharp maxima as function of v_E, and near-resonant CINATs have highest total rate constants. Vibronic states nearest to the initial ones are mainly populated in the near-resonant CINATs. The $\Delta J = \pm 1$ and 'ΔJ – large' propensity rules are valid in near-resonant and non-resonant CINATs, respectively;
- The $E, v_E \overset{I_2(X)}{\longleftarrow} D, v_D$ and $D, v_E \overset{I_2(X)}{\longleftarrow} E, v_D$ CINAT rate constants are similar;
- There are no distinct correlations between vibrational distributions (VDs) of a state populated in CINATs and *FCF*s of initial and final vibronic levels, though VDs 'try' to follow the trend in the *FCF*s with the initial states, but energy gap law, $P \sim \exp(-\frac{|\Delta E|}{\beta})$ (β—is a variable parameter, usually, $\beta = kT$) 'prevents' this trend.
- Rate constants of non-resonant processes correspond to small impact parameters. In these cases, the CINATs are provided by orbiting of the $I_2(E$ or $D)$ and $I_2(X)$ colliding partners due to dispersion,—C_6/R^6, interaction. The orbiting leads to population of several vibronic states in non-adiabatic transitions from the initial to final I_2 states.
- For M = N_2, CO_2, the $I_2\left(E0_g^+ \overset{M}{\longrightarrow} D0_u^+, D'2_g, \beta 1_g, \delta 2_u, \gamma 1_u\right)$ CINATs are also occurred.

Molecules possessing permanent electric dipole moments.

- In this case, long-range interaction between the *E-D* transition electric dipole and the permanent electric dipole of M, ~ R^{-3}, is dominant, and the $I_2\left(0_g^+ \overset{M}{\leftrightarrow} 0_u^+\right)$ CINATs have to be the most probable. For these collision partners, rate constants of quasi-resonant CINATs have to be huge, and VDs have to be narrow, if energy mismatches are compensated by rotational transitions in a colliding partner.

Molecules possessing transition electric dipole moments.

- In this case, the leading term of electrostatic expansion corresponds to E –D transition electric dipole moment–vibrational transitions electric dipole moment, $\sim R^{-3}$, interactions. Collisions with these partners produce very efficient electronic energy transfer with the total rate constant of the order of 10^{-9} cm^3/s. The total CINAT probability is predominantly associated with the transitions that lead to dipole allowed excitations of the partner. As a result, these partners are efficient for I_2 vibrational energy loss.

5.5.3.2 Non-Adiabatic Transitions Between Chlorine Ion-Pair States

The dependences of the $Cl_2\left(D0_u^+, v_D, J_D \xrightarrow{Cl_2(X)} E0_g^+, v_E, J_E\right)$ CINAT rate constants on the vibrational v_E, v_D, rotational J_E quantum numbers, energy gaps, and *FCF*s of the levels combined in the CINAT have been studied in the range of v_D = 1–5 [38] (see [39], also) (see Fig. 5.16).

A single rovibronic level of the D state was populated via accidental resonance in consecutive one- plus two-photon coherent transitions:

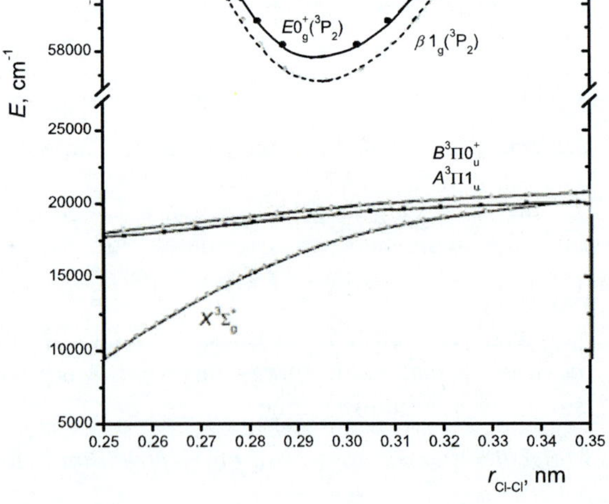

Fig. 5.16 *RKR* potential energy curves of three Cl_2 IP states (top) as well as the relevant portions of the covalent Cl_2(X, A, and B) states (bottom). The energy scale is given relative to the bottom of the Cl_2(X) state potential well [38] (Reprinted from Chemical Physics, Vol. 277, N. K. Bibinov, M. A. Nikitin, A. M. Pravilov, A. A. Zakharova, The collision-induced $Cl_2\left(D0_u^+ \xrightarrow{Cl_2(X)} E0_g^+\right)$ transition. p.p. 191–199 (2002) with permission from Elsevier)

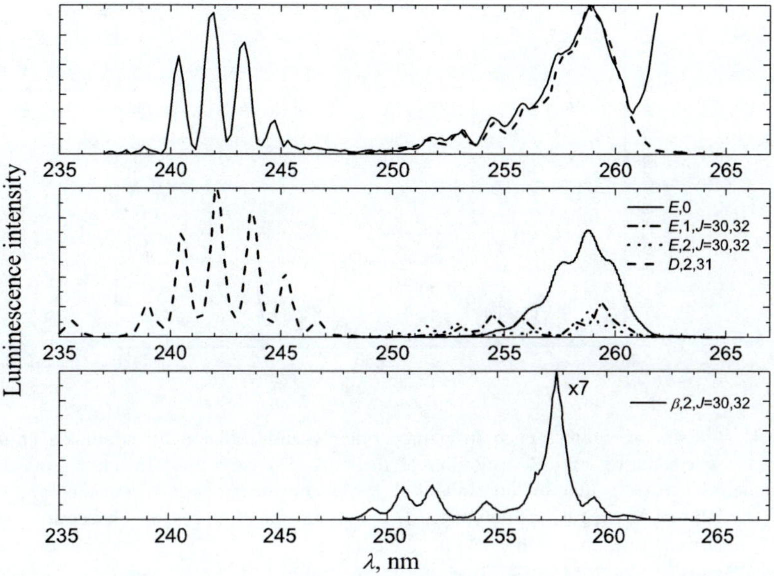

Fig. 5.17 Luminescence spectrum (solid lines) from the optically and collisionally populated chlorine IP rovibronic levels during optical excitation of the $v_D = 2, J_D = 31$ ($^{35}Cl^{37}Cl$) rovibronic levels and portions of the total simulated spectra (dotted lines), upper part of the figure. Simulations of partial spectra from optically and collisionally populated rovibronic levels are shown at lower parts of the figure. The simulations were made under assumption that $v_E = 1$–2, $J_E = 30{,}32$, $v_E = 0$, rotational Boltzmann distribution, T = 300 K and $\beta, v_\beta = 2$, $J_\beta = 30{,}32$ ($^{35}Cl^{37}Cl$) rovibronic levels are populated in the CINATs. Relative population of the E and β state levels, populated in the $Cl_2\left(D, v_D = 2, J_D = 31 \xrightarrow{Cl_2(X)} E, v_E = 0-2, \beta, v_\beta = 2\right)$ CINAT are equal to 1:0.29:0.21:0.12. Spectral resolutions are $FWHM = 0.8$ nm; $p_{Cl2} = 10$ Torr [38]. The increasing of the luminescence intensity in the long-wave part of the spectrum is due to $Cl_2\left(D, v_D \rightarrow 20_g^+\right)$ transition [40] (Reprinted from Chemical Physics, Vol. 277, N. K. Bibinov, M. A. Nikitin, A. M. Pravilov, A. A. Zakharova, The collision-induced $Cl_2\left(D0_u^+ \xrightarrow{Cl_2(X)} E0_g^+\right)$ transition. pp. 191–199 (2002) with permission from Elsevier)

$$Cl_2\left(D0_u^+, v_D, J_D \xleftarrow{h\nu_{exc} + h\nu_{exc}} B0_u^+, v_B, J_B \xleftarrow{h\nu_{exc}} X0_g^+, v_X, J_X\right) \tag{5.5.33}$$

The principal data obtained in [38, 39] are the following (Figs. 5.17 and 5.18):

- All the CINATs observed are non-resonant and correspond to loss of up to 1100 cm^{-1} $Cl_2(D0_u^+, v_D, J_D)$ vibrational energy.
- The $u \leftrightarrow g$ and $\Delta\Omega = 0$ propensity rules are valid in the $Cl_2(D \overset{Cl_2(X)}{\rightarrow}$ IP states).
- The vibrational and rotational distributions of the E state levels depend on v_D number. For $v_D = 3$–5, the E state vibrational level corresponding to maximum

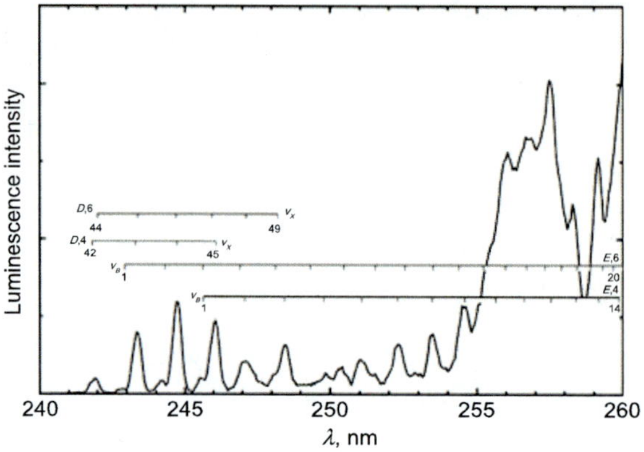

Fig. 5.18 Spectra of luminescence from the optically and collisionally populated chlorine IP rovibronic levels during optical excitation of the $^{35}Cl_2(D,v_D = 4,J_D = 28$ and $v_D = 6,J_D = 13)$ rovibronic levels. Assignment of unresolved *P, R* branches of transitions from the *D* and *E* state levels are shown. Spectral resolution is 0.2 nm. In the $\lambda > 254$ nm spectral range, the $E \to B$ spectrum overlaps with that of $^{35}Cl_2(D,v_D = 6 \to X,\ 20_g^+)$ [38] (Reprinted from Chemical Physics, Vol. 277, N. K. Bibinov, M. A. Nikitin, A. M. Pravilov, A. A. Zakharova, The collision-induced $Cl_2\left(D0_u^+ \xrightarrow{Cl_2(X)} E0_g^+\right)$ transition. pp. 191–199 (2002) with permission from Elsevier)

Table 5.4 Franck–Condon factors and energy differences of the vibronic levels combined in the CINATs observed (see [38])

D, v_D, J_D	E, v_E, J_E	*FCF*	$\Delta E = E_D,\ v_D,\ J_D - E_E,\ v_E,\ J_E$, cm^{-1}	Relative populations of the E, v_E vibronic levels
5.5	5.5	0.832	536.5	≥ 90%
4.28	4.28	0.831	565.3	≥ 90%
3.10	3.10	0.83	591.4	≥ 80%
2.31	0.31	0.210^{-3}	1114.3	61%
	1.31	0.0634	865.0	18%
	2.31	0.863	617.8	13%
	β, 2.31	0.839	864.1	8%
1.1	0.1	0.0501	889.0	≥ 80%

FCF with the initial one is mainly populated (Table 5.4). These CINATs are accompanied by resonant populations of the $Cl_2(X,v_X = 1,J_X)$ rovibronic states:

$$Cl_2\left(D, v_D, J_D\right) + Cl_2(X, v_X = 0, J_X) \to Cl_2\left(E, v_E, J_E\right) + Cl_2(X, v_X = 1, J_X), \quad (5.5.34)$$

$\omega_e^X = 559.72\ \mathrm{cm}^{-1}$ [8]. The (5.3.34) CINATs follow so-called 'the combined Franck–Condon—energy gap model' (see Sect. 5.7).

- The $E, v_E = 0,1,2$, $\beta 1_g, v_\beta = 2$ and $E, v_E = 0$ levels are populated from the D, $v_D = 2$ and 1, respectively. The 'ΔJ low' propensity rule is valid in the $Cl_2\left(D, v_D = 1-5, J_D \xrightarrow{Cl_2(X)} E, v_E = 1-5, J_E\right.$ and $\beta, v_\beta = 2, J_\beta)$ CINATs.

- Rotational distribution of the $v_E = 0$ levels populated from the $v_D = 1, 2$ can be described as the Boltzmann one, T = 300 K.
- The nonadiabatic transitions between potential energy surface corresponding to input $Cl_2(D) + Cl_2(X)$ and output $Cl_2(E) + Cl_2(X)$ channels occur in relatively short-lived collision complex, and the CINAT rate constants are in the (0.4–3.2) 10^{-11} cm^3/s range.

5.5.4 Collision-Induced Predissociation

Selection rules for collision-induced predissociation (CIP) are the same as those for CINATs, but $\Delta J = 0$ rule is satisfied automatically since one of the combining states is repulsive. This process is well kinetically studied using the example of the $I_2(B0_u^+, v_B)$ CIP (see [5] and references).

$$I_2(B, v_B, J_B) \xrightarrow{M} I_2(RS) \rightarrow I(^2P_{3/2}) + I(^2P_{3/2}) \qquad (5.5.35)$$

(RS means repulsive state (states), see Fig. 4.4) electronic deactivation of the $I_2(B, v_B, J_B)$ molecules in collisions with He, e.g., is impossible (see the next section). The CIP rate constant for $M = I_2(X)$, $k_{5.35}(I_2(X)) \approx 4 \cdot 10^{-10}$ cm^3/s, and is independent of v_B. The $k_{5.35}$(Rg) values smoothly increase with v_B. For Rg = He, e.g., $k_{5.35}(\mathrm{He}) \approx 5 \cdot 10^{-12}$ cm^3/s ($v_B = 6$) – 10^{-10} cm^3/s ($v_B = 50$), see [5] and references. The CIP rate constant for each RS has to oscillate with v_B due to the oscillation of Franck–Condon densities (FCDs) for B/RS states (see [41, 42] and references). Therefore, it is very likely that RS are *gerade* $a1_g, a'0_g^+$ and $(1)2_g$ states correlating with the *aa* limit (see Fig. 4.4). Oscillations of FCFs for the RS states are out of phase, so total $FCD(B/RS) = f(v_B)$ function can be smooth (see [28] and references). As to *ungerade* states, the $C1_u$ is the only state which intersects the B state. The $FCD(B/C) = f(v_B)$ function oscillates (see Fig. 4.26, and Sect. 4.2.1 in [28]). Therefore, process (5.5.35) cannot occur due to the $B \sim C$ collision-induced coupling.

5.6 Vibrational Relaxation Via a Complex of Electronic States

The rate constant of vibrational relaxation of the lower vibrational levels $N_2\left(X^1\Sigma_g^+\right)$ is small; for $v_X = 1$, it is of the order of 10^{-18} cm^3/s (see Sect. 5.3). What can it be equal for electronically excited states, where collision-induced non-adiabatic transitions can take place? Let us discussed the $N_2^* + N_2\left(X^1\Sigma_g^+\right)$ collisions (Fig. 5.19).

It has been shown in [26] (see [43], also) that the cross-section of the transitions $N_2(W^3\Delta_u, v'' \xrightarrow{N_2(X)} B^3\Pi_g, v'$ and $A^3\Sigma_u^+, v'' \xrightarrow{N_2(X)} B, v'$ are equal to $\sigma = 7.4$ exp (- $|\Delta E|$/ 665) and 3.5 exp ($-|\Delta E|$/380), respectively, (units, cross-sections Å^2, ΔE cm^{-1}). Cross-sections of reverse transitions can be obtained from the detailed balance principle (see Sect. 2.2). They differ by 2–3 times for the same ΔE due to differences in the statistical weights; ΔE influence can be much stronger. It is seen that the collision-induced transition rate constant of the process (5.6.1)

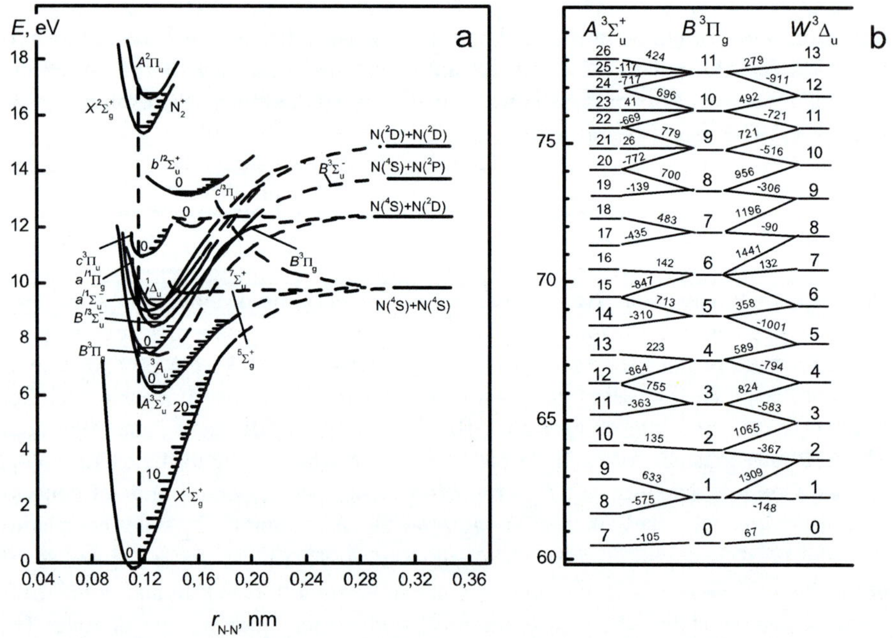

Fig. 5.19 The potential energy curves (a) and the diagram of vibrational levels (b) of the lower excited states of the N_2 molecule (see [26]) (Reproduced from R. Bachmann, Ch. Ottinger, A. F. Vilesov, J. Chem. Phys. **96**, 5151–5164 (1992). https://doi.org/10.1063/1.462756 with the permission of AIP Publishing)

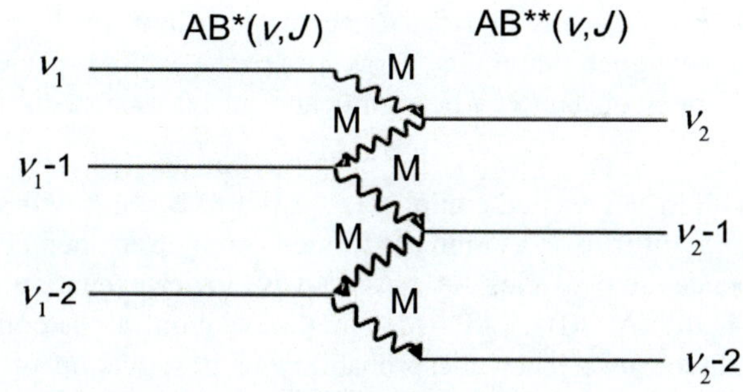

Fig. 5.20 Vibrational relaxation in a complex of electronic states

$$N_2\left(W^3\Delta_u, v_W\right)\xrightarrow{M}N_2\left(B^3\Pi_g, v_B\right)\xrightarrow{M}N_2\left(W^3\Delta_u, v_{W-1}\right), \tag{5.6.1}$$

equivalent vibrational relaxation in the sense of the $N_2(W^3\Delta_u, v_W)$ energy loss (Fig. 5.20) is many orders of magnitude higher than the corresponding value for $N_2(X, v_X)$.

For example, for ΔE = 665 cm^{-1}, $\sigma_{6.1}$ = 1/2 ·7.4 *exp*(-1) = 1.4 10^{-16} cm^2 ($k_{6.1}$ = 1.5 10^{-11} cm^3/s) (see [43, 44], also). Vibrational relaxation within $N_2(A, v_A)$ for $v_A < 6$, where a process similar to (5.6.1) is impossible, since the singlet $X^1\Sigma_g^+, v_X$ and triplet, $A^3\Sigma_u^+, v_A$ states combine poorly, and the bottom of the lower of the triplet states B, v_B is higher, occur via 'classical vibrational relaxation' within one state

$$N_2(A, v_A)\xrightarrow{M}N_2(A, v_A - 1), \tag{5.6.2}$$

Here, for M = N_2, for example, $k_{6.2}$ = 2.4 10^{-16} cm^3/s (v_A = 1) − 5.1 10^{-13} cm^3/s (v_A = 6) [45], and 6.3 10^{-16} cm^3/s (v_A = 1) − 1.2 10^{-15} cm^3/s (v_A = 6) for M = He [45] are much less than those of process (5.6.1).This is what can be expected for electronically excited states, however for the ground ones, if they are isoenergetic with an excited one. These excited states can also be one of the components of the multiplet of the ground state, as in the case of $NO(X^2\Pi_{3/2,1/2})$ (see Sect. 5.4).

5.7 Electronic Deactivation

Electronic deactivation is a collision-induced non-radiative transition to another electronic state that occurs with a large energy loss (often greater than 1 eV), that excludes a reverse transition to the state with which the species has just 'broken up' (see Sect. 4.1). The chemical composition of the interacting species remains unchanged. The mechanisms of such processes have not yet been fully studied. However, it can be argued that if the energy loss of the excited species is large, then the deactivation rate constant can be significant only in two cases:

1. In the collision of this species, for example, an excited atom A* or a AB* molecule with M, an excited complex (A···M)*, (AB···M)* with energy not less than the energy of its dissociation are formed. A complex then either dissociates or predissociates into species A**, AB** with lower excitation energy, and M (M*, M#). If the (A···M)*, (AB···M)* predissociation rate is comparable to the rate of its dissociation, then the probability of deactivation of A*, AB* in a collision with M is high. This process does not differ from those that we considered in Sect. 5.6. It should be noted that a deactivation rate constant to be large, a matrix element of the non-adiabatic interaction of the complex states that formed in the collision and the one through which the predissociation occurs has to be large as well. Otherwise, the complex should have a sufficiently long lifetime, i.e., be triatomic at least (see below).
2. An almost resonant transfer of excitation energy A*, AB* to M,

$$A^*(AB^*) + M \rightarrow A^{**}(AB^{**}) + M^*\left(M^{\#}\right) + \Delta E \qquad (5.7.1)$$

and the kinetic energy of the scattering partners $\Delta E = E_{A^*}(AB^*) - E_{A^{**}}(AB^{**})^+ E_{M^*(M^{\#})}$ is small, comparable to kT.

Both types of these processes are relatively well understood. The first of them, once called the complex-predissociation mechanism of deactivation (the term has not taken root), can be illustrated by the example of deactivation of O(^{1}D) by Xe atoms and CO($X^1\Sigma^+$) molecules.

The deactivation rate constant

$$O(^1D) + Xe \rightarrow O(^3P) + Xe \qquad (5.7.2)$$

is $k_{7.2}$ = 7 10^{-11} cm^3/s, and that of

$$O(^1D) + CO \rightarrow O(^3P) + CO \qquad (5.7.3)$$

$k_{7.3}$ = 4 10^{-11} cm^3/s. The O(^{1}D) atom excitation energy is E = 1.97 eV. The spin-orbit interaction in the XeO excimer is so large that the XeO complex with a probability of about 0.3 predissociates during one vibration, although this process is forbidden due to the Wigner rule [46], and with a probability of 0.7 dissociates (Fig. 5.21) [47].

In the complex O…CO, i.e., in a $CO_2(^1B_2, {}^1A_2)$ molecule with excitation energy equal to the dissociation energy of O(^{1}D) + CO($X^1\Sigma^+$) (Fig. 5.22), the spin-orbit interaction is orders of magnitude lower, but the lifetime is orders of magnitude longer with respect to dissociation. During this time, an image point repeatedly

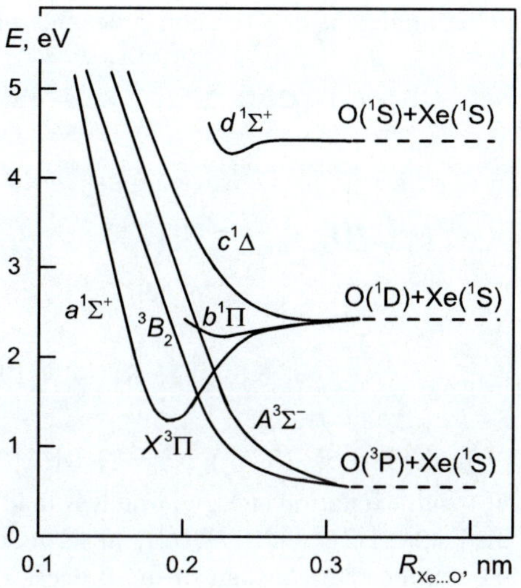

Fig. 5.21 The potential energy curves of the XeO excimer [47]

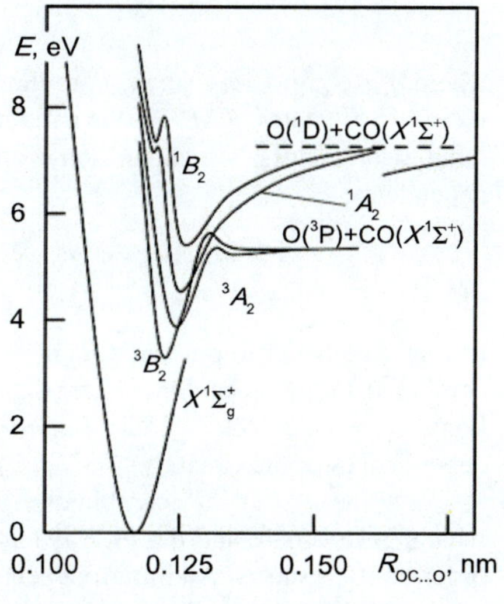

Fig. 5.22 Schematic arrangement of potential energy curves of the lower states of a CO_2 molecule as a function of the OC-O distance [48]

approaches the seam of intersections of the singlet $^1B_2(^1A_2)$ and triplet $^3B_2(^3A_2)$ states with a nonzero velocity component perpendicular to the seam, and despite the low probability of hopping, the probability of $CO_2(^1B_2(^1A_2))$ predissociation to O (^{3}P) + CO is high.

Examples of deactivation processes of the second type are:

$$N_2(A^3\Sigma_u^+, v_A) + CO(X^1\Sigma^+, v_X = 0) \rightarrow N_2\left(X^1\Sigma_g^+, v_X > 0\right) + CO(a^3\Pi, v_\alpha > 0) + \Delta E \tag{5.7.4}$$

$$\begin{aligned} {}^{15}N_2(B^3\Pi_g, v'_B) + {}^{14}N_2\left(X^1\Sigma_g^+, v_X = 0\right) &\rightarrow {}^{15}N_2\left(X^1\Sigma_g^+, v_X > 0\right) \\ &+ {}^{14}N_2(B^3\Pi_g, v''_B < v'_B) + \Delta E \end{aligned} \tag{5.7.5}$$

$$N_2^* + Xe(^1S) \rightarrow N_2\left(X^1\Sigma_g^+, Vx > 0\right) + Xe^* \tag{5.7.6}$$

In the process (5.7.6), $N_2^* = N_2(A^3\Sigma_u^+, B'^3\Sigma_u^-, B^3\Pi_g$ and, possibly, $X^1\Sigma_g^+$, high $v_X)$ with excitation energy of no less than 8.36 eV, Xe* = Xe ($^3P_{2,1,0}$, $^1P_0$ – [1/2,1/2]$_{0,1}$, [3/2,1/2]$_2$ and [3/2,1/2]$_1$ in terms of *J-J* bond). According to a very simple model, the rate constant of the process (5.7.4) can be relatively well estimated as [49]:

$$k_{7.4} = q_{N_2}(v_A, v_X) q_{CO}(v_X, v_a) f(\Delta E, T\ldots), \tag{5.7.7}$$

where $q_{N_2}(v_A, v_X)$, $q_{CO}(v_X, v_a)$ are the *FCF*s for the $N_2(A, v_A, X, v_X > 0)$ and CO(*X*, v_X, $a, v_a > 0$) states, $f(\Delta E, T\ldots)$ is a function of the energy gap ΔE, temperature and a number of factors, which, in some approximation, can be represented using the dependence

$$f(\Delta E, T\ldots) = \exp(-|\Delta E|/kT) \tag{5.7.8}$$

(it might be better to put some E_0 instead of *k*T [50]). *Equation* (5.7.8) presents the combined Franck–Condon—energy gap model. For CINATs, it is valid in exceptional cases (see Sect. 5.5.3.2). *Equation* (5.7.8) describes experimentally observed dependences approximately.

For some v', J', v'', J'' combinations of the donor and the acceptor of energy, at which ΔE is small, and the *FCF* is large, the rate constant can reach a value of the order of 10^{-11} cm^3/s. An initially excited molecule may not wholly lose energy, but remain with several vibrational–rotational energy quanta of the ground or other electronic states. This circumstance should provide large process rate constants. Since the density of the rotational levels in the molecules is quite high, the resonance $\Delta E \rightarrow 0$ in processes of the type (5.7.4–5.7.7) is ensured "almost automatically".

References

1. Kondratiev, V.N., Nikitin, E.E.: Gas-phase Reactions: Kinetics and Mechanism. Springer, Berlin Heidelberg New York (1981). https://doi.org/10.1007/978-3-642-67608-6
2. Kondratiev, V.N., Nikitin, E.E.: Kinetika i Mekhanism Gazofaznykh Reaktzij (Kinetics and Mechanism of Gas-phase Reactions). Nauka, Moscow (1974).(in Russian)
3. Herzberg, G.: Molecular Spectra and Molecular Structure I. Spectra of Diatomic Molecules. D. van Nostrand Co., Inc, Second Edition (1950)
4. Lordi, J.A., Mates, R.E.: Rotational relaxation in nonpolar diatomic gases. Phys. Fluids **13**, 291–308 (1970). https://doi.org/10.1063/1.1692920
5. Steinfeld, J.I.: Rate data for inelastic collision processes in the diatomic halogen molecules. J. Phys. Chem. Ref. Data. **13**, 445–553 (1984). https://doi.org/10.1063/1.555711
6. Parker, J.G.: Comparison of experimental and theoretical vibrational relaxation times for diatomic gases. J. Chem. Phys. **41**, 1600–1609 (1964). https://doi.org/10.1063/1.1726130
7. Billing, G.D., Fisher, E.R.: VV and VT rate coefficients in N_2 by a quantum-classical model. Chem. Phys. **43**, 395–401 (1979). https://doi.org/10.1016/0301-0104(79)85207-6
8. Huber, K.P., Herzberg, G.: Molecular Spectra and Molecular Structure IV. Constants of Diatomic Molecules. Springer (1979). https://doi.org/10.1007/978-1-4757-0961-2
9. Hancock, G., Smith, I.W.M.: Quenching of infrared chemiluminescence. 1: The rates of deexcitation of CO ($4 < v < 13$) by He, CO, NO, N_2, O_2, OCS, N_20, and CO_2. Appl. Opt. **10**, 1827–1842 (1971). https://doi.org/10.1364/AO.10.001827
10. Lambert, J.D., Salter, R.: Vibrational relaxation in gases. Proc. Roy. Soc. **A253**, 277–288 (1959). https://doi.org/10.1098/rspa.1959.0193
11. Powell, H.T.: Vibrational relaxation of carbon monoxide using a pulsed discharge. J. Chem. Phys. **59**, 4937–4941 (1973). https://doi.org/10.1063/1.1680709
12. Sackett, P.B., Hordvick, J.K., Schlossenberg, H.: Measurement of the V-V energy transfer rate from CO($v = 2$) using tunable parametric oscillator excitation. Appl. Phys. Letts. **22**, 367–368 (1973). https://doi.org/10.1063/1.1654676
13. Starr, D.F., Hancock, G., Green, W.H.: Vibrational deactivation of carbon monoxide by hydrogen and nitrogen from 100 to 650° K. J. Chem. Phys. **61**, 5421–5425 (1974). https://doi.org/10.1063/1.1681897
14. Wilson, G.J., Turnidge, M.L., Simpson, C.J.S.M.: Vibrational energy transfer between $^{12}C^{16}0$ and $^{15}N_2$ in the gas phase down to 45 K and in liquid Ne solution. Chem. Phys. Letts. **227**, 299–304 (1994). https://doi.org/10.1016/0009-2614(94)00833-7
15. Capitelli, M.: Nonequilibrium vibrational kinetics. Springer, Berlin e.a (1986)
16. Callear, A.B., Lambert, J.D.: The transfer of energy between chemical species. In Comprehensive Chemical Kinetics, eds. Bamford, C.H., Tipper, C.F.H., Vol.3. The formation and Decay of Excited Species , Elsivier (1969)
17. Poulsen, L.L., Billing, G.D.: Vibrational deactivation of CO($v = 1$) by p-H_2, and o-H_2. Chem. Phys. **73**, 313–322 (1982). https://doi.org/10.1016/0301-0104(82)85171-9
18. Millikan, R.C., Osburg, L.A.: Vibrational relaxation of carbon monoxide by ortho- and parahydrogen. J. Chem. Phys. **41**, 2196–2197 (1964). https://doi.org/10.1063/1.1726226
19. Dreiling, T.D., Setser, D.W.: State-to-state relaxation processes for XeCI(*B*, C). J. Chem. Phys. **75**, 4360–4378 (1981). https://doi.org/10.1063/1.442599
20. Steinfeld, J.I., Klemperer, W.: Energy-transfer processes in monochromatically excited iodine molecules. I. Experimental Results. J. Chem. Phys. **42**, 3475–3497 (1965). https://doi.org/10.1063/1.1695750
21. Lefebvre-Brion, H., Field, R.W.: Perturbations in the Spectra of Diatomic Molecules, Academic Press, Inc. (1986)
22. Lefebvre-Brion, H., Field, R.W.: The Spectra and Dynamics of Diatomic Molecules, Elsevier (2004)
23. Dagdigian, P.J.: State-resolved collision-induced electronic transitions. Annu. Rev. Phys. Chem. **48**, 95–123 (1997). https://doi.org/10.1146/annurev.physchem.48.1.95

24. Ottinger, Ch., Vilesov, A.F.: Intramolecular collisional transfer in NO($a^4\Pi \rightarrow B^2\Pi$, $b^4\Sigma$): Gateway-type, resonant versus direct, nonresonant mechanisms. J. Chem. Phys. **100**, 1805–1814 (1994). https://doi.org/10.1063/1.466533
25. Ottinger, Ch., Vilesov, A.F.: Lifetime measurements on perturbed levels of NO($B^2\Pi$) and precise term energy of the NO($a^4\Pi$) state. J. Chem. Phys. **100**, 1815–1822 (1994). https://doi.org/10.1063/1.466534
26. Bachmann, R., Ottinger, Ch., Vilesov, A.F.: Molecular-beam study of the collisional intramolecular coupling of N2(B3Πg) with the N2(A3 Σu+) and N2(W3Δu) states. J. Chem. Phys. **96**, 5151–5164 (1992). https://doi.org/10.1063/1.462756
27. Selwyn, J.E., Steinfeld, J.I.: Collision-induced predissociation by van der Waals interaction. Chem. Phys. Letts. **4**, 217–220 (1969). https://doi.org/10.1016/0009-2614(69)80106-5
28. Lukashov, S., Petrov, A., Pravilov, A.: The Iodine Molecule: Insights into Intra- and Intermolecular Perturbation in Diatomic Molecules. Springer (2018). https://doi.org/10.1007/978-3-319-70072-4
29. Furio, N., Ali, A., Dagdigian, P.J.: State-resolved study of collisional energy transfer between $A\ ^2\Pi\ v = 7$ and $X\ ^2\Sigma^+\ v = 11$ rotational levels of CN. J. Chem. Phys. **85**, 3860–3868 (1986). https://doi.org/10.1063/1.450906
30. Furio, N., Ali, A., Dagdigian, P.J.: State-to-state collisional interelectronic and intraelectronic energy transfer involving CN $A\ ^2\Pi\ v = 3$ and $X\ ^2\Sigma^+\ v = 7$ rotational levels. J. Chem. Phys. **85**, 7098–7105 (1986). https://doi.org/10.1063/1.451396
31. Ali, A., Jihua, G., Dagdigian, P.J.: State-resolved inelastic cross sections from CN $A\ ^2\Pi\ v = 8$ to $X\ ^2\Sigma^+\ v = 12$: Quenching of the even-odd alternation in the final rotational state populations. J. Chem. Phys. **87**, 2045–2050 (1987). https://doi.org/10.1063/1.453179
32. Dagdigian, P.J., Patel-Misra, D., Berning, A., Werner, H.-J., Alexander, M.H.: A joint experimental and theoretical study of $A\ ^2\Pi$ - $^2\Sigma^+$ electronic energy transfer in the CN molecule induced by collisions with helium. J. Chem. Phys. **98**, 8580–8592 (1993). https://doi.org/10.1063/1.464518
33. Katayama, D.H.: Collision induced electronic energy transfer between the A 2Πui (v = 4) and X 2 Σg+ (v = 8) rotational manifolds of N2+. J. Chem. Phys. **81**, 3495–3499 (1984). https://doi.org/10.1063/1.448076
34. Katayama, D.H., Dentamaro, A.V.: Propensities for collision Induced electronic transitions in a diatomic molecule. J. Chem. Phys. **81**, 2595–2598 (1986). https://doi.org/10.1063/1.451067
35. Katayama, D.H., Dentamaro, A.V., Welsh, J.A.: State specific electronic quenching rates for the $N_2\ a\ ^1\Pi_g(v{=}O)$ level from collisions with He, Ar, and N_2. J. Chem. Phys. **101**, 9422–9428 (1994). https://doi.org/10.1063/1.467973
36. Werner, H.-J., Follmeg, B., Alexander, M.H.: Adiabatic and diabatic potential energy surfaces for collisions of CN ($X\ ^2\Sigma^+$, $A\ ^2\Pi$) with He. J. Chem. Phys. **89**, 3139–3151 (1988). https://doi.org/10.1063/1.454971
37. Berning, A., Werner, H.-J.: Quantum scattering studies of electronically inelastic collisions of N2+(X 2 Σg+, A 2Πu) with He. J. Chem. Phys. **100**, 1953–1967 (1994). https://doi.org/10.1063/1.466548
38. Bibinov, N.K., Nikitin, M.A., Pravilov, A.M., Zakharova, A.A.: The collision-induced $D\mathrm{Cl}_2\left(D0_u^+ \overset{\mathrm{Cl}_2(X)}{\longrightarrow} E0_g^+\right)$ transition. Chem. Phys. **277**, 191–199 (2002). https://doi.org/10.1016/S0301-0104(02)00314-2
39. Bibinov, N.K., Davydov, V.G., Fateev, A.A., Kokh, D.B., Lugovoj, E.V., Ottinger, Ch., Pravilov, A.M.: The transition dipole moment of the chlorine $E0_g^+\left(^3\mathrm{P}_2\right) - B^3\Pi\left(0_u^+\right)$ system. J. Chem. Phys. **109**, 10864–10872 (1998). https://doi.org/10.1063/1.477783
40. Bibinov, N.K., Fateev, A.A., Kokh, D.B., Lugovoj, E.V., Pravilov, A.M.: Optical transitions from the chlorine 0+u (3P2) ion-pair state. Chem. Phys. 254, 89-98 (2000). https://doi.org/10.1016/S0301-0104(00)00019-7
41. Tellinghuisen, J.: Potentials for weakly bound states in I_2 from diffuse spectra and predissociation data. J. Chem. Phys. **82**, 4012–4016 (1985). https://doi.org/10.1063/1.448841

42. Buchachenko, A., Halberstadt, N., Lepetit, B., Roncero, O.: Ar ··· I2: A model system for complex dynamics. Int. Rev. Phys. Chem. **22**, 153–202 (2003). https://doi.org/10.1080/0144235031000075726
43. Sadeghi, N., Setser, D.W.: Collisional coupling of $N_2(B^3\Pi_g)$ and $N_2(W^3\Delta_u)$ states studied by laser-induced fluorescence. Chem. Phys. Letts. **77**, 304–308 (1981). https://doi.org/10.1016/0009-2614(81)80151-0
44. Slanger, T.G., Cosby, P.S.: Oxygen spectroscopy below 5.1 eV. J. Phys. Chem. **92**, 267–282 (1988). https://doi.org/10.1021/j100313a008
45. Dreyer, J.W., Perner, D.: Deactivation of by ground state nitrogen, ethane, and ethylene measured by kinetic absorption spectroscopy $N_2\left(A^3\Sigma_u^+; v = 0 - 7\right)$ by ground state nitrogen, ethane, and ethylene measured by kinetic absorption spectroscopy. J. Chem. Phys. **58**, 1195–1201 (1973). https://doi.org/10.1063/1.1679301
46. IUPAC. Compendium of Chemical Terminology, 2nd ed. (the Gold book). Complied by A.D. McNaught and A. Wilkinson, Blackwell Scientific Publications, Oxford (1997). https://doi.org/10.1351/goldbook
47. Pravilov, A.M., Sidorov, I.I., Skoroknodov, V.A.: Mechanism and kinetics of photorecombination of $O(^1D)$ with Xe. Sov. J. Chem. Phys. **3**, 793–807 (1985)
48. Pravilov, A.M., Protopopov, S.V., Sidorov, I.I., Skoroknodov, V.A.: Photorecombination of $O(^1D)$ with CO. Sov. J. Chem. Phys. **3**, 2010–2033 (1985)
49. Deperasinska, I., Beswick, J.A., Tramer, A.: A distorted wave calculation for electronic energy transfer in molecular collisions. Application to the $N_2\left(A^3\Sigma_u^+\right) + CO(X^1\Sigma^+) - + N_2\left(X^1\Sigma_g^+ + CO(a^3\Pi)\right.$ system. J. Chem. Phys. **71**, 2477–2487 (1979). https://doi.org/ https://doi.org/10.1063/1.438655
50. Fecko, C.J., Freedman, M.A., Stephenson, T.A.: Collision-induced electronic energy transfer from v=0 of the E(0g+)ion-pair state in I2: Collisions with I2(X). J. Chem. Phys. **115**, 4132–4138 (2001). https://doi.org/10.1063/1.1391264

Chapter 6
Weakly-Bound Complexes and Clusters

Abstract This chapter deals with a comprehensive class of weakly bound complexes and clusters. This book is devoted to gas-phase photoprocesses, that is, electronically excited states only those weakly-bound complexes that are weakly-bound in electronically excited states, as well, are considered. The complexes with so-called hydrogen and halogen bonds in the ground state will be considered briefly. The spectroscopic characteristics of the $R_gX_2(X0_g^+, B0_u^+, E0_g^+)$, X = Cl, Br, I, and RgICl ($X0^+$, $A1$, $B0^+$, $E0^+$, $\beta1$, $D'2$) states and the complex electronically-excited states dynamics are discussed very thoroughly. The ArICl($X0^+$, $A1$, $E0^+$, $\beta1$, $D'2$), NeIBr ($X0^+$, $A1$, $E0^+$, $\beta1$) spectroscopic characteristics, and dynamics of these complexes in IP states are discussed, also. The data, concerning the vdW complexes of iodine molecules with diatomic molecules, in the IP states, in particular, are considered. The Rg_nX_2, $n > 1$, and $HeNeI_2$($X0^+$, $A1$, $E0^+$, $\beta1$) spectral parameters, and their excited state dynamics, are analyzed. And finally, the vdW complexes of Rg and metal atoms are briefly considered.

6.1 Definitions

As it was mentioned in 1973 [1], 'A precise definition of the term *molecular complex* is not easy; and Person attempted to provide one as follows:'

> A *molecular complex* between two molecules is an association, somewhat stronger than ordinary van der Waals associations, of definite stoichiometry .... The partners are often already *closed-shell* (saturated valence) electronic structures. In loose complexes, the identities of the original molecules are, to a large extent, preserved [2].

The definitions of even van der Waals (vdW) complexes and forces differed from each other earlier:

> Van der Waals molecules are weakly bound complexes of small atoms or molecules held together, not by chemical bonds, but by intermolecular attractions [3].

A. Pravilov, *Gas-Phase Photoprocesses*, Springer Series in Chemical Physics 123,
https://doi.org/10.1007/978-3-030-65570-9_6

Van der Waals forces are 'The attractive or repulsive forces between molecular entities (or between groups within the same molecular entity) other than those due to bond formation or to the electrostatic interaction of ions or of ionic groups with one another or with neutral molecules. The term includes *dipole–dipole, dipole-induced dipole,* and London (instantaneous induced dipole-induced dipole) forces. The term is sometimes used loosely for the totality of nonspecific attractive or repulsive intermolecular forces' [4] (italics of the author).

In this chapter, the direct electrostatic interactions (dipole-dipole, $\sim 1/R^3$, quadrupole-quadrupole, $\sim 1/R^4$, etc. (see Sect. 6.3.3.2), will not be considered as van der Waals. Nevertheless, for transition electric dipole-permanent electric dipole, and transition electric dipole-permanent electric quadrupole interactions, they are long-range, $\sim 1/R^3$ and $\sim 1/R^4$, respectively (see Sect. 6.3.3.1).

Recently, two types of bonds (complexes), namely, *hydrogen* and *halogen* bonds, began to be considered independently.

'The hydrogen bond is an attractive interaction between a hydrogen atom from a molecule or a molecular fragment X–H in which X is more electronegative than H, and an atom or a group of atoms in the same or a different molecule, in which there is evidence of bond formation.

A typical hydrogen bond may be depicted as X–H···Y–Z, where the three centered dots denote the bond. X–H represents the hydrogen bond donor. The acceptor may be an atom or an anion Y, or a fragment or a molecule Y–Z, where Y is bonded to Z' [5].

'A halogen bond occurs when there is evidence of a net attractive interaction between an electrophilic region associated with a halogen atom in a molecular entity and a nucleophilic region in another, or the same, molecular entity. The halogen bond donor (R-X) are halogen-containing molecules (electrophillic region), homonuclear or heteronuclear halogens, in particular. The halogen bond is represented conventionally by the three centered dots in R-X···Y, where the halogen atom X of the molecule X–R interacts with a nucleophilic acceptor atom/center Y' [6]. In the gas phase, the halogen bond acceptor (electrophilic region) are Lewis bases B, such as N_2, CO, HCN, H_2O, C_2H_2, H_2S, NH_3. In this case, the general notation of halogen bonds would be B···XY (XY are a homohalogen or heterohalogen) [7].

This book is devoted to gas-phase photoprocesses, that is, electronically excited states, in particular. Therefore, in this chapter, we consider only those weakly bound complexes *that are weakly bound in electronically excited states*, as well. As far as the author knows, there are no data in the literature on experimental studies of electronically excited states of complexes with hydrogen and halogen bonds. For this reason, complexes with hydrogen and halogen bonds in the ground states are considered briefly.

6.2 Hydrogen- and Halogen-Bonded Complexes

Systematic experimental studies of the hydrogen- and halogen-bond complexes in the gas phase are carried out by the A.C. Legon's and J.W. Bevan's scientific groups. An analysis of rotational spectra of isolated weakly-bound complexes formed in a B/XY/Rg supersonic molecular beam in conjunction with ab initio quantum-chemical calculations gives a clue to the symmetry and geometry of the complexes. Besides, it allows determining other spectroscopic characteristics of the complexes (parameters of the one-dimensional potential energy function to the dissociation coordinate, the intermolecular dissociation energy, D_σ, the dipole moment, and some others, see [7–15] and references).

The quadratic intermolecular stretching force constant k_σ provides a measure of the force required for an infinitesimal unit increase in the separation of the B and XY subunits along the intermolecular B···HX or B···XY bonds. It has been shown (see [7, 8] and references) that

$$k_\sigma = cNE, \tag{6.2.1}$$

where c is a constant, N and E are numbers representing the B *nucleophilicities* and HX (XY) *electrophilicities* [16, 17]. The intermolecular dissociation energy D_σ is directly proportional to k_σ

$$D_\sigma = C_\sigma k_\sigma, \tag{6.2.2}$$

C_σ= 1.50(3)·10^3 m^2mol^{-1} for both, hydrogen- and halogen- bonds. The electrophilicity of the H and halogen atoms depends on their *electronegativities* [4]. It increases with electronegativity, which is equal to 2.20 (H), 3.96 (F), 3.16 (Cl), 2.96 (Br), 2.66 (I) in the Pauling scale [16]. Therefore, the D_σ increases in the HI–HBr–HCl–HF raw for the hydrogen-bonded complexes, HI (1007 cm^{-1})–HF (2958 cm^{-1}) for B = H_2O, for example, and in the F_2–Cl_2–Br_2–ClF–BrCl–ICl raw for halogen-bonded complexes, F_2 (387 cm^{-1})– ICl (2059 cm^{-1}) for B = H_2O, for example. Nucleophilities and D_σ increase in the N_2, CO, HCN, H_2O, C_2H_2, H_2S, NH_3 raw [7], N_2 (591 cm^{-1})– NH_3 (3905 cm^{-1}) for XY = ICl, for example (see [8]).

6.3 Van der Waals Complexes

6.3.1 General

In this book, the author uses the following definition of the vdW complexes: 'Van der Waals complexes are weakly-bound molecular species in which the dominant attractive contribution are weak intermolecular attractions'. The weak intermolecular interactions are polarization (induction and dispersion interactions, see Sects. 6.3.3, 6.3.4 as well as Sects. 5.3.2 and 6.4 in [18]).

The population and decay of vdW complexes in the gas phase play a fundamental role in the kinetics and dynamics of electronically-excited molecular states. They are responsible for radiative emission and energy transfer in the atmosphere, energy pooling and conversion in laser media, relaxation phenomena in chemiluminescence processes, plasma formation, and in many other situations, where electronically-excited states are involved. The description of these processes is significant for understanding the mechanisms of non-adiabatic transitions in weakly-bound complexes and clusters. An understanding of population and decay of vdW complexes is also essential for an interpretation and a modeling of various photoinitiated processes in liquids and solids, where the role of intermolecular interaction is greatly magnified through the formation of solvation shell(s) and multiple collisions between a molecule and a solvent. One should note that decay of vdW complexes has much in common with vibrational relaxation and electronic predissociation induced by collisions with atoms and molecules, which have been discussed in Chap. 3.

An energy of a vdW bond is small, less than several hundred cm^{-1} (see below). Therefore, to observe and study vdW complexes, it is required cooling atoms or molecules to sufficiently low temperatures by utilizing the supersonic molecular beam technique (see Sect. 6.2 in [18]).

The first pioneering work devoted to studies vdW complexes in electronically-excited states using the supersonic molecular beam was published in 1976 by D. H. Levy and his team [19]. Heretofore, rotational constants of the RgHX (X, $v_X = 0$) and ROCS(X, $v_X = 0$) by means of molecular beam electric resonance spectroscopy were determined by W. Klemperer's team (see [20, 21] and references as well as [22]). This field of chemical physics and molecular spectroscopy is trendy at present. One of the most thoroughly studied groups of vdW complexes are complexes consisting of homonuclear halogens (hereafter, X_2) or interhalogens (XY) and rare gas atoms, RgX_2 ($X0_g^+$, $B0_u^+$, $E0_g^+$), RgXY($X0^+$, $B0^+$, $A1$, $E0^+$, $\beta 1$, $D'2$). Decades of experimental and theoretical papers devoted to vdW complexes have been published (see [23–31] and references). Bound excited X_2, XY valence states are perturbed by repulsive states (see Sect. 4.2.3 and Fig. 4.4), and predissociation of the complexes to Rg and halogen atoms occurs. Therefore, the only methods of studies spectroscopic characteristics and dynamics of these complexes at the supersonic molecular bands are the following (see Sect. 6.2 in [18]):

- *Intracavity laser spectroscopy.* Supersonic jet is placed within the optical resonator cavity of an amplitude modulated cw dye laser. The laser output was dispersed with a spectrometer. Absorption features appear as wavelength-dependent losses in the spectral profile of the laser output (see [32] and references for details);
- *Measurements of luminescence spectra* of halogens (interhalogens) which are the products of vibrational predissociation of the complexes; a loss of several vibrational quanta can occur in these processes;

- *The pump-probe method* when $RgHal_2(B, v_B)$ complex vibrational predissociation (VP) products, *e.g.*, produced due to absorption of a pump laser photon by a complex are excited to the gerade (*g*) ion-pair (IP) states, Hal_2 $(E0_g^+)$, for example, and luminescence of this IP state are monitored during measurement of excitation spectra. Action spectra can also be measured: wavelength of probe radiation is fixed and that of pump radiation scan;
- *Luminescence depletion techniques*. The two dye laser beams are counter-propagated through the jet expansion chamber and spatially overlapped. The depletion laser is fired before the probe laser, and depopulate the lower vdW state from which the transition studied occurs. Therefore, this transition is depleted (see [33] for details);
- *Velocity-map imaging method*. A supersonic molecular beam is produced by pulsed supersonic jet expansion generated by a pulse valve into a source chamber evacuated with high pumping speed. A central part of the gas jet passes through a skimmer into an electrostatic lens chamber where it crosses with a focused pulse laser beam. The energy of the laser pulse has to be enough to excite ground state molecule to Rydberg state and then ionize it in the (2 + 1) schemes. Ions produced at the laser beam spot are extracted by a three-electrode electrostatic lens to time-of-flight (TOF) tube. After passing it, ions hit an imaging detector consisting of a microchannel plate and phosphor screen. A 2D image of the phosphor screen is recorded by a charge-coupled device. The ring structures in the VMI images reflect the distribution of the ions as a function of kinetic energy. The sum of the kinetic energies of the fragment ions formed in the dissociation or predissociation processes is termed as Kinetic Energy Release (KER). Analysis of the KER spectra allows assigning the channels of Rydberg state predissociation (see Sect. 2.8 in [18]).

A large amount of experimental and theoretical data have been obtained, and spectroscopic characteristics of the $RgX_2(X, B)$ and $RgXY(X, B, A)$ states and the complex electronically-excited states dynamics are studied. Three conformers of RgX_2 complexes have been observed and studied. These are *T*-shaped (C_{2v} symmetry group), linear ($C_{\infty v}$ symmetry group) $RgX_2(X, B)$ complexes which vibrational wave functions are localized at Θ = 90°, 0 and 180°, respectively, as well as 'free-rotor' or bending complexes which vibrational wave functions are delocalized over Θ = 0–360° (Fig. 6.1).

The vast majority of works are devoted to the research of the RgX_2 and RgXY ground states. As to excited valence states, there are experimental and theoretical data on the I-containing, $RgI_2(B)$, and, to a lesser extent, $RgXY(B, A)$ states (see Sects. 6.3.2 and 6.3.3). As to RgI_2(IP) and RgXY(IP) complexes, to the best of the author's knowledge, six works devoted to the $HeICl(E0^+, \beta1)$ [28], $NeICl(E, \beta)$ [34–36] and $HeI_2(E)$ [37, 38] had been published up to 2016, when the first paper on the $RgI_2(E)$ complexes was published by the author's team [39].

As it has been mentioned in Sect. 6.1, in this chapter, we consider only those weakly bound complexes *that are weakly bound in electronically excited states*, as well. An analysis of the papers, mentioned in [22, 40] shows that the RgMe,

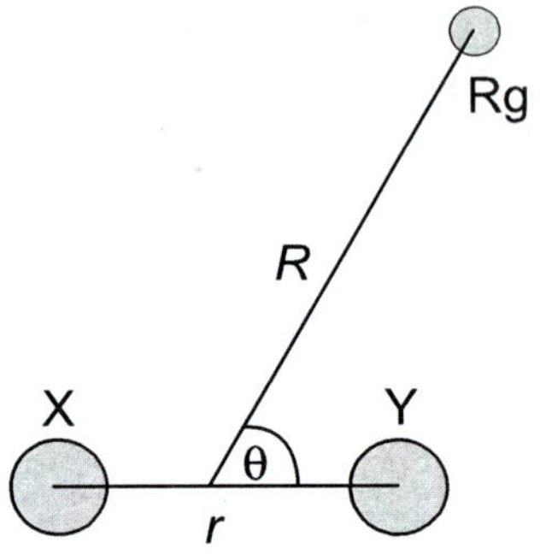

Fig. 6.1 *T*-shaped ($\Theta = 90^{\circ}$), free-rotor ($360^{\circ} > \Theta > 0^{\circ}$) and linear ($\Theta = 0, 180^{\circ}$), RgXY vdW complex

Me = Al [41], Cd, Zn, Hg (see [40–43] and references), and Ni [44], vdW complexes have weakly bound electronically-excited states. Electronically-excited state of the RgX_2 and RgXY vdW complexes have been studied most detailed. Therefore, we start with the RgX_2 complexes.

6.3.2 RgX_2 vdW Complexes

The lowest electronically-excited states of the X_2 molecules is $^1\Pi_u$ and $^3\Pi_u$ (in terms of Hund's cases (*a*), (*b*), see Sect. 4.2.3). The perpendicular $X_2(^1\Pi_u - X^1\Sigma_g^+)$ transitions are weak (see [45] and references), and the $X_2(^3\Pi_u - X^1\Sigma_g^+)$ electric dipole transitions can be allowed due to spin-orbit coupling which increases as $\sim Z^4$ (*Z* is X nuclear charge, see Sect. 4.6.1and Table 4.1). Therefore, maximal cross-sections of the X_2 *A* absorption bands are $2.4{\cdot}10^{-20}$ cm^2, λ_{max} = 290 nm (F_2), $2.4{\cdot}10^{-19}$ cm^2, $\lambda_{max} \approx 330$ nm (Cl_2), $6.1{\cdot}10^{-19}$ cm^2, $\lambda_{max} \approx 410$ nm (Br_2) and $2.0{\cdot}10^{-18}$ cm^2, $\lambda_{max} \approx 500$ nm (I_2), see [46, 47] and references). The electronically-excited valence state RgX_2 complexes can be populated in optical transitions from the $RgX_2(X)$ state if transitions occur to bound parts of the upper states.

6.3.2.1 The RgF_2 vdW Complexes

Both $F_2(^3\Pi(0_u^+)$ and $^1\Pi_u)$ states are repulsive in the Franck-Condon zone of transitions from the $F_2(X, v_X = 0)$ state (see [48] and references), and $RgF_2(^3\Pi(0_u^+)$, $^1\Pi_u)$ complexes cannot be populated in optical transitions. There are theoretical data on some spectroscopic parameters of the $RgF_2(X)$ state in the literature [24, 49] (Table 6.1).

One sees, that binding energy of lowest, $n^{str} = 0$, $n^b = 0$ vdW levels, $D_0 = D_e - \frac{1}{2}\left(\omega_e^{str} + 2\omega_e^b\right)\left(\Theta = 0^{\circ}\right)$ and $D_0 = D_e - \frac{1}{2}\left(\omega_e^{str} + \omega_e^b\right)\left(\Theta = 90^{\circ}\right)$ is negative,

Table 6.1 Equilibrium distances R_e(Å), dissociation, D_e, and binding, $D_0^{(1)}$, energies (units, cm^{-1}) and vdW stretching (ω_e^s) and bending (ω_e^b) wavenumbers of the $RgF_2(X,v_X = 0)$ linear ($\Theta = 0°$) and *T*-shaped ($\Theta = 90°$) complexes, $r_{F\text{-}F} = 1.142$ Å [24] (see Fig. 6.1)

$RgF_2(X)$	Θ	R_e	D_e	D_0	ω_e^s	ω_e^b
HeF_2	0	3.47	35.9	−12.0	54.1	20.8
	90	3.00	31.9	−5.4	47.3	27.2
NeF_2	0	3.59	61.5	23.7	36.8	19.4
	90	3.08	61.4	30.0	34.4	28.5
ArF_2	0	3.88	122.8	76.9	38.3	26.8
	90	3.44	110.0	76.8	34.4	32.1

(1)—Dissociation energy of lowest vdW levels

so the reliability of the calculation results is low (n^{str}, n^b are the vdW quantum numbers).

6.3.2.2 The $RgCl_2$ vdW Complexes

The bound-bound $Cl_2\left(A^3\Pi(1_u) \leftarrow X^1\Sigma\left(0_g^+\right)\right)$ transition takes place, but there is no evidence in the absorption spectrum of Cl_2 for bands which could be assigned to the $A \leftarrow X$ transition. The dipole moment of the $Cl_2(A\text{–}X)$ transition is much less than that of the $Cl_2\left(B^3\Pi(0_u^+) - X^1\Sigma\left(0_g^+\right)\right)$, radiative lifetime of the $Cl_2(A)$ state is (0.02–0.08) s [50], whereas that of $Cl_2(B)$ is $3.05 \cdot 10^{-4}$ s [51], and *FCF*s of the bound-bound $Cl_2(A \leftarrow X,0)$ transitions are extremely small. Therefore, the information on spectroscopic characteristics of the $Cl_2(A)$ state has been obtained after analysis of the $Cl_2(\beta 1_g \rightarrow A)$ transitions and recombination of Cl atoms accompanied by the $Cl_2(A,v_A \rightarrow X,v_X)$ radiation (see [50, 52] and references).

The $RgCl_2(B \leftarrow X)$, Rg = He, Ne, Ar transitions have been observed, and some spectroscopic characteristics of the $RgCl_2(B, X)$ complexes as well as the $RgCl_2(B)$ dynamics have been studied.

There are $^{35}Cl_2$ (57%), $^{35}Cl^{37}Cl$ (37%), and $^{37}Cl_2$ (7%) isotopes in the nature [51].

The $HeCl_2$ Complexes. Experimental data on spectroscopic characteristics of the $HeCl_2(X)$ complexes is scarce. It has been shown that the $HeCl_2(X)$ is *T*-shaped, and $R_X = (3.8 \pm 0.4)$ Å [53]. The binding energies of the $n^b = 0\text{–}4$ vdW levels are $(12.83 \div 0.77)$ cm^{-1}. It has also been shown that there are no *bound* levels in which the vdW *stretch* is excited for the *X* states [54]. Some spectroscopic characteristics of the $RgCl_2(X, v_X = 0)$ linear and *T*-shaped complexes are calculated in [55] (see [23, 49, 56] and references, also).

One sees, that calculated binding energy of lowest, $n^{str} = 0$, $n^b = 0$ vdW levels, $D_0 = D_e - \frac{1}{2}\omega_e^b$ (see above) for *T*-shaped $HeCl_2(X,v_X = 0)$ is 29.3 cm^{-1}, more than two times higher than the experimental value, 12.83 cm^{-1}.

The spectroscopic characteristics and dynamics of $He^{35}Cl_2(B0_u^+)$ VP are studied relatively good. It has been shown that the binding energies of the lowest vdW modes of the $He^{35}Cl_2(B, v_B = 7\text{-}12)$ complexes are ~ 3.7 cm^{-1} less than that of $He^{35}Cl_2(X, v_X = 0)$ one, i.e., equal to (12.83–3.7) ≈ 9.1 cm^{-1} (see above). The $He^{35}Cl_2(B, v_B \leftarrow X, v_X = 0)$ transitions occur in the *T*-shaped configuration, R_B = (3.9 ± 0.4) Å, and $He^{35}Cl_2(B, v_B = 8\text{–}24)$ lifetime is 506 ps (v_B= 8)– 52.37 ps (v_B= 12) [53], 5.4 ps (v_B= 20), 2.5 ps (v_B= 24) [54]. The $He^{35}Cl_2(B, v_B = 8)$ binding energies of the n^b = 0–4 vdW levels are (9.39−1.31) cm^{-1}. There are no *bound* levels in which the vdW *stretch* is excited for the *B* states. The $He^{35}Cl_2(B, v_B) \rightarrow He + Cl_2(B, v_B - 1)$ VP channel dominates, and bimodal rotational distributions of the $Cl_2(B, v_B - 1, J_B)$ VP products takes place [54] (see [57] and references, also). The $HeCl_2(B, v_B)$ binding energy is small, and $\Delta v_B = 1$ VP channel is energetically possible even at $v_B = 20$. The rotational distributions of the $Cl_2(B, v_B)$ VP shows that $\Delta v_B =$ 2, 3 VP channel should occur even at $v_B = 12$. The signals due to VP become unobservable above $He^{35}Cl_2(B, v_B > 24)$ [54].

The $NeCl_2$ Complexes. Experimental and calculated data on spectroscopic characteristics of the $NeCl_2(X)$ are given in [23, 25, 55, 56, 58–61]. According to experimental data, the *T*-shaped $NeCl_2(X, v_X = 0)$ binding energy is D_0 = (60 ± 2) cm^{-1} [59], and R_0 = 3.57(4) Å see [60] and references). Calculations show [55, 56, 58] that linear $NeCl_2(X,0)$ conformer is also exist, its binding energy is ~10 cm^{-1} less and equilibrium distances is ~0.8 Å larger (Table 6.2), see [56, 58], also. The $NeCl_2(X, v_X = 1, 2)$ vdW complexes (lifetime, $\tau \geq 10^{-5}$ s) were observed in [61].

The $Ne^{35}Cl_2(B, v_B \leftarrow X, v_X = 0)$ transitions occur in the *T*-shaped configuration [60]. It has been shown that the binding energies of the lowest vdW modes of the $Ne^{35}Cl_2(B, v_B = 6\text{–}13)$ complexes are ~6 cm^{-1} less than that of $Ne^{35}Cl_2(X, v_X = 0)$ one [60, 61]. It is equal to D_0 = (55 ± 2) cm^{-1} (v_B= 16, 17, exp.) and (54.4–51.0) cm^{-1} (v_B= 11–21, theor.) [25]. The R_0 value is R_0 = 3.52(5) Å for v_B= 7– 12 (see [60] and references). The $Ne^{35}Cl_2(B, v_B)$ lifetimes are (258 ± 42) ps (v_B= 9) ÷ (33 ± 2) ps (v_B= 13), 11 ps (v_B= 16) [60, 61], (11 ± 2 and 6 ± 1) ps (v_B= 16 and

Table 6.2 Equilibrium distances R_e (Å), dissociation, D_e, and binding, $D_0^{(1)}$, energies (units, cm^{-1}) and vdW stretching (ω_e^s) and bending (ω_e^b) wavenumbers of the $RgCl_2(X, v_X = 0)$ linear (Θ = 0°) and *T*-shaped (Θ = 90°) complexes calculated in the aug-cc-pVTZ1(3*s*3*p*2*d*1*f*1*g*) basis set, $r_{Cl\text{-}Cl}$ = 1.988 Å [55] (see Fig. 6.1)

$RgCl_2(X)$	Θ	R_e	D_e	D_0^a	ω_e^{str}	ω_e^b
$HeCl_2$	0	4.17	45.6	0.1	53.2	18.9
	90	3.40	43.5	5.8	46.9	28.5
$NeCl_2$	0	4.21	87.3	49.9	35.6	16.0
	90	3.45	88.0	59.1	33.0	24.7
$ArCl_2$	0	4.45	223.7	178.5	41.5	24.4
	90	3.68	215.4	179.6	38.0	33.6

(1)—$D_0 = D_e - \frac{1}{2}(\omega_e^{str} + 2\omega_e^b)(\Theta = 0^\circ)$, and $D_0 = D_e - \frac{1}{2}(\omega_e^{str} + \omega_e^b)(\Theta = 90^\circ)$

17, respectively). The populations of VP products become quite weak above $Ne^{35}Cl_2(B, v_B > 16)$ [25].

Calculations carried out in [25] show the following:

- The $NeCl_2(B, v_B \rightarrow {}^3\Pi(2_g)) \rightarrow Ne + 2Cl({}^2P_{3/2})$ electronic predissociation (EP) channel opens at $v_B = 13$ (Fig. 6.2);

The $Ne^{35}Cl_2(B, v_B) \rightarrow Ne + Cl_2(B, v_B\text{-}\Delta v_B)$ VP channel dominates at $Ne^{35}Cl_2(B$, low $v_B)$, but the $\Delta v_B = 1$ channel becomes energetically impossible with v_B increasing due to strong $Cl_2(B, v_B)$ anharmonicity. In this case, the $\Delta v_B = (2 \div 5)$ channels become appreciable due to intramolecular vibrational redistribution (relaxation) (IVR): the decay occurs predominantly in a stepwise fashion:

$$NeCl_2\left(B, v_B, n_B^i \rightarrow B, v_{B-1}, n_B^f\right) \rightarrow Ne + Cl_2(B, v_B - \Delta v_B), \qquad (6.3.1)$$

$\Delta v_B > 1$, n_B^f is more than the initial vdW quantum number, n_B^i. The IVR results from quasi-degeneracy between the initial excited $NeCl_2(B, v_B, n_B^i)$ state ('bright state') and intermediate zero-order $NeCl_2(B, v_B\text{-}1, n_B^f)$ 'dark states' with less $Cl_2(B)$ vibrational excitation and more vdW energy, $NeCl_2(B, v_B\text{-}1, n_B^f)$. The vibrational coupling induced by the neon atom mixes these zero-order quasi-degenerate states, which leads to a population to the dark state. It, in turn, decay to the $Ne + Cl_2(B, v_B\text{-} \Delta v_B)$, $\Delta v_B = (2 \div 5)$ (Fig. 6.3), see Sect. 6.4.3.2 in [18], [62, 63] and references.

IVR realized in so-called 'sparse', 'intermediate', and 'statistical' regimes for which half-widths of the bound $ArI_2(B, v_B, n_B)$ levels are much less, comparable,

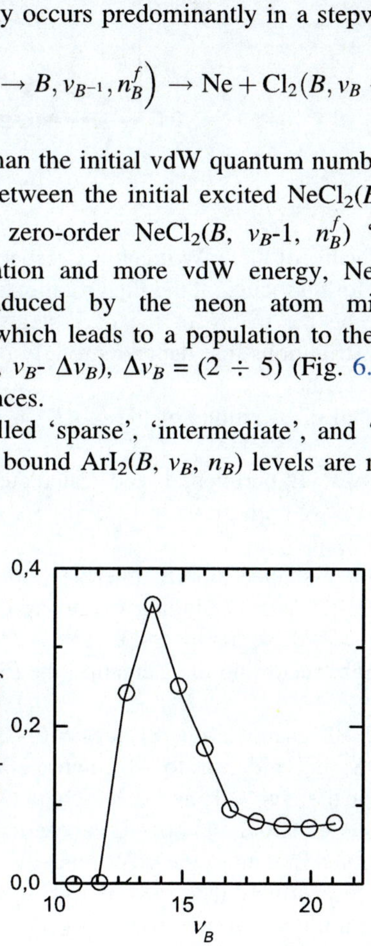

Fig. 6.2 The $NeCl_2(B,v_B)$ EP probability as a function of v_B (see [25]) (Reprinted with permission C. R. Bieler, K .C Janda, R. Hernández-Lamoneda, O. Roncero, $NeCl_2$ and $ArCl_2$: Transition from direct vibrational predissociation to intramolecular vibrational relaxation and electronic nonadiabatic effects. J. Phys. Chem. A. **114**, 3050–3059 (2010). https://doi.org/10.1021/jp906392m. Copyright 2010 American Chemical Society)

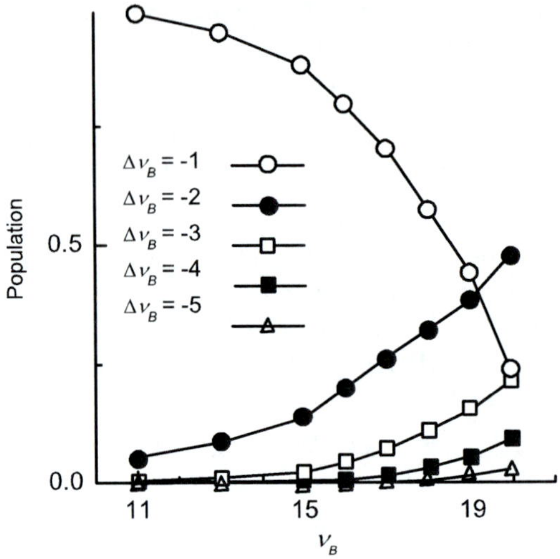

Fig. 6.3 Calculated channels of $Cl_2(B, \nu_B\text{-}\Delta\nu_B)$ VP from the initial $NeCl_2(B, \nu_B)$ state (see [25]) (Reprinted with permission C. R. Bieler, K. C Janda, R. Hernández-Lamoneda, O. Roncero, $NeCl_2$ and $ArCl_2$: Transition from direct vibrational predissociation to intramolecular vibrational relaxation and electronic nonadiabatic effects. J. Phys. Chem. A. **114**, 3050–3059 (2010). https://doi.org/10.1021/jp906392m. Copyright 2010 American Chemical Society)

and much larger than a value of the vdW quantum, respectively [63]. Since the dark state acts as a doorway for dissociation, the final rotational distribution of $Cl_2(B, \nu_B\text{-}\Delta\nu_B)$ depends strongly on the nature of the dark state. As a result, complicated oscillatory rotational distributions that depend strongly on the initial excitation are obtained in the sparse regime.

The bimodal rotational distributions of the $Cl_2(B, \nu_B{-}1, J_B)$ VP products takes place [59].

The $ArCl_2$ Complexes. Experimental and calculated data on spectroscopic characteristics of the $ArCl_2(X)$ are given in [23, 55, 56, 63–68]. They are given in Table 6.3 (see Table 6.2, also)

One sees that data on *T*-shaped binding energies, only, are consistent enough.

The *T*-shaped $Ar^{35}Cl_2(B, \nu_B{=}\ 7)$ binding energy is D_0 = (188 ± 1) cm^{-1}, and R_0 = (3.7 ± 0.1) Å, $\tau \approx$ 100 ps at ν_B= 6–11. The $ArCl_2(B, \nu_B)$ VP product disappear at ν_B > 12, though, according to calculation the EP probability less than 1up to ν_B= 21 (Fig. 6.4).

The $\Delta\nu_B$ = 1, and 2 VP channels open for ν_B = 6, 7, and ν_B > 7, respectively. Rather narrow, $\Delta J \leq 8$, and wide, up to 28, rotational distributions of the $Cl_2(B, \nu_B)$ are characteristic for the $\Delta\nu_B$ = 1, and 2 VP channels, respectively [67].

To break the *T*-shaped Ar-$Cl_2(B, \nu_B)$ bond it is necessary to transfer 1, 2, 3, 4 and so on $Cl_2(B, \nu_B)$ vibrational quanta to the vdW modes for the ν_B = 1–7, 8–16, 17–20, and 21–23 levels, respectively (see [48], p. 146, [68] and [69], p. 94). It has been shown that this feature is due to the IVR process (see the description of the IVR for $NeCl_2(B)$ above).

The $KrCl_2$ and $XeCl_2$ complexes. These complexes were studied in the K. C. Janda's group [70, 71] using the pump-probe method. It was shown that the binding energy of the ground state complexes are $D_0^X = (236.6 \pm 2.0)\,\text{cm}^{-1}$ ($KrCl_2(X, \nu_X{=}\ 0)$ and (268.3 ± 1.1) cm^{-1} ($XeCl_2(X, \nu_X{=}\ 0)$. The binding energy of

Table 6.3 Equilibrium distances R_e,[(1)], $R_0^{(2)}$ (Å), dissociation, D_e, and binding, D_0, energies (units, cm^{-1}) and vdW stretching (ω_e^s) and bending (ω_e^b) wavenumbers of the $Ar^{35}Cl_2(X, v_X = 0)$ linear ($\Theta = 0°$) and *T*-shaped ($\Theta = 90°$), $r_{Cl–Cl} = 1.988$ (see Fig. 6.1)

Θ	R_e [(1)]		D_e		R_0 [(2)]		D_0		ω_e^s	ω_e^b				
0	4.132 [56]	3.5 [63]	226.88 [56]	220.1 [63]					41.2 [65]			37.9 [65]		
90	3.657 [56]	3.9 [63]	220.82 [56]	183.6 [63]	3.58 [56]	3.71 [62]	3.719 [63]	188.4 [56]	188 [62]	180.8 [64]	34.5 [66]	41 [67]	29.9 [66]	12 [67]

(1)—The equilibrium distances corresponding to the potential well depths
(2)—The equilibrium distances corresponding to the $n_X = 0$ vdW modes

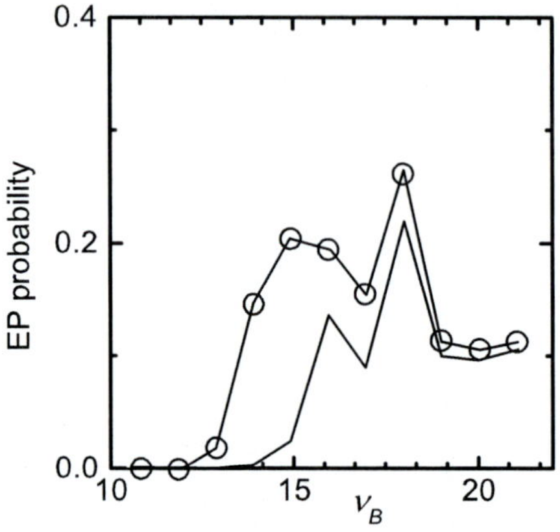

Fig. 6.4 The $ArCl_2(B,v_B)$ EP probability as a function of v_B (see [25]) (Reprinted with permission C.R. Bieler, K.C Janda, R. Hernández-Lamoneda, O. Roncero, $NeCl_2$ and $ArCl_2$: Transition from direct vibrational predissociation to intramolecular vibrational relaxation and electronic nonadiabatic effects. J. Phys. Chem. A. **114**, 3050–3059 (2010). https://doi.org/10.1021/jp906392m. Copyright 2010 American Chemical Society)

the $KrCl_2(B,v_B$= 8-10) complexes are (230.2 ± 4.4) cm^{-1} (v_B= 8) ÷ (223.8 ± 3.1) cm^{-1} (v_B= 10), and that of $XeCl_2(B$, v_B= 10,11) is 277(2) cm^{-1}. Rotational distributions of the VP products show that IVR could occur in the VP process. Lifetime of the $XeCl_2(B$, v_B= 10) complex is $\tau \geq 50$ ps [70].

6.3.2.3 The $RgBr_2$ vdW Complexes

The bound-bound $Br_2(A1_u(^3\Pi) \leftarrow X0_g^+(^1\Sigma))$ transition takes place, but its dipole moment and *FCF*s are low, $\tau_{rad} = 3.5{\cdot}10^{-4}$ s for $Br_2(A$, v_A= 11) [51]. To the best of the author's knowledge, there are no data on the $RgBr_2(A)$ complexes in the literature.

The $RgBr_2(B \leftarrow X)$, Rg = He, Ne, Ar transitions have been observed, and some spectroscopic characteristics of the $RgBr_2(B, X)$ complexes as well as the $RgBr_2(B)$ dynamics have been studied.

There are $^{79}Br_2$, $^{79}Br^{81}Br$, and $^{81}Br_2$ isotopes in the nature by natural occurrence ~ (1 : 2 : 1) [51]. Vibrational constants, ω_e, $\omega_e x_e$, of the isotopes differ in the third character.

The $HeBr_2$ complexes. Some spectroscopic characteristics and data on *PES* and vdW vibrational wavefunctions of the $HeBr_2(X,v_X$= $0,n_X)$ complexes are given in Table 6.4 and Figs. 6.5, 6.6 and 6.7.

According to calculations carried out in [74], the n_X = 0,1 vdW states are bound ($E = -16.46$ cm^{-1} relative to dissociation limit) and represent the symmetric and antisymmetric linear combinations of the two states that have the helium atom localized near one of the Br atoms, $\theta = 0°$, 180°. The n_X = 2 state is bound also

Table 6.4 The dissociation, D_e, and binding, D_0, energies (units, cm^{-1}) as well as equilibrium distances $R_e^{(1)}$, $R_0^{(2)}$ (Å) of the $RgBr_2(X, v_X = 0)$ linear ($\Theta = 0^{\circ}$) and *T*-shaped ($\Theta = 90^{\circ}$) conformers (see Fig. 6.1)

		Linear				*T*-shaped			
Rg	Reference	D_e	D_0	R_e	R_0	D_e	D_0	R_e	R_0
He	[72] (ab initio)	48.8	17.2	4.42		40.3	17.7	3.58	
	[73] (ab initio)		16.03				14.89		
	[74] (ab initio)	49.45	16.46	4.412	4.86	40.64	15.81	3.599	
	[74] (exp)		17.0(8)				16.6(8)		
	[75] (exp.)						17.0 ± 1.5		
	[76]								3.7(2)
Ne	[72] (ab initio)	93.6	68.0	4.49		83.9	67.3	3.60	
	[77] (semiemp. values)					92	71.2	3.52	3.68
	[78] (exp)						70.5 ± 2.0	3.67	
	[79] (exp)		71 ± 3						
	[80] (exp)						70.0 ± 1.1		
Ar	[72] (ab initio)	262.7	228.0	4.63		226.4	203.5	3.80	
	[81] (semiemp. values)	256.6	220.0	4.60		247.2	213.5	3.65	

(1)—The equilibrium distances corresponding to the potential well depths
(2)—The equilibrium distances corresponding to the $n_X = 0$ vdW modes

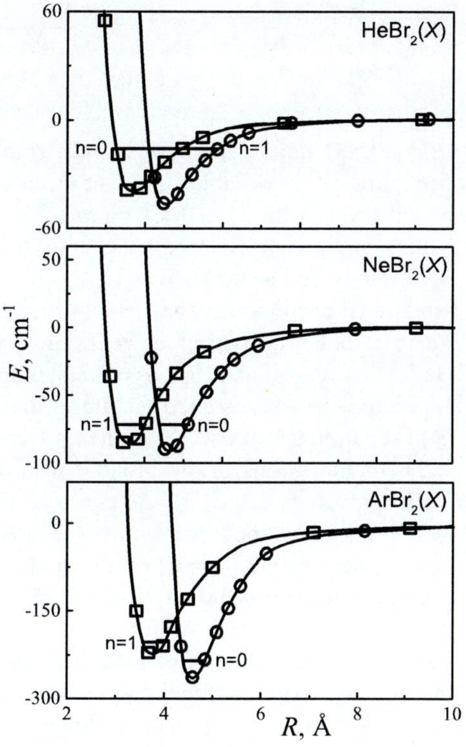

Fig. 6.5 Potential energy curves for $RgBr_2(X)$ complexes, Rg = He, Ne, Ar. Open circles indicate ab initio results for the linear, and open squares for the *T*-shaped configurations. Full lines are for the parametrized potential curves (see [72] and Fig. 6.1). (Potential energy curves for $RgBr_2(X)$ complexes measured in molecular beam scattering experiments are presented in [26], also)

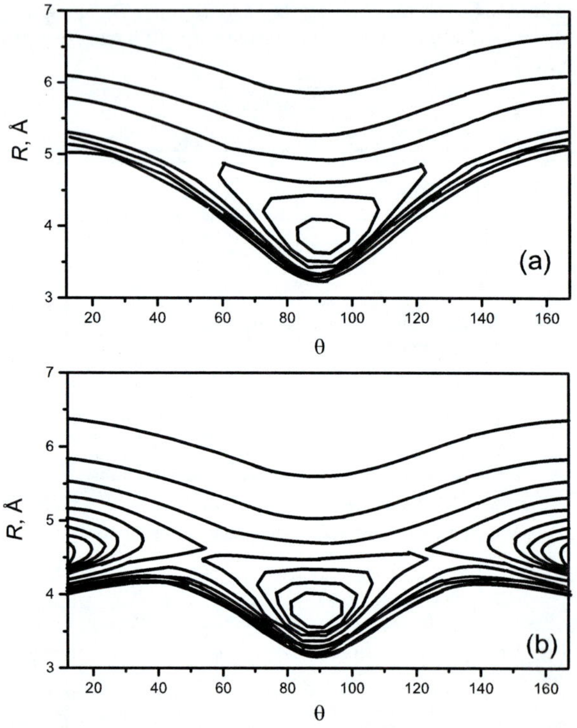

Fig. 6.6 The $HeBr_2(X, v_X = 0)$ (**a**) and $HeBr_2(B, v_B = 12)$ (**b**) *PES*s. The contours begin at -5 cm^{-1} and incrementally decrease by 5 cm^{-1} in each panel [74] (Reproduced from D. S. Boucher, D. B. Strasfeld, R. A. Loomis, J. M. Herbert, S. E. Ray, A. B. McCoy, J. Chem. Phys. **123**, 104312 (14 pp) (2005). https://doi.org/10.1063/1.2006675 with the permission of AIP Publishing)

($E = -15.81$ cm^{-1}) and corresponds to the helium atom localized in the *T*-shaped configuration (see Table 6.4). These data contradict to those in [72]. The $n_X > 2$ bound levels with $J_X = 0$ are close to or above the barrier for free rotation of the helium around the Br_2 and have amplitudes in the $\theta = (0–180)^{\circ}$ range. There are $n_X = 0–6$ bound states ($E = -(16.46– 3.12)$ cm^{-1}), and none of these states contain vibrational excitation in the He–Br_2 stretching coordinate (see [74] and references). There is potential barrier of ~ 10 cm^{-1} separating the $n_X = 0$, 1 states of linear $He^{79}Br_2(X, v_X = 0$ and the $n_X = 2$ states of the *T*-shaped conformer [74].

There is a well resolved feature in the $HeBr_2(B, v_B = 8, 12, 21, n_B \leftarrow X, v_X = 0, n_X)$ excitation spectra at ~ 2.5 and ~ 4 cm^{-1} (calculation and experiment) as well a series of transitions in the broader feature those are peaked at 10 cm^{-1} from the $HeBr_2(B, v_B = 8, 12, 21 \leftarrow X, v_X = 0)$ origins. The lower-energy feature is associated with transitions from the *T*-shaped $HeBr_2(X, v_X = 0)$ vdW levels to the lowest-energy state ($n_B = 0$) on the *B*-state potential. The broader feature at higher energy reflects transitions from all three lowest-energy levels on the *X* state, $n_X = 0–2$, to excited intermolecular vibrational levels on the *B* surface, $n_B \geq 1$. The lines attributed to transitions from the linear $n_X = 0$, 1 levels that appear within the higher-energy features tend to be more intense than those from the *T*-shaped, $n_X = 2$ level (see [74]). The binding energy of the $HeBr_2(B, v_B = 44)$ complex is $D_0 = (13.5 \pm 1.0)$ cm^{-1}.

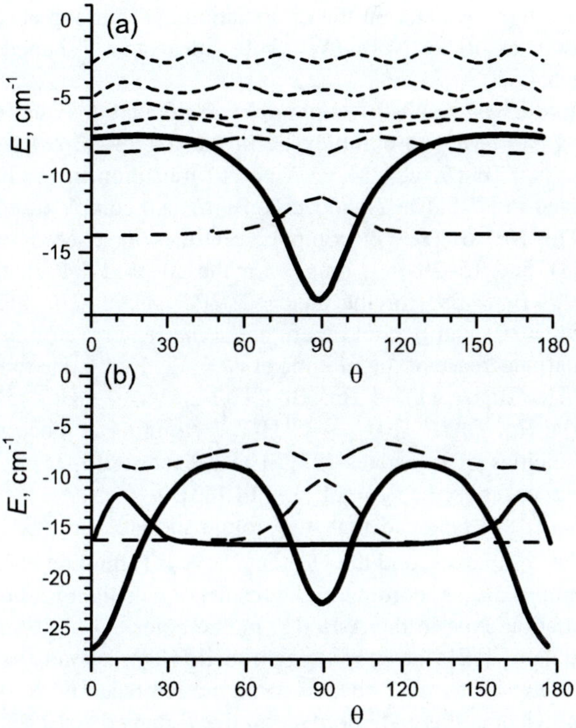

Fig. 6.7 The $He^{79}Br_2(B,v_B = 12)$ (**a**) and $He^{79}Br_2(X,v_X = 0)$ (**b**) adiabatic potentials (black lines), plotted as a function of θ (see Fig. 6.1). Superimposed on the potentials are the probability amplitudes of the lowest-energy vdW vibrational eigenstates. Each state is plotted so that zero amplitude corresponds to the energy of the level. The state localized in the linear wells are shown as solid lines, those localized in the *T*-shaped minimum as dashed lines, and the free rotor states are plotted as dotted lines (see [74]) (Reproduced from D.S. Boucher, D.B. Strasfeld, R.A. Loomis, J. M. Herbert, S.E. Ray, A.B. McCoy, J. Chem. Phys. **123**, 104312 (14 pp) (2005). https://doi.org/10.1063/1.2006675. https://doi.org/10.1063/1.2006675 with the permission of AIP Publishing)

The $HeBr_2(B, v_B)$ complex undergoes VP and EP [75, 76, 82, 83]. The $He^{79}Br_2(B, v_B)$ complex lifetimes have been measured in the v_B = 10-20 [83], 16–23 [76] and 34-48 [75] ranges. The lifetime decreases from 157 ps ($v_B = 10$) to 1.8 ps ($v_B = 44$) and there are some features in this dependency:

There are sharp decrease of the lifetime at v_B = 11, 13, 14 due to EP rate increase caused by the $He^{79}Br_2(B, v_B) \rightarrow He^{79}Br_2(^3\Pi(1_g, 2_g)) \rightarrow He + {}^{79}Br + {}^{79}Br$ predissociation. The EP rate is (0–15)% of total predissociation rate except those at v_B = 11, 13, 14 [83];

VP rate decrease at $v_B = 45$ due to the $\Delta v_B = 1$ VP channel [75].
Below $v_B = 43$, the VP product rotational distribution for the Δv_B = 1, 2 VP channels are similar [82].

The $NeBr_2$ complexes. Data on the dissociation and binding energies as well as equilibrium distances of the $NeBr_2(X, v_X= 0)$ linear and *T*-shaped conformers, is given in Table 6.4.

According to the best results of semiempirical calculations, the binding energies of the *T*-shaped $NeBr_2(B, v_B)$ complex lie in the (64.23–62.28) cm^{-1} range [77]. The bound-free $Ne^{79}Br_2(B, v_B= 14 \leftarrow X, v_X= 0)$ transition in the linear configuration was observed in [79]. The *T*-shaped $NeBr_2(B, v_B)$ complex undergoes VP and EP [83–85]. The $Ne^{79}Br_2(B, v_B)$ complex lifetimes have been measured in the v_B = 10–20 [83], and 15–29 [84] ranges for the $\Delta v_B = 1$ VP channel, and 19 ps ($v_B = 27$) −15 ps ($v_B = 29$) for the $\Delta v_B = 2$ VP channel. The lifetime decreases from 312 ps ($v_B = 10$) with v_B, and there are some features in this dependency:

There are sharp decrease of the lifetime at v_B = 11, 13, 14 due to EP rate increase caused by the $He^{79}Br_2(B, v_B) \rightarrow He^{79}Br_2(^3\Pi(1_g, 2_g)) \rightarrow He + {}^{79}Br + {}^{79}Br$ predissociation. The $He^{79}Br_2(B^3\Pi(0_u^+ \sim C^1\Pi(1_u))$ coupling is much weaker since to different multiplicities of the states [85]. The EP rate is (0–15)% of total predissociation rate except those at v_B = 11, 13, 14 [83].

The authors of [84] believed that they found the $He^{79}Br_2$ $(E0_g^+, v_E \leftarrow B, v_B)$ transitions in the complexes, and the $He^{79}Br_2(E, v_E= 3)$ binding energy is 82 cm^{-1}.

The $ArBr_2$ complexes. According to the results of calculations, both linear and *T*-shaped configuration exist in the $ArBr_2(X, v_X)$ complexes [72, 81] (see Table 6.4). According to data of [72], the $n_X= 0$ eigenfunction corresponds to linear configuration, while the $n_X = 1$ to the *T*-shaped configuration (see Fig. 6.5). To the best of the author's knowledge, there are no data on the *T*-shaped $ArBr_2(B, v_B)$ complexes in the literature. The bound-free $Ar^{79}Br_2(B, v_B= 14 \leftarrow X, v_X= 0)$ transition in the linear configuration was observed in [79].

6.3.2.4 The RgI_2 vdW Complexes

The data on spectroscopic characteristics of ground *X*, valence *B* and IP *E* state vdW complexes (dissociation energies, topologies of *PES*s, energies of vdW levels) and data on the RgI_2(*B* and *E*) decay published up to the mid-2017, have been analyzed in detail in Chapter 6 of [18]. Therefore, in this book, a short review of these data and an analysis of the data obtained after mid-2017 are presented. The spectroscopic parameters of the RgI_2 complexes, Rg = He, Ne, Ar, Kr are given in Table 6.5.

The HeI_2 complexes. High-level ab initio calculations of $HeI_2(X0_g^+, B0_u^+$ and $E0_g^+)$ complex *PES*s have been performed in [86–88] (see [89], also). The *T*-shaped, linear as well as 'free-rotor' complexes which vibrational wave functions are delocalized over θ = 0– 360° vdW HeI_2 complexes have been analyzed. It has been shown in particular that:

- the degenerate *X*, $v_X= 0$, $n_X = 0$, 1 vdW levels ($E = -15.72$ cm^{-1} relative to dissociation limit) are localized in the linear well, whereas $n_X = 2$ one

Table 6.5 Spectroscopic parameters of the RgI_2 complexes, Rg = He, Ne, Ar, Kr

HeI_2								
T-shaped					Linear (free-rotor)			
	D_e,$^{(1)}$ cm^{-1}	R_e, $^{(2)}$ Å	D_0,$^{(3)}$ cm^{-1}	R_0, $^{(4)}$ Å	D_e, cm^{-1}	R_e, Å	$D_0^{(3)}$, cm^{-1}	R_0, Å
X	38.92 [86]	3.84 [37]	15.51 [86] 16.6 [37]	4.36 [86]	44.28 [86]	4.83 [86] 4.89 [37]	15.72 [86] 16.3 37]	5.34 [86]
B	29.48 [87]	3.96 [87]	12.33 [87] 12.8 [37]	4.58 [87]			12.33 [87]	
E	33.2 [88]	4.1 [88]	16.85 [88] 16.7(6) [37] 14.0(1) [32] 13.9(1) [91, 92]				10.8 [89]	
NeI_2								
X	89 [93]	3.78 [93]	72.4-74.7 [94] 65.4 [33] 69.62 [93] 64 [95]		91 [93]	4.91 [93]	68 [93]	5.01 [93]
B			65.0-67.1 [94] 57.6 [33] 59(1) [95, 96]					
E			73 [95, 96]					
ArI_2								
X	235 [97]	3.96 [97]	212 [97] 240.5 [98]		268 [97]	5.05 [97]	237.8 [97]	
B		3.83 [44]	227 [99] 226.5 [98]				250.3 [98]	
E			see Fig. 6.27					
KrI_2								
X			320 [100]					
B			307 [100]					
E			663–683 [100]					

(1)—potential well depths
(2)—equilibrium distances corresponding to potential well depths
(3)—binding energies of the lowest vdW levels corresponding to the *T*-shaped or linear (free-rotor) complexes
(4)—equilibrium distances corresponding to $n = 0$ vdW levels

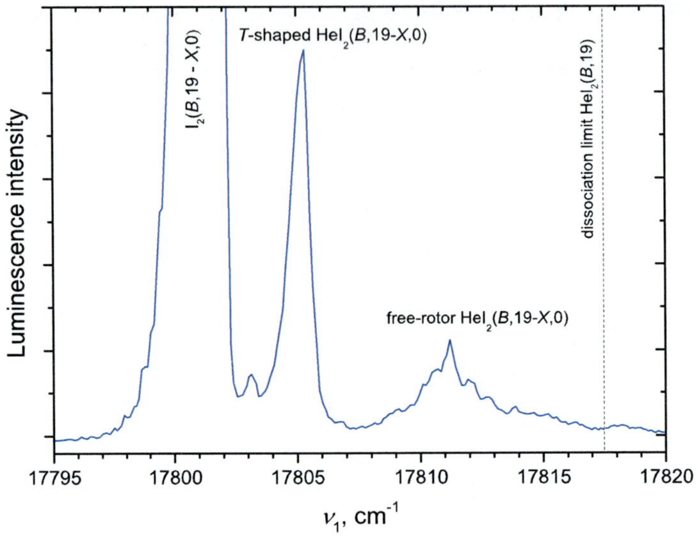

Fig. 6.8 Excitation spectrum of luminescence of the $I_2(B,19)$ and the $HeI_2(B,19)$ complex VP products (see [89]) (Baturo, V.V., Lukashov, S.S, Poretsky, S.A., Pravilov, A.M., Zhironkin A.I.: The HeI_2 van der Waals complexes in a 'free-rotor' configuration. J. Phys. B: At. Mol. Opt. Phys. **53**, 035101 (8 pp) (2020). https://doi.org/10.1088/1361-6455/ab582b. © IOP Publishing. Reproduced with permission. All rights reserved)

($E = -15.51$ cm^{-1}) is localized in the T-shaped well. The $n_X \geqslant 3$ levels ($E \geqslant -8.3$ cm^{-1}) have energies above the barrier between linear and T-shaped configurations and are spread over all θ values [86].

- the $B, v_B = 0, n_B = 0$ vdW level of the HeI_2 complex ($E = -12.33$ cm^{-1}) is localized in the T-shaped configuration, whereas $n_B = 1$–6, $E = -(8.36$–$2.34)$ cm^{-1}, wave functions are spread over all θ values [87]. An excitation spectrum of luminescence of $I_2(B,19)$ and products of the $HeI_2(B, 19)$ complexes VP is shown in Fig. 6.8, as an example.
- the binding energy of the E, $v_E = 0$, $n_E = 0$ vdW level is $D_0^{E,0} = 16.85\,\text{cm}^{-1}$, and this level is localized in the T-shaped configuration. The $n_E = 1$ level ($E = -10.68$ cm^{-1} relative to the dissociation limit) is localized around θ = 70° and 110°, and $n_E \geqslant 2$ levels ($E \geqslant -7.74$ cm^{-1}) show oscillatory character over all θ = 0–360° values [88] (see Fig. 6.9).

One sees in Fig. 6.9 that symmetry of the wave function of both, B and E, states to permutation of the nuclei are different for even and odd n: they are even for even n, and odd for odd n. Therefore the $HeI_2(B, v_B, n_B \leftarrow X, 0, n_X$ and $E, v_E, n_E \leftarrow B, v_B, n_B)$ transitions follow the $\Delta n = even$ selection rule.

T-shaped $HeI_2(E, v_E = 0$–$17, n_E)$ complexes. The spectroscopic characteristics and decay of the T-shaped $HeI_2(E, v_E = 0$–$17, n_E)$ complexes populated in two-step, two-color scheme

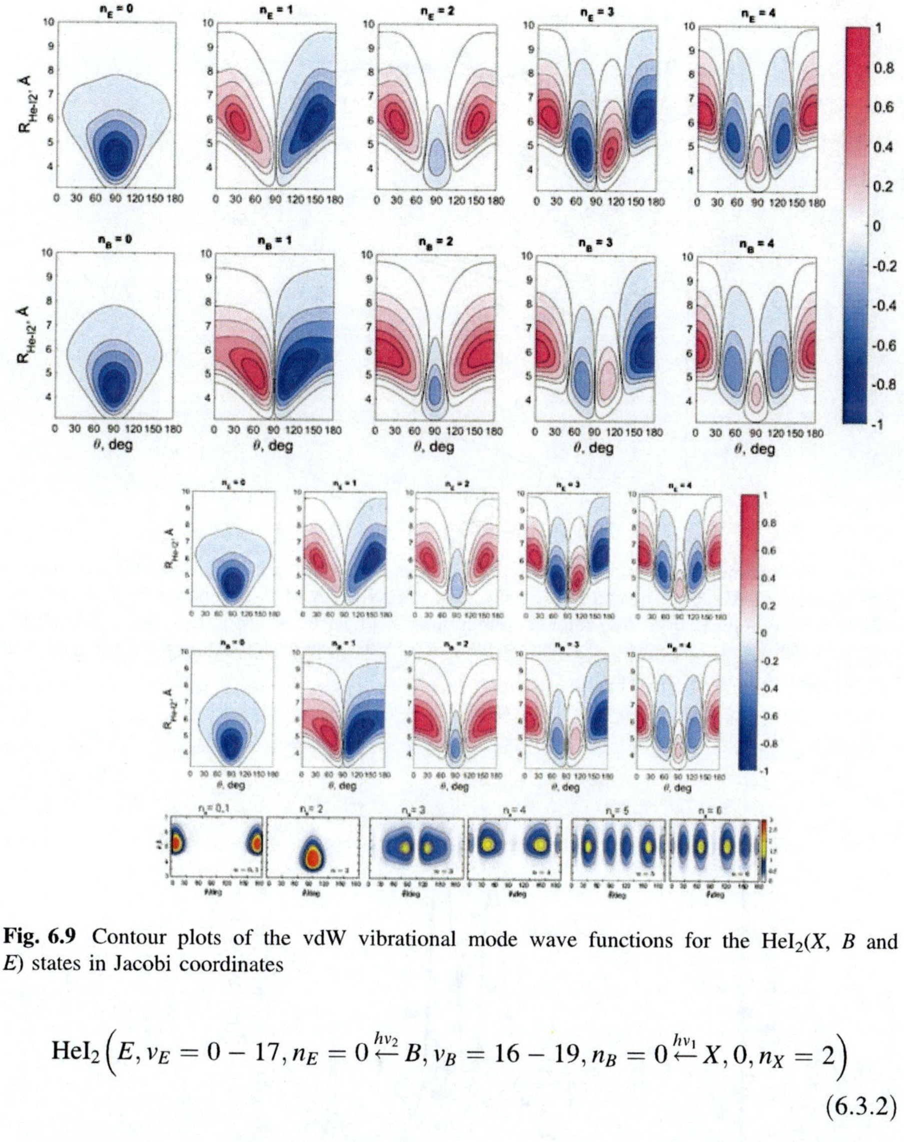

Fig. 6.9 Contour plots of the vdW vibrational mode wave functions for the HeI_2(X, B and E) states in Jacobi coordinates

$$HeI_2\left(E, v_E = 0 - 17, n_E = 0 \overset{h\nu_2}{\leftarrow} B, v_B = 16 - 19, n_B = 0 \overset{h\nu_1}{\leftarrow} X, 0, n_X = 2\right) \tag{6.3.2}$$

(see Fig. 6.10) have been studied in [92] (see Figs. 6.11, 6.12) as examples. The $h\nu_2$ laser radiation is produced by mixing of the TDL laser fundamental output and fundamental harmonic of the YG981C laser in KDP crystal, $1/\lambda_2 = 1/\lambda' + \nu_f^i$ (λ' and ν_f^i are generation wavelengths of the TDL laser and wavenumbers of fundamental harmonics of the YG981C laser, respectively). The fundamental YG981C laser harmonic consists of 4 spectral components: $\nu_f^1 = 9395.12 \pm 0.02\,\text{cm}^{-1}$, $\nu_f^2 = 9393.53 \pm 0.07\,\text{cm}^{-1}$ and $\nu_f^3 = 9396.67 \pm 0.10\,\text{cm}^{-1}$ with relative intensities of about 1 : 0.3 : 0.05, and a very weak line with the frequency of $\nu_f^4 = 9392.04\,\text{cm}^{-1}$. Therefore, the $h\nu_2$ laser radiation consists of 4 spectral components, also (see Sect. 6.2.1.2 in [18]).

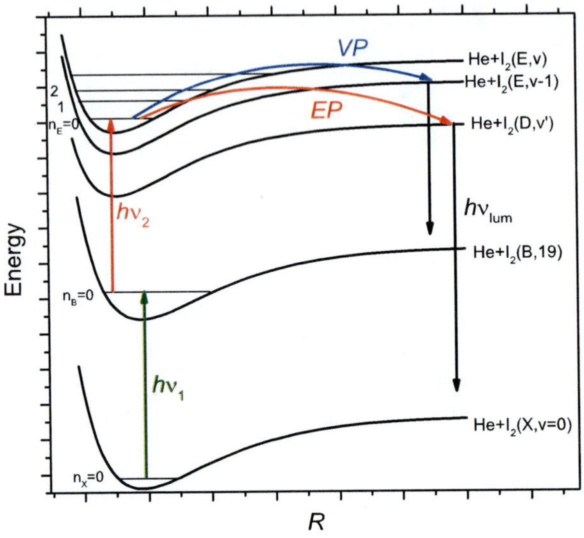

Fig. 6.10 Schematic diagram describing excitation as well as VP and EP of the $HeI_2(E)$ complex [89] (Baturo, V.V., Lukashov, S.S, Poretsky, S.A., Pravilov, A.M., Zhironkin A.I.: The HeI_2 van der Waals complexes in a 'free-rotor' configuration. J. Phys. B: At. Mol. Opt. Phys. **53**, 035101 (8 pp) (2020). https://doi.org/10.1088/1361-6455/ab582b © IOP Publishing. Reproduced with permission. All rights reserved)

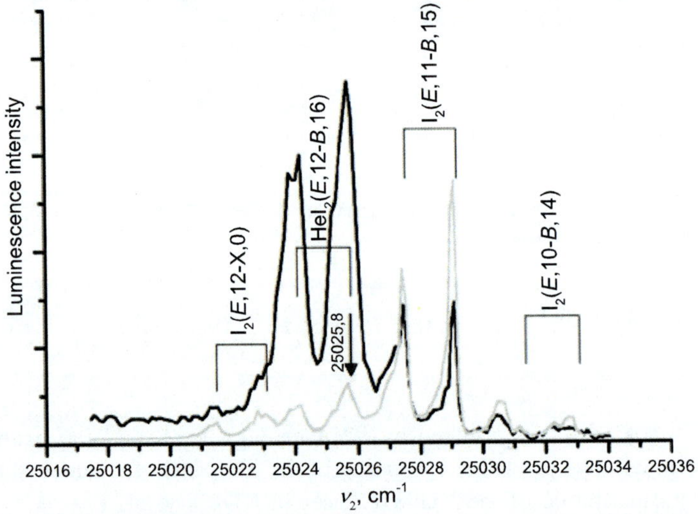

Fig. 6.11 Excitation spectra of luminescence in the $E \rightarrow B$, $\lambda_{lum} \approx 4284$ Å, (grey line) and UV spectral range, $\lambda_{lum} \approx 2600$–$3800$ Å (black line) measured at $\lambda_1 = 5707.94$ Å ($\nu_1 = 17519.45$ cm^{-1}, *T*-shaped $HeI_2\left(B, 16, n_B = 0 \overset{h\nu_1}{\leftarrow} X, 0, n_X = 2\right)$ transition)) in a vicinity of the $HeI_2\left(E, 12, n_E = 0 \overset{h\nu_2}{\leftarrow} B, 16, n_B = 0\right)$ transition. The transitions from the $HeI_2(B,16,n_B = 0)$ VP products, $I_2(B,v_B = 14,15)$, two-photon $I_2(E,12 \leftarrow X,0)$ and $HeI_2(E,12,n_E = 0 \leftarrow B,16,n_B = 0)$ transitions are shown. The $h\nu_2$ laser radiation, produced by mixing of the fundamental Nd: YAG laser harmonics and the TDL laser fundamental output in KDP crystal, consists of 2 spectral components (see Sect. 6.2.1.2 in [18] and [90–92] for details)

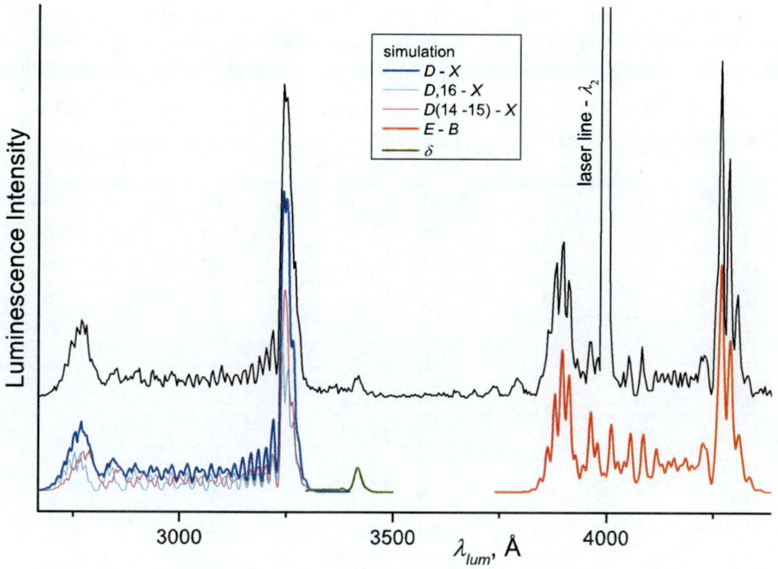

Fig. 6.12 Experimental (black line) and simulated luminescence spectra in the λ_{lum} = 2800– 4400 Å spectral range measured at the ν_2 = 25025.8 cm^{-1}, the $\mathrm{HeI_2}\left(E, 12, n_E = 0 \overset{h\nu_2}{\leftarrow} B, 16, n_B = 0\right)$ transition (see Fig. 6.11). Spectral resolution FWHM = 10 Å. The experimental spectrum is offset for clarity. The λ_2 laser line is marked (see Table 6.5)

The studies have shown the following:

1. No $\mathrm{HeI_2}(D, D' \leftarrow B)$ transitions in the excitation spectra have been observed though they are allowed at the C_{2v} symmetry group. It means that the propensity rules for the HeI_2(IP − B) transitions are the same as selection rules for the isolated I_2 molecule, and it may be deduced that the He atom behaves as a 'spectator' without significant perturbations of the 'pure' I_2(IP) states (see [62] and references). The same feature is observed in all the studied RgI_2(IP ← B) transitions in the T-shaped configuration.
2. The binding energy of the HeI_2(E, v_E= 0−17, n_E= 0) do not vary with v_E= 0 −17, and equal to $D_0^E = 13.9(1)\,\mathrm{cm}^{-1}$.
3. The

$$\mathrm{HeI_2}(E, v_E, n_E = 0) \rightarrow \mathrm{He} + \mathrm{I_2}(E, v_E - \Delta v_E), \tag{6.3.3}$$

$\Delta v_E = 1$, VP and

Table 6.6 Branching ratios, *br.r.*[(1)], of the $HeI_2(E, v_E, n_E)$ VP and EP product formation and maximal vibronic state (v_{IP}^{max}) of the products determined by simulation of the $I_2(E \rightarrow B$ and $D \rightarrow X)$ luminescence spectra measured at the $HeI_2(E, v_E, n_E = 0 \leftarrow B, v_B, n_B)$ excitation bands

E		D					D'	δ
v_E	*br.r.*	v_D^{max}	$\Delta E^{(2)}$, cm^{-1}	*br. r.*	*br.r.* (D, v_D^{max})	$D, v_D^{max}/D, v_D^{all}$		
0	0.15[(3)]	3	94	0.81	0.05	0.07	0.04	0
1	0.4	5	3.5	0.59	0.08	0.14	0.01	0
2	0.25	6	10.4	0.73	0.16	0.22	0.02	0
3	0.48	7	17.7	0.51	0.15	0.3	0.01	0
4	0.42	8	23.7	0.55	0.21	0.38		0.03
5	0.35	9	30.1	0.60	0.24	0.4		0.05
6	0.37	10	36.3	0.56	0.23	0.42		0.06
7	0.31	11	42.5	0.65	0.33	0.5		0.02
8	0.22	12	48.6	0.7	0.41	0.59		0.08
11	0.42	15	65.7	0.56	0.24	0.43		0.02
12	0.44	16	71.3	0.54	0.25	0.46		0.02
13	0.27	17	76.7	0.67	0.33	0.49		0.06
14	0.48	18	81.9	0.47	0.25	0.56		0.05
15	0.28	19	87.1	0.68	0.33	0.48		0.04
16	0.26	21	2.4	0.72	0.37	0.51		0.02
17	0.14	22	6.6	0.84	0.56	0.67		0.04

(1)—The probabilities of the decay channels (branching ratios, *br.r.*) are estimated as the ratios of integrated partial intensity of the selected IP state luminescence to the sum of all integrated intensities

(2)—Differences in the energy of the $HeI_2(E, v_E, n_E = 0)$ and nearest low I_2 (D, v_D^{max}) level

(3)—Luminescence of the $HeI_2(E, v_E = 0, n_E = 0)$ complex itself

$$HeI_2(E, v_E, n_E = 0) \rightarrow He + I_2\left(D0_u^+, v_D, D'2_g, v_{D'}, \delta 2_u, v_\delta\right) \tag{6.3.4}$$

EP occur. Only the $\Delta v_E = 1$ VP channel is observed at all, $v_E = 0–17$ levels. The $\Delta v_E = 0$ channels are energetically closed, and luminescence at the $\Delta v_E = 0$ for each $v_E = 0–17$ are the luminescence of the $HeI_2(E, v_E, n_E = 0)$ complexes themselves. The probabilities of the decay channels (branching ratios, *br.r.*) are estimated as the ratios of integrated partial intensity of the selected IP state luminescence to the sum of all integrated intensities. One sees in Table 6.6 that the probabilities of two last EP channels are low.

4. There are no dependences of $br.r._E$ and $br.r._D$ on v_E; $br.r._E = 0.37 \pm 0.08$, $b.r._D = 0.60 \pm 0.07$. The $v_E = 1$ and 16 levels in which the energy gap between $HeI_2(E, v_E, n_E)$ and the nearest lower level I_2 $(D, v_D^{max}, J_D = 0)$ is small could be the 'particular points'; however, these points are not too different from the surrounding.

5. The

$$HeI_2(E, v_E, n_E = 0) \rightarrow He + I_2\left(D, v_D^{max} \div \left(v_D^{max} - 3\right)\right), \quad (6.3.5)$$

EP are observed at $v_E > 1$. Four $I_2(D,\ v_D)$ vibronic states are populated, and the energy gap between them is $\sim$300 cm^{-1}.

6. The energy gaps between the initial state of the complex, $HeI_2(E,\ v_E= 1–17,\ n_E= 0)$, and the final state $I_2(D,\ v_D)$ are $\sim$(20–400) cm^{-1}; the depth of the $HeI_2(E,\ v_E,\ n_E= 0)$ is $\sim$14 cm^{-1} (see above).
7. In EP, there is a mixing of the wave functions corresponding to the bound part of the $HeI_2(E)$ *PES* and the repulsive part of the $HeI_2(D)$ *PES*. It is this circumstance that explains the absence of correlations of the channel (6.3.5) *br. r* with the energy gaps. The predissociation rate is

$$k_{pr} = 1/\tau_{pr} = \hbar V_{ij}^2 s^{-1}, \quad (6.3.6)$$

where

$$V_{ij} = \Psi_i^0|\hat{V}|\Psi_j^0 \approx \Phi_i^0\chi_i^0|\hat{V}_e + \hat{V}_Q|\Phi_j^0\chi_j^0 \approx A_{el}\chi_i^0|\chi_j^0, \quad (6.3.7)$$

A_{el} is the matrix element of the electron interaction of the states i, j, $\langle\chi_i^0|\chi_i^0\rangle$ is the overlap integral of the vibrational wave functions of these states, equal to the square root of the Frank-Condon factor for these states (see Sect. 3.6). Since the EP rate (non-adiabatic process) is almost 2 times higher than that of VP (adiabatic process), one has to admit that both A_{el} and $\chi_i^0|\chi_j^0$ are large. It is possible that the A_{el} value and the VP rate change weakly and monotonously with v_E. There is a non-adiabatic transition from the bound $HeI_2(E)$ *PES* to the repulsive part of the $HeI_2(D)$ *PES*, and then the image point slides down to the pelvis of the $He + I_2\left(D, v_D = v_D^{max} - 0\right)$ dissociation channel, and the $I_2\left(D, v_D = v_D^{max} - v_D^{min}\right)$ vibronic states are populated.

Free-rotor $HeI_2(E,\ v_E= 0–6,\ n_E)$ complexes. The spectroscopic characteristics and decay of the free-rotor $HeI_2(E,\ v_E= 0–6,\ n_E)$ complexes populated in two-step, two-color scheme

$$HeI_2\left(E, v_E = 0-6, n_E = 0 \overset{h\nu_2}{\leftarrow} B, v_B = 19, n_B \overset{h\nu_1}{\leftarrow} X, 0, n_X = 0, 1\right) \quad (6.3.8)$$

have been studied in [89].

According to the literature data [37, 101], the binding energy of the $HeI_2(X,0,\ n_X= 0,1)$ complex is $D_0^X \approx 16\,\text{cm}^{-1}$. The $HeI_2(E)$ complex term energy (upper x-axis in Fig. 6.13) relative to that of the $I_2(X,\ v_X= 0,\ J_X= 0)$ is $\nu_1 + \nu_2 - D_0^X = 17795 + \nu_2$. The binding energies of the $HeI_2(E,\ v_E= 0–6, n_E)$ complexes can be determined as the energy gaps, ΔE, between (17795 + ν_2) of the transitions corresponding to the $\nu_f^1 = 9395.12\,\text{cm}^{-1}$ component (see caption to

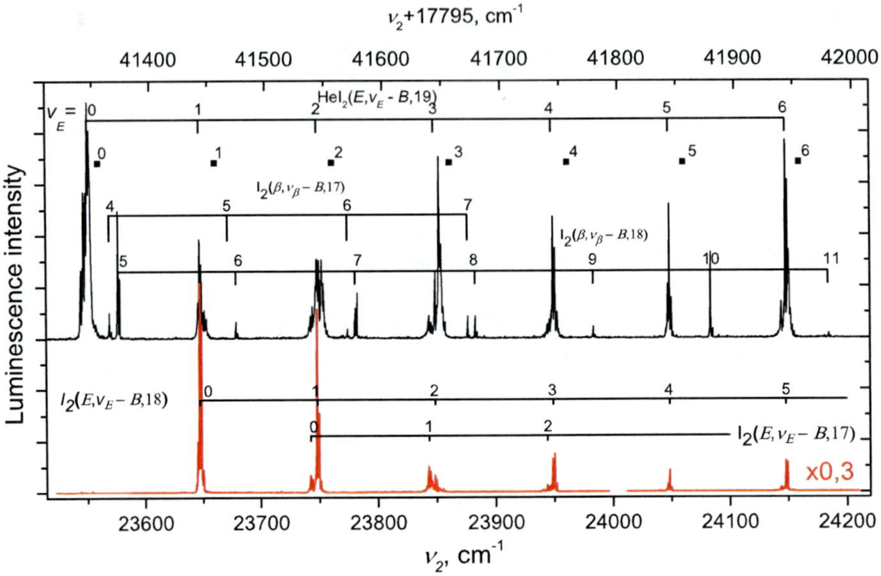

Fig. 6.13 Survey excitation spectra of luminescence in the $E \rightarrow B$, $\lambda_{lum} \approx 4284$ Å, (red (grey) line) and UV spectral range, $\lambda_{lum} \approx 2600$–$3800$ Å (black line) measured at $\lambda_1 = 5614.47$ Å ($\nu_1 = 17811.1$ cm^{-1}, free-rotor HeI$_2$(B,19,$n_B \approx$ 3 ← X,0,n_X= 0,1) transition, see Fig. 6.8). Energy of transitions to the I$_2$(E,v_E= 0–6,J_E= 0) states corresponding to the most intense spectral ν_f^1 = 9395.12 cm^{-1} component (dissociation limits of the HeI$_2$(E,v_E= 0-6,n_E) complexes) are marked out as black squares. The bands corresponding to the I$_2$(E,v_E= 0-5 ← B,v_B= 17,18 and $\beta 1_g$, v_β= 5-7 ← B,v_B= 17,18) transitions are shown. Groups of excitation bands of the I$_2$(IP → valence states) luminescence corresponding to the HeI$_2$(E,v_E = 0-6,n_E) states and located lower than the corresponding dissociation limits of the HeI$_2$(E,v_E= 0-6,n_E) complexes are marked out as v_E = 0-6 (see [89]) (Baturo, V.V., Lukashov, S.S, Poretsky, S.A., Pravilov, A.M., Zhironkin A.I.: The HeI$_2$ van der Waals complexes in a 'free-rotor' configuration. J. Phys. B: At. Mol. Opt. Phys. **53**, 035101 (8 pp) (2020) https://doi.org/10.1088/1361-6455/ab582b. © IOP Publishing. Reproduced with permission. All rights reserved)

Fig. 6.13) and energy of transitions to the I$_2$(E, v_E= 0–6, J_E= 0) states. Excitation spectra of the UV luminescence measured in the ν_1 = 17819.0–17812.6 cm^{-1} spectral range are given in Fig. 6.14.

Spectrum (1) measured at ν_1 = 17819.0 cm^{-1} corresponds to a *bound-free* B f← X transition in the complex with subsequent dissociation to the I$_2$(B, 19) + He. Another (2)–(5) spectra are broad and correspond to the *bound-bound* $B \leftarrow X$ transition in the complex (see Fig. 6.8) and [37]). The bands in the ν_2 = 23549–23553 cm^{-1} range become more and more distinct with ν_1 decreasing and in the ν_1 < 17814.2 cm^{-1} spectral range positions of the band maxima independent of ν_1 value [89].

The I$_2$($D \rightarrow X$, $E \rightarrow B$) and very weak luminescence, $\lambda_{lum}^{max} \approx 342$ nm, *br. r.* ≈ 0.03 and 0.02 at v_E = 0 and 6 bands, respectively, occur. The HeI$_2$(E,v_E,

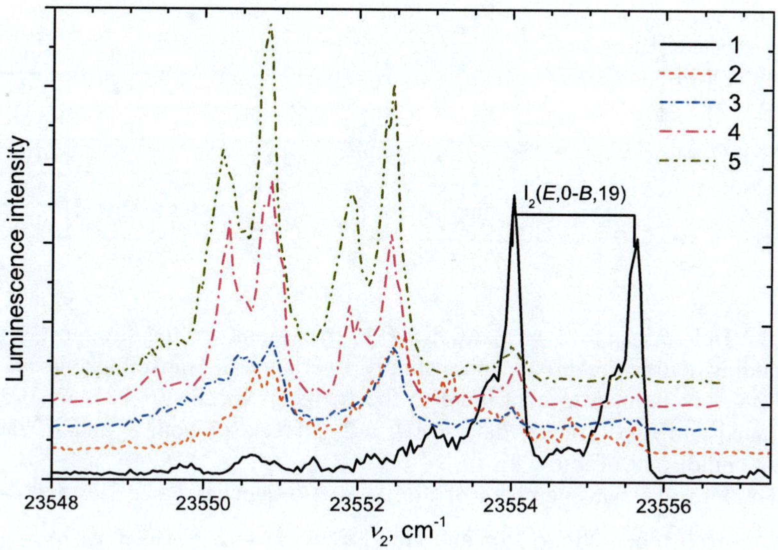

Fig. 6.14 Excitation spectra of the UV luminescence in the ν_2 = 23548 −23560 cm^{-1} (the $v_E = 0$ group) corresponding to the UV spectral range, $\lambda_{\text{lum}} \approx$ 2600–3800 Å measured at ν_1 = 17819.0 (1), 17817.4 (2), 17815.8 (3), 17814.2 (4), 17812.6 cm^{-1} (5) (see [89]) (Baturo, V.V., Lukashov, S.S, Poretsky, S.A., Pravilov, A.M., Zhironkin A.I.: The HeI_2 van der Waals complexes in a 'free-rotor' configuration. J. Phys. B: At. Mol. Opt. Phys. **53**, 035101 (8 pp) (2020) https://doi.org/10.1088/1361-6455/ab582b. © IOP Publishing. Reproduced with permission. All rights reserved)

n_E) → He + $I_2(E,v_E\text{-}\Delta v_E)$, Δv_E= 0, VP channels are energetically closed, as it occurs for the *T*-shaped $HeI_2(E,v_E,n_E)$ complexes (see above). A weak luminescence, *br.r.* ≈ 0.1 in the λ_{lum} = 4000–4400 Å spectral range at the $v_E = 0$ band belongs to the $HeI_2(E,v_E= 0,n_E)$ complex itself. The VP probabilities (if they are energetically allowed) are ~2–3 times higher than those of EP (Table 6.7) as opposed to what is observed in *T*-shaped complexes (see Table 6.6).

As it has been mentioned above (see Fig. 6.9), two degenerated $HeI_2(X,0, n_X= 0,1)$ states are localized at the linear geometry. As to the *B* state, $n_B > 0$ states are delocalized. The n_B states of different parity can be populated from the $n_X = 0$, 1 states in the Δn = *even* transitions. According to experimental data obtained in [37,

Table 6.7 Branching ratios, *br.r.*, of the $HeI_2(E, v_E, n_E)$ VP and EP product formation and maximal vibronic state determined by simulation of the $I_2(E \rightarrow B$ and $D \rightarrow X)$ luminescence spectra measured at the $HeI_2(E, v_E, n_E = 0 \leftarrow B, 19, n_B)$ bands (v_{IP}^{max}) [89]

v_E	$v_D^{max}/br.r.$	$v_E^{max}/br.r.$
0	4/0.97	–
2	6/0.33	1/0.67
3	7/0.27	2/0.73
6	10/0.21	5/0.77

Table 6.8 Experimental and calculated binding energies of the $HeI_2(E, 0, n_E$ and $B, v_B, n_B)$ vdW levels

n	B state		E state		
	[37]	[89]	[89] (exp.)	[38]	[89] (calc.)
0	12.8	12.3	13.9 [92]	16.8	13.9
1	–	8.4		14.1	10.9
2	7.9	7.6	10.8		10.7
3	6.8	6.8	9.8		9.3
4	5.7	5.6	8.3		7.7

89], the $HeI_2(B,v_B,n_B$= 2-4 ← $X,0,n_X$= 0,1) transitions are the strongest, and corresponding bands are broad and strongly overlapped. Transitions via all these, n_B = 2-4, levels have to contribute to the two-step spectra (6.3.8). According to calculated and experimental data, the n = 2–4 levels of both B and E states are almost equidistant (Table 6.8).

Therefore, the wavenumbers of the Δn = 0 sequences (2-2, 3-3, and 4-4) are close to each other (Fig. 6.14), and $HeI_2\left(E, v_E, n_E = 2-4 \overset{hv_2}{\leftarrow} B, v_B, n_B = 2-4\right)$ transitions are realized as the $\Delta n_{E-B} = |n_E - n_B|) = 0$ sequences.

The NeI_2 complexes. According to most recent theoretical data [93], the *PES* of the $NeI_2(X)$ complex state has two minima corresponding to *T*-shaped and linear isomers (see Table 6.5). The stretching n_X= 0, 3, 6 vdW levels (energies relative to the dissociation limit are −69, −58 and −51 cm^{-1}) of the *T*-shaped and the n_X = 1, 2, 4, 5 ones (energies relative the dissociation limit are −68.56 and −55.15 cm^{-1}) of the linear isomers are stable due to a high (the energy relative the dissociation limit is ~49 cm^{-1}) isomerization barrier between them. The n_X = 0–2 states are near-degenerated. The *T*-shaped, linear and delocalized $NeI_2(B, v_B$= 32–38) isomers can be populated in the $NeI_2(B, v_B \leftarrow X, 0)$ transitions from these near-degenerate n_X = 0–2 states [33].

The blue shift between the $NeI_2(B, 19, n_B \leftarrow X, 0, n_X$= 0) and $I_2(B, 19 \leftarrow X, 0)$ bands is 6.4 cm^{-1} (Fig. 6.15), so the binding energy for the lowest $NeI_2(B, 19)$ levels is $D_0^B = 65.4 - 6.4 = 59\,cm^{-1}$. It can be considered that the vdW bending mode is not excited in the *T*-shaped $NeI_2(B, 19, n_B \leftarrow X, 0, n_X$= 0) transition at the band assigned as $NeI_2(T)$ in Fig. 6.15 (v_1 = 17807.9 cm^{-1}), i.e., the $NeI_2(B, 19, n_B$= 0) state utilized as an intermediate one is populated at this band.

The binding energy of the $NeI_2(X, 0, n_X$= 0, 1) complex is $D_0^X = 65.4\,cm^{-1}$ [33]. The $NeI_2(E)$ complex term energy (upper *x*-axis in Fig. 6.16) relative to that of the $I_2(X,v_X$= 0,J_X= 0) is $v_1 + v_2 - D_0^X = 17742.5 + v_2$. The binding energies of the $NeI_2(E,v_E$= 0–6,n_E) complexes can be determined as the energy gaps, ΔE, between (17742.5 + v_2) of the transitions corresponding to the $v_f^1 = 9395.12\,cm^{-1}$ component and energy of transitions to the $I_2(E,v_E$= 0–6,J_E= 0) states.

Seven groups of the excitation bands located lower than the dissociation limits of the $NeI_2(E,v_E$= 0–6) complexes (see Figs. 6.16 and 6.17) are called v_E = 0–6 groups.

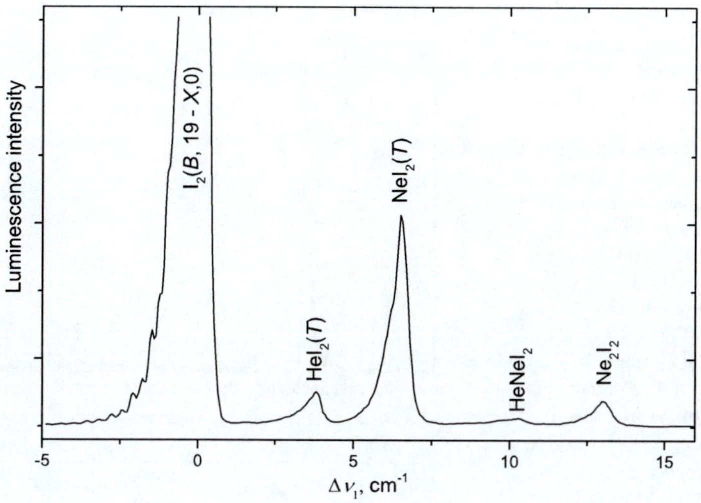

Fig. 6.15 Excitation spectrum of luminescence of the $I_2(B,19)$ and products of VP of the $HeI_2(B,19)$, $NeI_2(B,19)$ complexes, and $HeNeI_2(B)$, $Ne_2I_2(B,19)$ clusters plotted relative to the $I_2(B,19 \leftarrow X,0)$ origin (see [96]) (Baturo, V.V., Lukashov, S.S., Poretsky, S.A., Pravilov, A.M.: The *T*-shaped NeI_2 ($E0_g^+$) van der Waals complex. J. Phys. B: At. Mol. Opt. Phys. **52**, 145101 (9 pp) (2019). https://doi.org/10.1088/1361-6455/ab582b. © IOP Publishing. Reproduced with permission. All rights reserved)

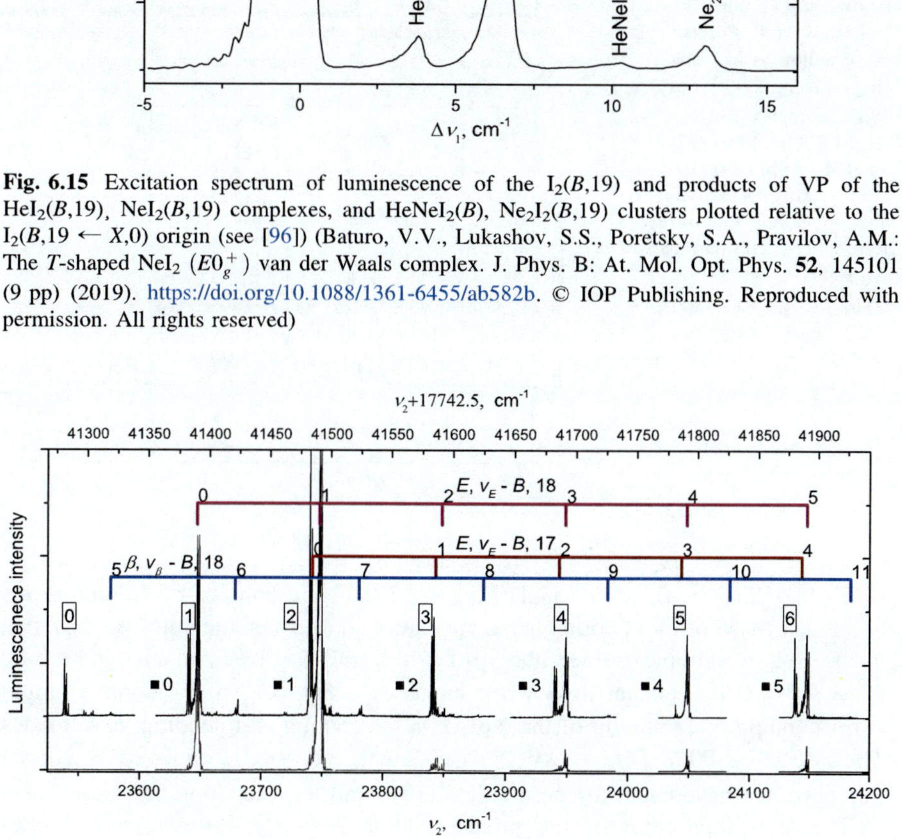

Fig. 6.16 Excitation spectra of luminescence in the $E \rightarrow B$, $\lambda_{lum} \approx 4284$ Å, (lower line) and UV spectral range, $\lambda_{lum} \approx 2600$–$3800$ Å (upper line) measured at $\nu_1 = 17807.9\ cm^{-1}$ (*T*-shaped $NeI_2(B,19,n_B{=}\ 0 \leftarrow X,0,n_X{=}\ 0)$ transition). Dissociation limits of the $NeI_2(E,\nu_E{=}\ 0\text{-}5,n_E)$ complexes are marked out by filled squares. Assignments of the $I_2(E,\nu_E \leftarrow B,\nu_B{=}\ 17,18$ and β, $\nu_\beta \leftarrow B,\nu_B{=}\ 17)$ transitions are shown. Groups of excitation bands of the I_2(IP $\rightarrow$ valence states) luminescence corresponding to the $NeI_2(E,\nu_E{=}\ 0\text{-}6,n_E)$ states are marked out as 0-6 in frames (see Fig. 6.18 in [18])

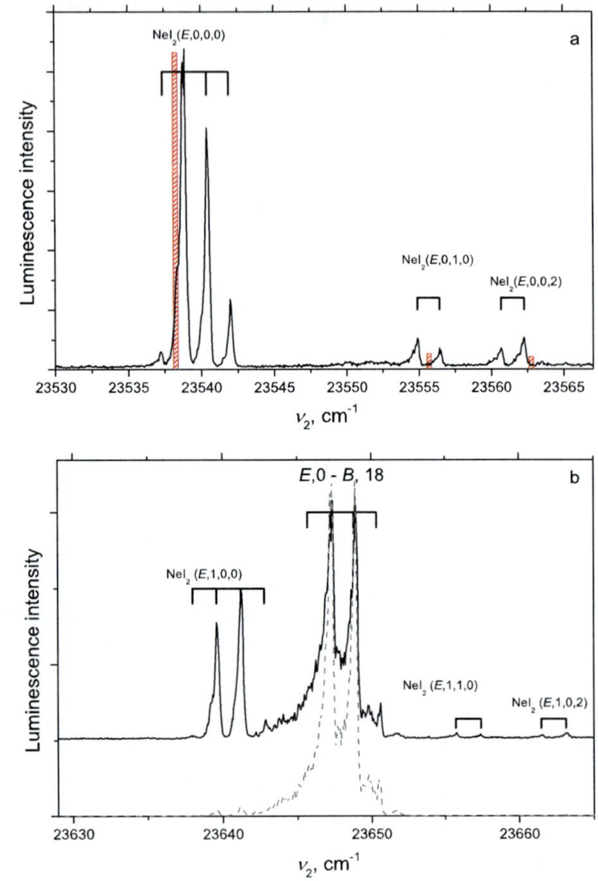

Fig. 6.17 Excitation spectra of luminescence in the in the ν_2 = 23530-23567 cm^{-1} (the $\nu_E = 0$ group) (**a**) and 23529–23665 cm^{-1} (the $\nu_E = 1$ group) (**b**) corresponding to the $\lambda_{lum} \approx 4284$ Å, (the $E \to B$ transition, lower line) and UV spectral range, $\lambda_{lum} \approx 2600$–3800 Å (upper line) at ν_1 = 17807.9 cm^{-1}. Simulated spectrum is shown by bars in Fig. 6.17 a (see below) [96] (Baturo, V.V., Lukashov, S.S., Poretsky, S. A., Pravilov, A.M.: The *T*-shaped NeI_2 ($E0_g^+$) van der Waals complex. J. Phys. B: At. Mol. Opt. Phys. **52**, 145101 (9 pp) (2019). https://doi.org/10.1088/1361-6455/ab2496.

Each $I_2(E,\nu_E \leftarrow B,\nu_B)$ and $NeI_2(E,\nu_E \leftarrow B,19)$ band consists of 4 components, and at least two of these components are saturated (the same feature was observed in the $HeI_2(E)$ complexes, see above). Each ν_E = 0–6 group contains one strong, $E,\nu_E,n_E^{str}=0,n_E^b=0$ and two weak bands (see Fig. 6.17 a, b, as an example) corresponding to excitation of the $NeI_2(E,\nu_E)$ stretching and bending vdW modes. The $I_2(E0_g^+ \to B0_u^+$, $D0_u^+ \to X0_g^+$, $\beta 1_g \to A1_u$, $\gamma 1_u \to a1_g$ and $\delta 2_u \to 2_g(aa))$ luminescence have been observed (Fig. 6.18) at all the excitation bands.

The branching ratios of the predissociation channels observed, and highest vibrational levels the predissociation products are given in Table 6.9.

Energies of the $I_2(\mathrm{IP},\nu_{\mathrm{IP}},J_{\mathrm{IP}} = 5)$ rovibronic levels of the first-tier iodine molecule IP states, IP = β, D, E, γ, δ and calculated relative to that of $I_2(X,\nu_X= 0,J_X= 4)$ as well as energies of the $NeI_2(E,\nu_E,n_E= 0)$ states are given in Fig. 6.19.

One sees in the figure what $I_2(\mathrm{IP},\nu_{\mathrm{IP}})$ vibronic states can be populated in the $NeI_2(E,\nu_E, n_E^{str}, n_E^b)$ VP and EP.

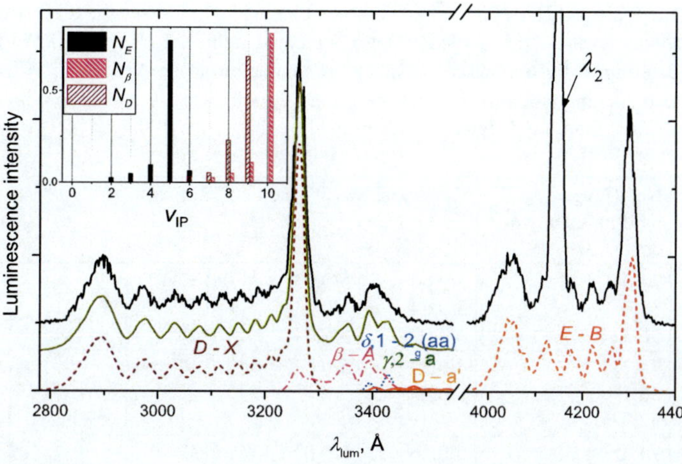

Fig. 6.18 Experimental (solid line) and simulated (gray line) luminescence spectra in the λ_{lum} = 2900–3550 Å and 4000–4400 Å spectral ranges at the $NeI_2(E,6,0,0)$ band, ν_2 = 24138.08 cm^{-1}. Spectral resolution FWHM = 20 Å. Experimental spectrum is offset for clarity. Partial simulated luminescence spectra are also shown. Populations of the E,ν_E, D,ν_D and β, ν_β vibronic states are given in the inset. The λ_2 laser line is marked [96] (Baturo, V.V., Lukashov, S.S., Poretsky, S.A., Pravilov, A.M.: The *T*-shaped NeI_2 ($E0_g^+$) van der Waals complex. J. Phys. B: At. Mol. Opt. Phys. **52**, 145101 (9 pp) (2019). https://doi.org/10.1088/1361-6455/ab2496. © IOP Publishing. Reproduced with permission. All rights reserved)

To assign observed transitions in excitation spectra to specific vdW bending and stretching modes, the *PES*s of both, $NeI_2(B, E)$, states have been constructed, and the energies of the vdW states have been calculated in [97]. The pairwise potential energy curves within the intermolecular DIM PT1 model (see details in [62, 96]) have been utilized.

The DIM model predicts one minimum of the *T*-shaped configurations for both *E*, *B* states. The *PES*s of the *B* and *E* states are similar and have almost the same equilibrium intermolecular distances, $R_e^B = 3.69\,Å$, $R_e^E = 3.7Å$, that explains a small number of bands in all the ν_E groups observed in the experiment. The model predicts no linear isomer for both states.

The calculated *FCF*s for the ν_E= 0 group and the energies of the $NeI_2(E,\nu_E= 0, n_E^{str}, n_E^b \overset{h\nu_2}{\leftarrow} B,19, n_B= 0)$ transitions are in excellent agreement with experimental data (Fig. 6.17a).

It has been found that three states ($n_E^{str} = 0$, $n_E^b = 0$), (1,0) and (0,2) are populated from the lowest ($n_E^{str} = 0$, $n_E^b = 0$) vdW level of the *B* state. The assignment is given in Fig. 6.17a. The species of the *T*-shaped $NeI_2(B, E)$ states (C_{2v} symmetry point group) are B_2 and A_1, respectively, and those of the stretching and bending modes are A_1 and B_1. The species of the transition moments are A_1 (M_z), B_1 (M_x), and B_2 (M_z). A transition is electric dipole allowed if $\mu = B|\hat{\boldsymbol{\mu}}|E \neq 0$, that is valid if the product $B|\hat{\boldsymbol{\mu}}|E$ is totally symmetric for at least one orientation of $\hat{\boldsymbol{\mu}}$, i.e., the

Table 6.9 The *T*-shaped $NeI_2(E,v_E= 0–6)$ complex binding energies (see the text for details). The highest vibrational levels (v_{IP}^{max}) of a I_2(IP) states populated at the $NeI_2(E,v_E, n_E^{str}, n_E^b)$ VP and EP as it has been determined by luminescence spectrum simulation at the $NeI_2(E,v_E, n_E^{str}, n_E^b \leftarrow B,19)$ bands and the branching ratios (*br.r.*)[(1)] of the $NeI_2(E,v_E, n_E^{str}, n_E^b)$ VP and EP product formation. Uncertainties of the binding energies are determined by that of the *T*-shaped $NeI_2(B,v_B)$ complex, 1 cm^{-1} [33] (see [96])

v_E, n_E^{str}, $n_E^{b\,(2)}$	$NeI_2(E,v_E,n_E)$ binding energies, cm^{-1}	$v_D^{max}/br.r.$	$v_\beta^{max}/br.r.$	$v_E^{max}/br.r.$	$v_\gamma^{max}/br.r.$	$v_\delta^{max}/br.r.$
0,0,0 0,1,0 0,0,2	74.0 58.0 52.3	3/0.60 3/0.64 3/0.68	5/0.03 5/0.03 5/0.04	0/0.37[(3)] 0/0.33[(3)] 0/0.28[(3)]	–	–
1,0,0 1,1,0 1,0,2	74.3 58.1 52.3	4/0.34	5/0.02	1/0.64	–	–
2,0,0 2,1,0 2,0,2	74.4 58.2 52.3	5/0.46	6/0.07	2/0.47	–	–
3,0,0 3,1,0 3,0,2	74.5 58.3 51.5	6/0.35	7/0.05	3/0.51	0/0.09	–
4,0,0	74.6	7/0.40	8/0.07	4/0.46	1/0.07	–
5,0,0	74.7	7/0.38	9/0.07	5/0.48	1/0.05	0/~ 0.02
6,0,0	74.8	9/0.51	10/0.09	6/0.36	2/0.02	1/~ 0.02

(1)—*br.r.* is a ratio of integrated partial intensity from selected IP state to the sum of all integrated intensities
(2—Approximate assignments based on the nodal patterns of the wavefunctions (see below)
(3)—The spectra observed can be simulated by the $I_2(E,\ v_E= 0 \rightarrow B)$ transitions. They belong to the $NeI_2(E,\ v_E= 0,\ n_E^{str},\ n_E^b)$ complexes themselves

direct product of the species, corresponding to the *B*, *E* state vdW vibrational modes and transition moments has a totally symmetric component, A_1 (see Sect. 4.2.1). It is valid for the transitions to the NeI_2(*E*,0,0,0; *E*,0,1,0, and *E*,0,0,2) states shown in Fig. 6.17a, while those to the NeI_2(*E*,0,0,1 and *E*,0,1,1) states are symmetry forbidden ($v_2 \approx 23551.2$ and 23565.2 cm^{-1}, respectively).

These results contradict to those obtained later [102]

The wavenumbers of the bands corresponding to the stretching and bending modes in Fig. 6.20 are described as following:

$$v_2^{str} = 23632.2 + 24 \cdot \left(v_E^{str} + 1/2\right) - 1.6 \cdot \left(v_E^{str} + 1/2\right)^2 (cm^{-1}) \tag{6.3.9}$$

$$v_2^b = 23632.2 + 9.8 \cdot \left(v_E^b + 1/2\right) - 0.9 \cdot \left(v_E^b + 1/2\right)^2 (cm^{-1}) \tag{6.3.10}$$

(see Fig. 6.21).

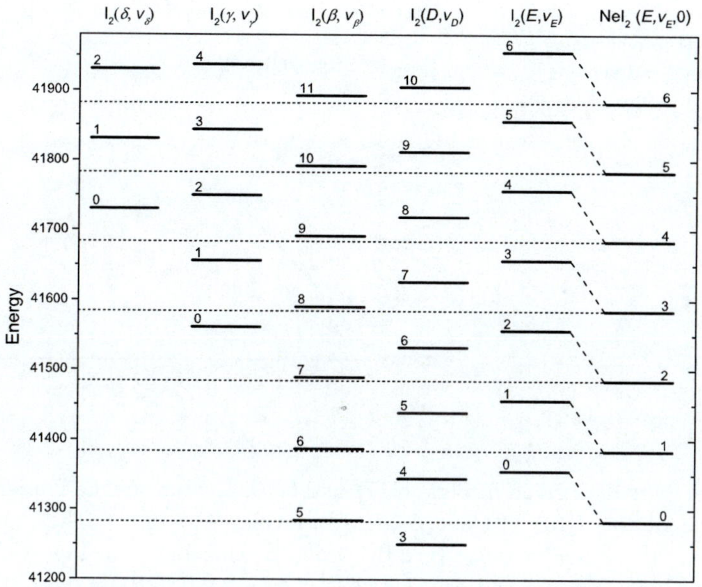

Fig. 6.19 Energies of the $I_2(IP,v_{IP},J_{IP} = 5)$ rovibronic levels of the first-tier iodine molecule IP states, IP = β, D, E, γ, δ calculated relative to that of $I_2(X,v_X= 0,J_X= 4)$ and $NeI_2(E,v_E= 0–6,n_E= 0)$ energies (the last column). Vibrational quantum numbers of the states are indicated. It is assumed that the $NeI_2(X,0,0)$ binding energy is $D_0^X = 65.4\,\text{cm}^{-1}$ (see Fig. 6.20 in [18])

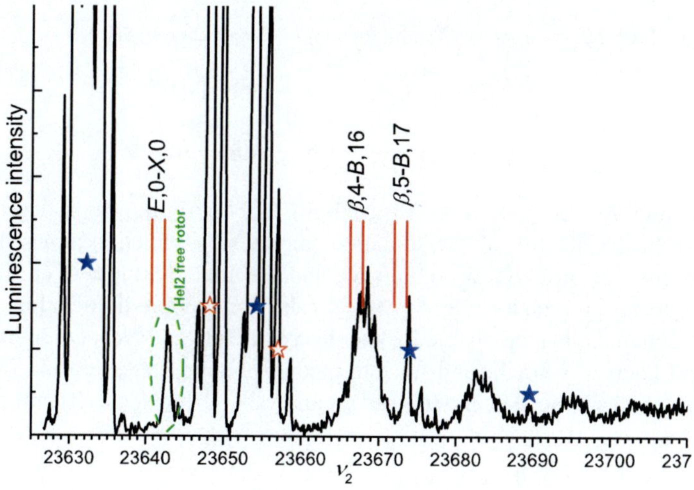

Fig. 6.20 Excitation spectra of the $I_2(D,v_D \rightarrow X)$ luminescence measured at v_1 = 17714.56 cm^{-1} ($NeI_2(B,18,n_{str}= 0,n_b= 0 \leftarrow X,0,n_{str}= 0,n_b= 0)$ transition). The bands corresponding to the n_{str} = 0-3 and n_b = 0, 2, 4 vdW modes are shown as full blue stars and open red stars, respectively. Transitions in the free iodine molecules and HeI_2 free-rotor complexes are shown [102]

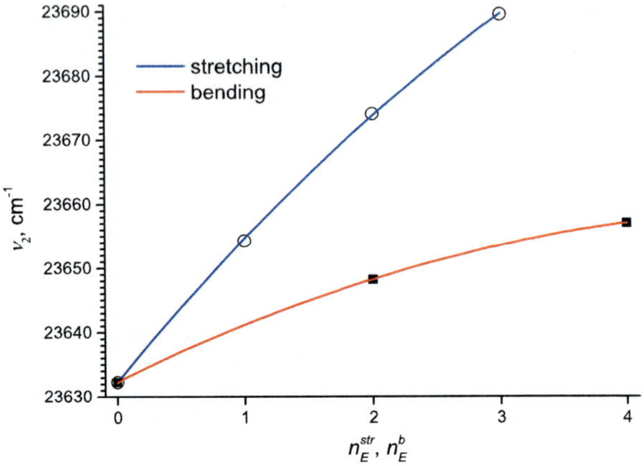

Fig. 6.21 The wavenumbers of the bands corresponding to the stretching and bending modes as functions of the n_{str} and n_b values

To compare data given in Figs. 6.17a and 6.20, one has to take into account the difference of the $I_2(B,\ 19)$ and $I_2(B,\ 18)$ levels, 93.3 cm^{-1}. The experimental energies of the E,0,n_{str}= 0,n_b= 0, E,0,1,0 and E,0,0,2 bands in Fig. 6.17 a correspond to 23631, 23648.2 and 23653.9 cm^{-1}, i.e., E,0,0,0, E,0,0,2 and E,0,1,0 bands in Fig. 6.20. The band assigned as E,0,1,0 in Fig. 6.17 a is assigned as E,0,0,2 in Fig. 6.20 and the band assigned as E,0,0,2 in Fig. 6.17 a is assigned as E,0,1,0 in Fig. 6.20. One sees that calculated and measured data have to be checked.

Vibrational and electronic predissociations of the NeI$_2$(E) complexes. The following $NeI_2(E,v_E,n_E)$ vdW complex decay channels are energetically allowed:

$$NeI_2(E, v_E, n_E) \rightarrow Ne + I_2(E, v_E - \Delta v_E), \Delta v_E \neq 0 \tag{6.3.11}$$

$$NeI_2(E, v_E, n_E) \rightarrow Ne + \text{other } I_2(\text{IP}) \tag{6.3.12}$$

$$NeI_2(E, v_E, n_E \rightarrow B, v_B, n_B) + hv \tag{6.3.13}$$

It is commonly believed that VP of light $RgI_2(B)$ complex is described adequately in the framework of the so-called 'direct VP', which assumes direct coupling between the initial quasi-bound state and the final continua. According to [95, 103], the strong propensity rule $\Delta v = 1$ is valid for this predissociation.

This mechanism is responsible for the decay of HeI_2 and NeI_2 complexes, which vdW bond energy is small, and no intermediate bound states are involved to the dynamical process (the IVR model, see an analysis of the $NeCl_2(B)$ complex decay above).

The

$$NeI_2(E, v_E) \rightarrow I_2(E, v_E - 1) + Ne \tag{6.3.14}$$

VP channels are energetically open for the $NeI_2(E,v_E \geq 1)$ states (see Figs. 6.16, 6.19), and relative probabilities of both VP and EP become comparable in these cases (Table 6.9). The $\Delta v_E = 1$ propensity rule is valid in these cases, and the luminescence at $\Delta v_E = 0$ VP is ascribed to the $NeI_2(E,v_E,n_E)$ complexes themselves (6.3.13) (see below).

The $RgI_2(E)$ state couples with other *bound* RgI_2(IP) states, and EP mechanism has to be sufficiently different from that of the $RgI_2(B)$ state.

Two possible mechanisms of EP may occur after $RgI_2(E,v_E,n_E)$ population: direct coupling of the initially excited bound $RgI_2(E,v_E,n_E)$ with a quasi-continuum Rg + $I_2(\mathrm{IP},v_{\mathrm{IP}})$ ('direct EP') or a sequential non-adiabatic $RgI_2(E \to \mathrm{IP})$ transition with the following VP of the RgI_2(IP) including IVR ('EP + IVR').

The principal channel for the $HeI_2(E)$ and $NeI_2(E)$ EP is $RgI_2(E) \to$ Rg + $I_2(D, v_D)$. Taking into account only pairwise *u–g* interactions between IP states in DIM PT1 model, the $RgI_2(E \to D)$ EP channel is forbidden in the *T*-shaped configuration [105]. However, a contribution of this channel among all EP channels is > 70% in the experiments (see Table 6.9). Non-adiabatic coupling between all the IP states is allowed in the bent configuration (C_s symmetry group). Averaging of the interaction matrix elements over bending θ angle gives non-zero $NeI_2(E \sim D)$ coupling. Therefore, one can believe that the DIM PT1 model is capable of describing NeI_2 *PES*s and energies of vdW levels, but unable to describe nonadiabatic processes.

The experimental vibrational distributions of the $I_2(D,v_D)$ EP products are described in the energy-gap law [97]

$$P(v_D) \sim e^{-\alpha \Delta E}, \tag{6.3.15}$$

where $P(v_D)$ is the probability of a population of the D,v_D vibronic state, α is a fitting parameter, and ΔE is an energy gap between the initial $NeI_2(E,v_E,n_E)$ and final $I_2(D,v_D)$ levels (see Fig. 6.18).

Luminescence of the $NeI_2(E)$ complexes. Luminescence similar to that of $I_2(E,v_E)$ corresponding to $\Delta v_E = 0$ occurs at the v_E = 0-2 groups (see Fig. 6.22, as an example), though the $NeI_2(E,v_E,n_E) \to$ Ne + $I_2(E,v_E)$ ($\Delta v_E = 0$) VP channels are energetically closed. The lifetimes of a luminescence species at λ_{lum} = 4297 Å corresponding to the v_E = 0-2 groups differ significantly: it is similar to the $I_2(E,0 \to B)$ transition radiative lifetime, $\tau \approx 26$ ns for the v_E = 1, 2, and much less for v_E = 0 one, $\tau = (8.0 \pm 0.5)$ ns (Fig. 6.23).

The only open channels of the $NeI_2(E,0,n_E)$ decay are EP (6.3.12), and luminescence of the complex (6.3.13). Therefore, luminescence, λ_{lum} = 4000–4400 Å, has to be assigned to luminescence of the complex itself so the lifetime of the $NeI_2(E,0,n_E)$ complex is 8 ns, and the total rate of the complex decay is equal to $12.5 \cdot 10^7\ \mathrm{s}^{-1}$. The total rate of the EP channels (6.3.12), is $k_{3.12} = (7.8–8.8) \cdot 10^7\ \mathrm{s}^{-1}$, $\tau_{\mathrm{EP}} \approx 12$ ns (see data on $NeI_2(E,0,n_E)$ in Table 6.9), and the radiative decay rate of the $NeI_2(E,0,0)$ vdW complex is $k_{3.13} = (3.8–4.8) \cdot 10^7\ \mathrm{s}^{-1}$, similar to that of $I_2(E, 0)$. The $NeI_2(E,0,n_E)$ EP proceeds slowly, since light Ne atom perturbs a I_2 molecule weakly. The maximum of the $I_2(D, v_D \to X)$ luminescence temporal

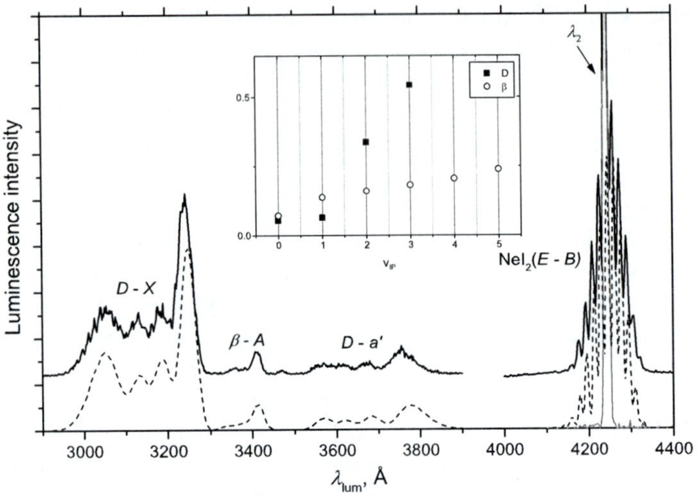

Fig. 6.22 Experimental (solid line) and simulated (broken line) luminescence spectra in the λ_{lum} = 2900–3550 Å and 4000–4400 Å spectral ranges at the $v_E = 0$ band of the $NeI_2(E,v_E, n_E= 0 \leftarrow B,19,n_B= 0)$ transitions. Spectral resolution, FWHM = 10 Å. Experimental spectrum is offset for clarity. Populations of the D,v_D and β,v_β vibronic states are given in the inset. The λ_2 laser line is marked [95] (Reprinted from Chemical Physics Letters, Vol. 696, V.V. Baturo, S.S. Lukashov, S.A. Poretsky, A.M. Pravilov, Luminescence of the NeI_2 van der Waals complex., p. p. 26-30 (2018) with permission from Elsevier)

history for $NeI_2(E, 0,n_E= 0)$ complex lies at larger time than those for the v_E = 1–3 groups (Fig. 6.24)

since the former is populated from the rather long-lived $NeI_2(E, 0, 0)$ complex (τ = 8 ns) and the latter from the short-lived one ($\tau \leq 2$) ns (see below).

The VP channels (6.3.11) open for the $NeI_2(E,v_E\geq 1,n_E)$ complexes (see Figs. 6.16, 6.19), and the complex decay becomes much faster. The origins of the luminescence in the $\lambda_{lum} \approx$ 3900–4400 Å spectral range are both VP products, $I_2(E)$ ($\Delta v_E \neq 0$ VP channel), and the complex (Δv_E = 0 VP channel). Luminescence of the complex or the VP products in this spectral range is simulated as luminescence from the optically populated (v_E^{max}) or lower vibrational levels, respectively. Simulations of the luminescence spectra show that the relative contribution of the *complex luminescence* is estimated as ~ 6%. The branching ratio of VP increases two times at $NeI_2(E,v_E= 1,n_E)$ complex relative to that of $NeI_2(E,v_E= 0,n_E)$ one, and then decreases smoothly due to a competition of the VP and EP channels (see Table 6.9). Assuming that the radiative decay rate of the complex, $k_{3.13}$, depends weakly on v_E, one can estimate the rate of the VP channel, $k_{3.11}$, as $5{\cdot}10^8$ s^{-1}. Comparison of branching ratios of VP and EP channels gives $k_{3.12} \approx 5{\cdot}10^8$ s^{-1}, and lifetime of the $NeI_2(E,v_E> 1,n_E)$ complex is < 1 ns, much less than that of the $NeI_2(E,0,n_E)$ one (8 ns). Time-resolved measurements confirm that both $I_2(E,v_E$ and $D,v_D)$ states are populated in these cases from the short-lived complex (Figs. 6.23,

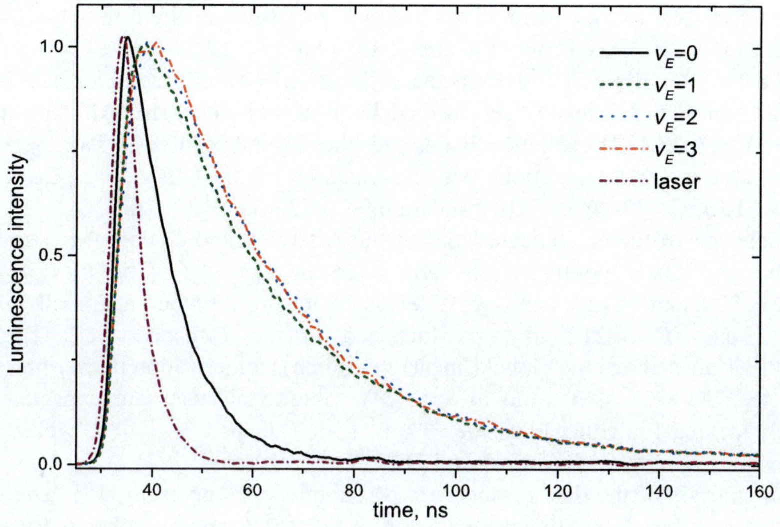

Fig. 6.23 Experimental temporal histories of the luminescence intensities at λ_{lum} = 4297 Å, where the $I_2(E \rightarrow B)$ transition can occur, measured at the v_E = 0-3 groups (see Fig. 6.16). The temporal behavior of the $h\nu_2$ laser pulse is also shown [95] (Reprinted from Chemical Physics Letters, Vol. 696, V.V. Baturo, S.S. Lukashov, S.A. Poretsky, A.M. Pravilov, Luminescence of the NeI_2 van der Waals complex., p.p. 26–30 (2018) with permission from Elsevier)

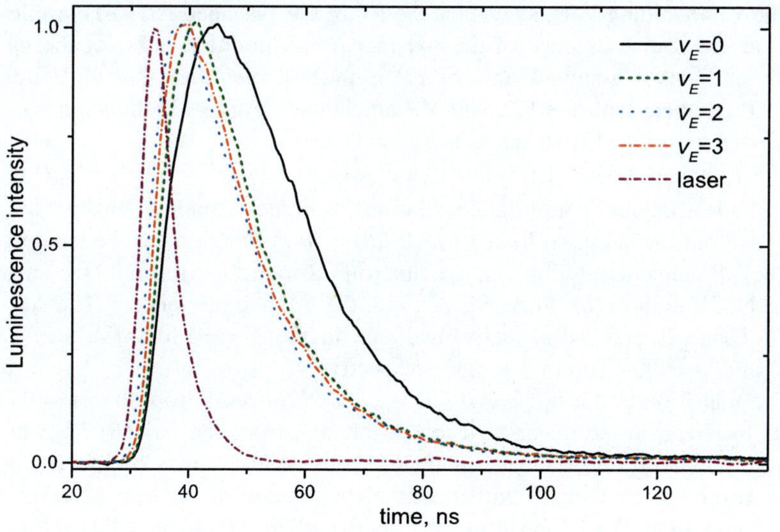

Fig. 6.24 Experimental temporal histories of the luminescence intensities at λ_{lum} = 3240 Å, $I_2(D \rightarrow X)$ transition, measured at the v_E = 0–3 groups (see Fig. 6.16). The temporal behavior of the $h\nu_2$ laser pulse is also shown [95] (Reprinted from Chemical Physics Letters, Vol. 696, V.V. Baturo, S.S. Lukashov, S.A. Poretsky, A.M. Pravilov, Luminescence of the NeI_2 van der Waals complex., p.p. 26–30 (2018) with permission from Elsevier)

6.24). The $NeI_2(E,v_E = 1,n_E)$ state has an intermediate lifetime ($\tau \approx$ 1–2 ns), although the VP decay channel is already available.

The ArI_2 complexes. The *T*-shaped and linear $ArI_2(X)$ vdW complexes as well as $Ar_nI_2(X)$ (n = 2, 3) clusters are formed in a supersonic beam. All the $Ar_nI_2(B, v_B \leftarrow X,0)$ (n = 1–3) excitation bands are blue-shifted relative to $I_2(B,v_B \leftarrow X,0)$ bands, and spectral separations between maxima of the $I_2(B,v_B = 13\text{-}26 \leftarrow X,0)$ and $Ar_nI_2(B,v_B = 13\text{-}26 \leftarrow X,0)$ bands are $\sim n{\cdot}13.4$ cm^{-1} [32, 62].

Some spectroscopic characteristics of the $ArI_2(X, B$ and $E)$ complexes in the *T*-shaped and linear configurations are given in Table 6.5. The $n_X = 0$ and 1 ($E = -237.8$ and -213.7 cm^{-1}) vdW levels are localized in the linear well, whereas $n_X = 2$ one ($E = -212$ cm^{-1}) is localized in the *T*-shaped well. The high (E = 132 cm^{-1} above the global (linear) minimum) isomerization barrier presents at $R \approx 4.92$ Å, $\Theta \approx 52.3°$, and lowest vdW vibrational states are expected to be mostly localized in either linear (n_X = = 0, 1) or *T*-shaped (n_X = = 2) wells; the ω_e values are equal to 24.0 (linear) and 26.5 (*T*-shaped) cm^{-1} [97].

An analysis of the data on the $ArI_2(B)$ complexes is given in [18]. The lowest vdW $ArI_2(B,v_B = 21,n_B)$ levels with $E < -$ 112 cm^{-1} energy relative to $ArI_2(B, v_B = 21,n_B)$ dissociation limit belongs to the *T*-shaped configuration. The levels with $E \approx -(10\text{–}88)$ cm^{-1} belongs to the linear configuration, and those of $E \approx -(0\text{–}10)$ cm^{-1}, to the free-rotor, probably. Since both $ArI_2(X$ and $B)$ *PES*s are similar near the bottom of the *T*-shaped well, the Franck–Condon principle implies that the transition probability from the $n_X = 2$ level has to be largest for $n_B \approx 0$ and rapidly decreases with n_B [99].

The VP including IVR, as well as the EP of the *T*-shaped $ArI_2(B)$ complexes are well studied. Luminescence of the $ArI_2(B,v_B)$ VP products lacks for the $v_B \leq 11$ levels since, as it assumed, rate of EP is much larger than that of VP at them. The VP compete with the EP, and VP efficiencies, $\Delta v_B \geqslant 3$, show strong oscillations over the $v_B \approx 12\text{–}26$ range for n_B = 0-2.

The *T*-shaped $ArI_2(B)$ EP is due to the $ArI_2(B\text{–}RS)$, $RS = a'0_g^+$, $a1_g$, $(1)2_g$ (and $C1_u$, in a less degree) coupling. Oscillations of Franck-Condon densities for the *RS* states are out of phase, so total $FCD(B/RS) = f(v_B)$ function can be smooth.

The VP efficiency oscillations are due to IVR in the sparse limit (see an analysis of the $NeCl_2$ complexes, Sect. 6.3.2.2, [18, 62, 63] and references). The $\Delta v_B \geqslant 3$, 2 and 1 channels are energetically available in the $E < -180$ cm^{-1}, $\approx -(180\text{–}90)$ cm^{-1} and $\approx -(90\text{–}10)$ cm^{-1} ranges, respectively.

The transition to the linear $ArI_2(B)$ isomer *PES* occurs from the n_X = 0, 1 vdW levels localized in the linear well to the left branch of the $ArI_2(B)$ *PES*, and continuum excitation spectra correspond to them. Wide ranges of the $I_2(B,v_B\text{-}\Delta v_B)$ states (from v_B = 12 to 23 with maximal population at the v_B= 18 level for the $ArI_2(B,v_B = 26 \leftarrow X,0)$ transition) are populated. Rotationally cold $I_2(B,v_B)$ fragments, $T_{rot} \approx 5$ K, consistent with direct dissociation from a near-linear geometry are observed.

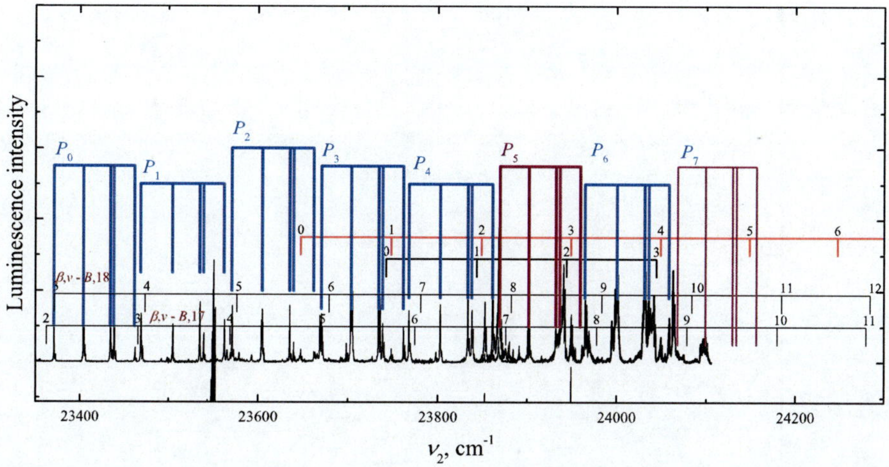

Fig. 6.25 Excitation spectra of the $ArI_2(E)$ EP product luminescence in the $\lambda_{lum} \approx 2600–3800$ Å spectral range measured at T-shaped $ArI_2(B, v_B = 17, 18)$ intermediate. The position of the $I_2(\beta, v_\beta \overset{h\nu_2}{\leftarrow} B, 17$ and $\beta, v_\beta \overset{h\nu_2}{\leftarrow} B, 18)$ transitions as well as dissociation limits of the $ArI_2(E, v_E = 0–6)$ complexes for the transitions via $ArI_2(B, v_B = 18)$ intermediate are shown

Very recently, spectroscopic characteristics, as well as VP and EP of the $ArI_2(E, v_E, n_E)$ complexes were studied in wide $v_E = 0–16$, range [106]. The T-shaped $ArI_2(E, v_E = 0–16, n_E)$ complexes were populated in two-step, two-color scheme

$$ArI_2\left(E, v_E = 0 - 16, n_E \overset{h\nu_2}{\leftarrow} B, v_B = 16 - 19, n_B = 0 \overset{h\nu_1}{\leftarrow} X, 0, n_1 = 2\right) \quad (6.3.16)$$

An analysis of excitation spectra allowed to determine positions of the progression terms, $P_{vE}(n_E)$, $n_E \leq 4$ (see Fig. 6.25 as an example and Fig. 6.26) and the $ArI_2(E, v_E = 0\text{-}16, n_E = 0)$ binding energies (Fig. 6.27)

The VP, EP branching ratios (Fig. 6.28), and vibrational populations of the VP, EP products were determined after luminescence spectra simulations (see Fig. 6.29 as an example).

One sees in Fig. 6.27 (at least at low v_E) that $D_0^{v_E}$ are higher on odd v_E than those on the even v_E. At first glance, this can only be explained by perturbations of the $ArI_2(E, v_E, n_E)$ vdW states by nearest $ArI_2(IP, v_{IP}, n_{IP})$, IP = D', β, D). One should note that the energy gaps of the nearest vibronic levels of the $I_2(E, D, \beta, D')$ states, i.e., between the dissociationlimit of $ArI_2(IP, v_{IP}, n_{IP})$ complexes change monotonously.

The population ratio $I_2(D', \beta)$ at $P_0(1)$ is distinguished from that observed on other progression terms, namely, the population of $I_2(\beta)$ is 3.2 times higher than that of I_2 (D'). Further, there is a tendency toward a decrease in the populations of $I_2(D', \beta)$ states, apparently, due to the appearance of other decay channels of the $ArI_2(E,$

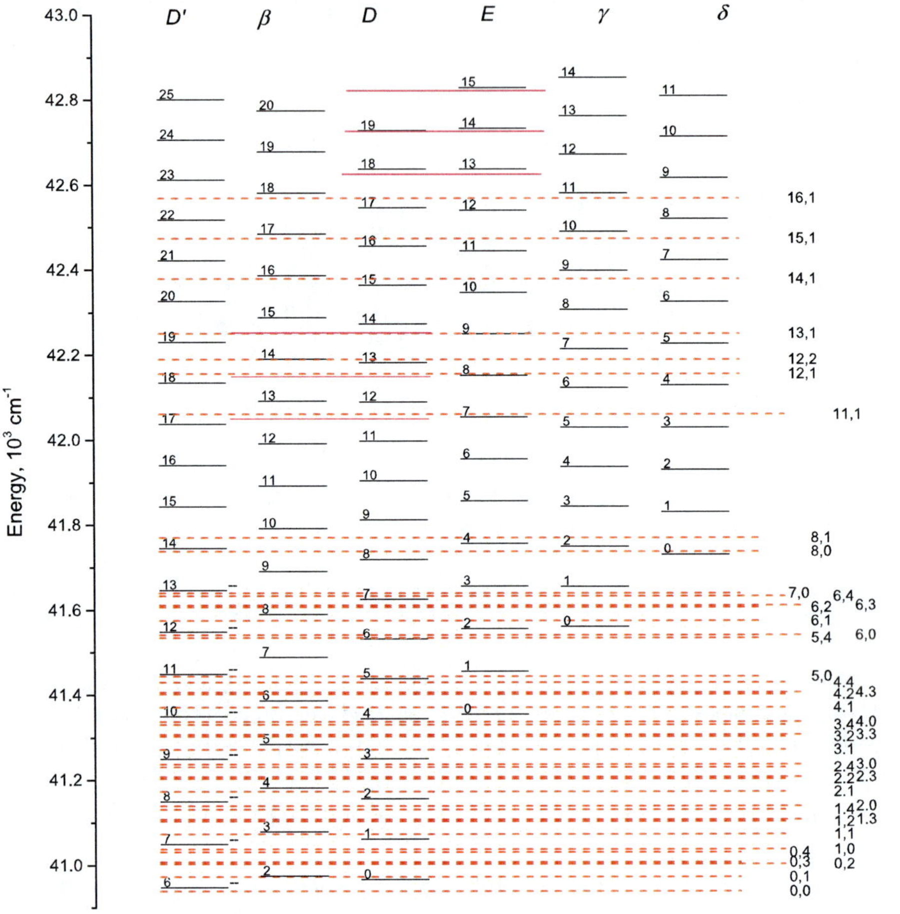

Fig. 6.26 Energies of the $I_2(IP, v_{IP}, J_{IP} = 5)$ rovibronic levels of the first-tier iodine molecule IP states, IP = β, D, E, γ, δ calculated relative to that of $I_2(X, v_X = 0, J_X = 4)$, and the $ArI_2(E, v_E = 0\text{-}16, n_E)$ level energies (shown as horizontal dotted lines). Positions of the progression terms $P_{vE}(n_E)$ are indicated as v_E, n_E (the last column). It is assumed that the $ArI_2(X,0,0)$ binding energy is $D_0^X = 240.5\,\text{cm}^{-1}$ [105]

v_E, n_E) complex. At all the $P_{vE}(n_E)$ terms except $P_{16}(1)$, the decay channel to Ar + $I_2(\beta)$ is prevailing (Fig. 6.28). The VP relative probability increases with v_E.

It follows from the data given in Fig. 6.26 that there are resonances on some terms of the progressions and the $I_2(E, D, \beta, D')$ state vibronic levels. They are: $P_0(1)/\beta,2$; $P_5(0)/D,5$; $P_5(4)/D,6$; $P_6(4)/D,7$; $P_8(0)/\delta,0$; $P_{11}(1)/E,7$; $P_{12}(1)/E,8$; $P_{12}(2)/D,13$, $\beta,14$; $P_{13}(1)/E,9$. The presence of these resonances does not affect the values of the branching ratios; moreover, resonances do not affect the population of the 'resonance' level of a free iodine molecule. The reason for this 'phenomenon' is explained in the 'T-shaped $HeI_2(E, v_E = 0\text{-}17, n_E)$ complexes' Sect (see above).

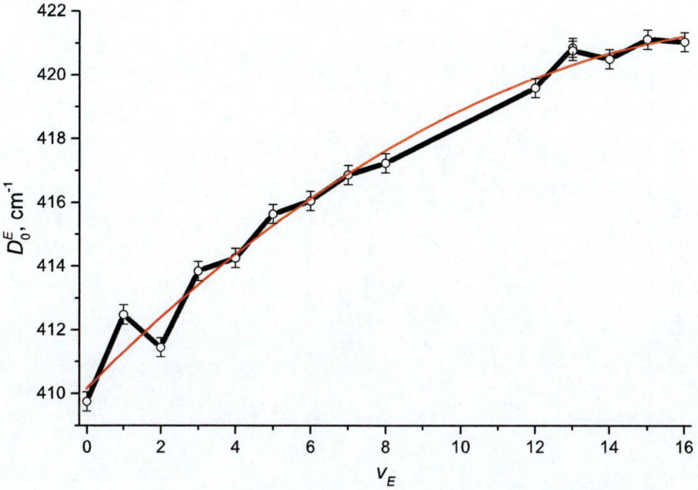

Fig. 6.27 The $ArI_2(E,v_E,n_E = 0)$ binding energies as a function of v_E quantum number

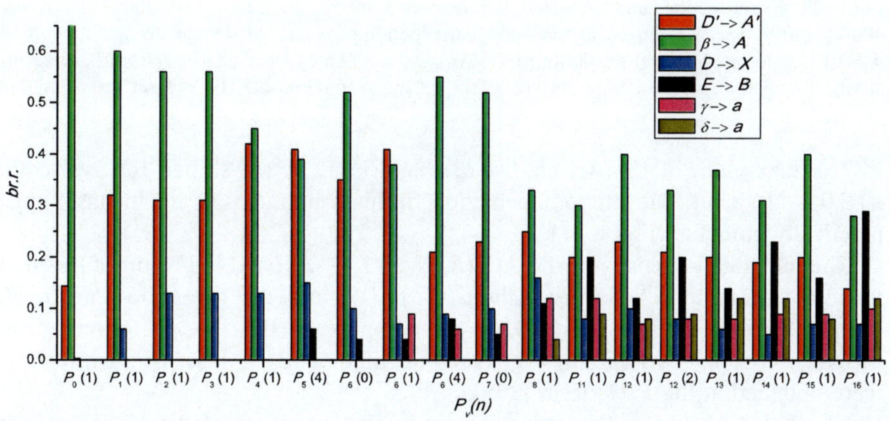

Fig. 6.28 Branching ratios, *br.r.*, of the $ArI_2(E,v_E = 0\text{-}16,n_E)$ VP and EP product formation determined by simulation of the I_2(IP → valence state) luminescence spectra measured at the $ArI_2(E,v_E,n_E \leftarrow B,v_B,n_B)$ excitation bands shown in Fig. 6.26

There is a non-adiabatic transition from the bound $ArI_2(E)$ *PES* to the repulsive part of the ArI_2(IP) *PES*s, and then the image point slides down to the pelvis of the $Ar + I_2\left(\text{IP}, v_{\text{IP}} = v_{\text{IP}}^{max} - 0\right)$ dissociation channel, and the $I_2\left(\text{IP}, v_{\text{IP}} = v_{\text{IP}}^{max} - v_{\text{IP}}^{min}\right)$ vibronic states are populated.

Luminescence of the $ArI_2(E,v_E,n_E)$ complexes. In the $I_2(E \to B)$ transition spectral range, the luminescence band corresponding to the $ArI_2(E,0 \to B)$ transition is observed [105] (see Fig. 6.30).

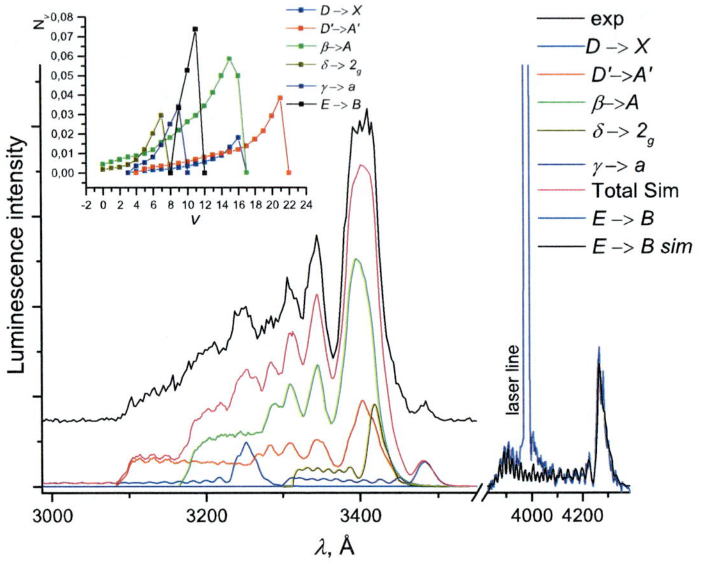

Fig. 6.29 Experimental and simulated luminescence spectra in the λ_{lum} = 2900–3550 Å and 4000–4400 Å spectral ranges at the band corresponding to the $ArI_2(E,v_E= 15,n_E= 1 \leftarrow B,17, n_B= 0)$ transitions. Spectral resolution, FWHM = 10 Å. The experimental spectrum is offset for clarity. Populations of the IP,v_{IP} vibronic states are given in the inset. The λ_2 laser line is marked

The maximum of the $ArI_2(E,0 \rightarrow B)$ band is 32 Å red-shifted relative to the $I_2(E,0 \rightarrow B)$ band. The temporal behaviors of this band intensity and the laser pulse practically coincide (Fig. 6.31).

The integrated intensity of the $ArI_2(E,0,n_E= 1 \rightarrow B)$ band is 110 times less than the total luminescence intensity of the $I_2(D', \beta, D)$ states, that gives a lower estimate of the lifetime of the $ArI_2(E,0,n_E= 1)$ complex, ~ 0.22 ns. It follows from Fig. 6.31 that the $ArI_2(E,0,n_E= 1)$ complex lifetime is ~ 1 ns. Similar features were obtained at the $P_1(1)$ term [105].

The KrI$_2$ complexes. Some spectroscopic characteristics of the $KrI_2(X, B$ and $E)$ complexes in the *T*-shaped configuration are given in Table 6.5. One can see the data obtained at these studies in [18, 91, 100] and references.

The XeI$_2$ complexes. Information on the XeI_2 vdW complexes is very scarce (see [18] and references).

The author's team tried to 'catch' $XeI_2(E)$ vdW complexes by the way which was successfully utilized for a study of the KrI_2 vdW complexes but failed. Maybe, it is due to very fast EP of the $XeI_2(B)$ complexes.

The binding energies of the $RgI_2(X, B$ and $E)$ complexes, D_0, increase from He to Kr, but trends of D_0^B and D_0^E are different (Fig. 6.32). So, one can conclude that the nature of potential interaction for *B* and *E* states is different.

One sees in Fig. 6.32 that the D_0^E value is directly proportional to Rg polarizability, α_{Rg} (see [106, 107]. This feature can be explained as follows

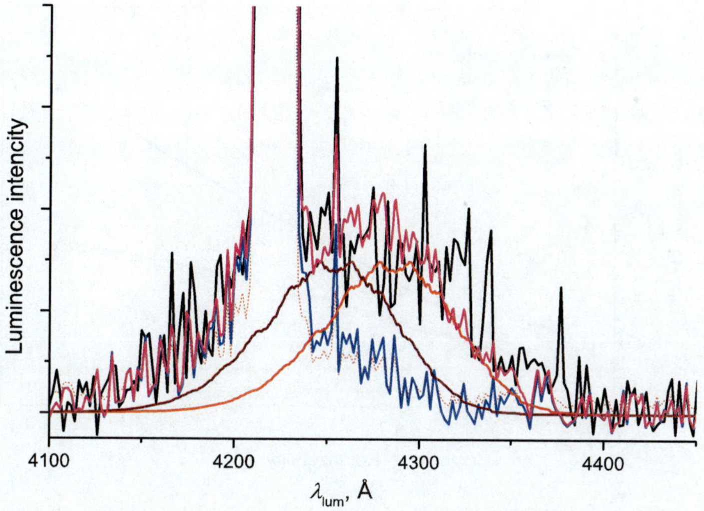

Fig. 6.30 Experimental and simulated luminescence spectra in the λ_{lum} = 4100–4450 Å spectral ranges at the $P_0(1)$ term. Spectral resolution, FWHM = 10 Å. Experimental spectrum is shown as a black line, the laser λ_2 scattered line measured with blocked λ_1 laser beam is shown as a blue line. Simulation of the $I_2(E,0 \rightarrow B)$ transition is shown as a brown line, and simulation of the $ArI_2(E,0 \rightarrow B)$ transition is shown as a red line, the sum of the scattered light with the λ_1 laser beam and the simulation of the $ArI_2(E,0 \rightarrow B)$ transition is shown as a magenta line

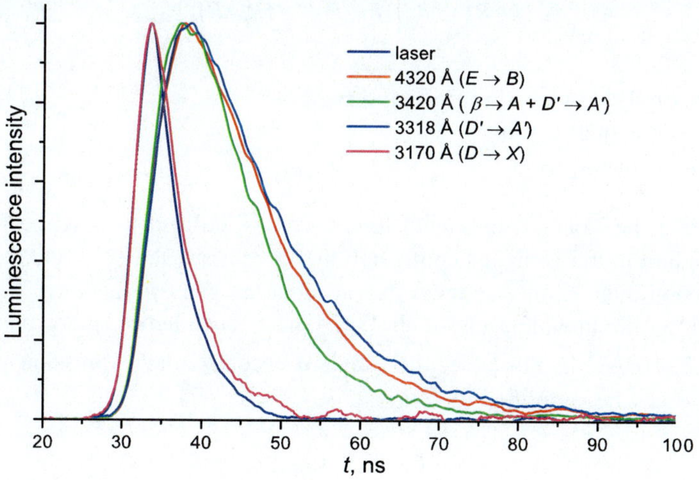

Fig. 6.31 Temporal behaviors of the luminescence intensity measured at the $P_0(1)$ term. The λ_2 laser pulse is also shown

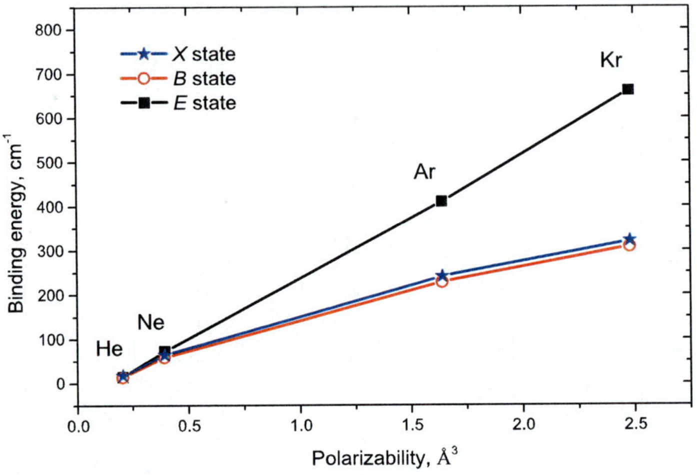

Fig. 6.32 Experimental binding energies of the RgI_2(X, B and E) complexes, as functions of the Rg polarizability (see Fig. 6.29 in [18])

(see Sects. 3.3.3, 3.4, [106–109] and references): Homonuclear I_2 has no permanent dipole moment, but the $I_2(E)$ state has giant polarizability due to a huge $I_2(E0_g^+ - D0_u^+)$ transition dipole moment μ_{ED}. The Lennard-Jones (6-12) potential

$$U(R) = 4\varepsilon\left[(R_0/R)^{12} - (R_0/R)^6\right], \tag{6.3.17}$$

R is the I_2(IP) center of mass– Rg distances,
ε is the potential depth,
R_0 is the R distances at which $U(R) = 0$,

representing the London dispersion forces, can be utilized to describe Rg– I_2(IP) vdW bonding in the T-shaped configuration. To estimate the $C_6 = 4\varepsilon R_0^6$ coefficient of dispersion interaction, one takes into account the giant $\mu_{ED} = e \cdot r_{II}$, $\sim 18 \cdot 10^{-18}$ *esu* for low vibrational levels of the $I_2(E)$ state, and, hence, giant polarizability $\alpha_{I_2(E)} \approx 2 \cdot 10^5 \text{Å}^3$. In this case, the standard second-order expression for the C_6 coefficient can be simplified to

$$C_6 \approx \alpha_{Rg}|\mu_{ED}|^2. \tag{6.3.18}$$

There is a correlation between a vdW complex dissociation energy D_e and a C_6 dispersion coefficient

$$D_e = A\frac{C_6}{R_e^6}, \tag{6.3.19}$$

A is the correlation coefficient, $A = 0.72$, usually, and R_e is the equilibrium Rg–I_2(IP) distance (see [97] and references).

6.3.3 The RgXY vdW Complexes

The low-lying valence electronic states of the interhalogens are analogous to those of the homonuclear halogens, but there are two essential differences; the *u*/*g* parities for the permutation of identical nuclei does not exist, and there are four dissociation limits, I($^2P_{3/2}$ + Br($^2P_{3/2}$), I($^2P_{3/2}$ + Br($^2P_{1/2}$), I($^2P_{1/2}$ + Br($^2P_{3/2}$), I($^2P_{1/2}$ + Br($^2P_{1/2}$) for IBr molecule, e.g. The lowest XY($X0^+$, $A'2$ and $A1$) states correlate with the lowest X($^2P_{3/2}$) + Y($^2P_{3/2}$) dissociation limits, and there are no avoided crosses for them. The characters of a XY($B0^+$) states correlating with a X($^2P_{3/2}$) + Y($^2P_{1/2}$) dissociation limit (spin-orbit couplings are lower for the Y atom than for the X ones) differ significantly from those of the homonuclear halogens since a XY($B0^+$) state is perturbed by a repulsive 0^+ state, usually labeled as YO^+ which correlates diabatically with the lowest X($^2P_{3/2}$) + Y($^2P_{3/2}$) dissociation limit (see Fig. 6.33).

To the best of the author's knowledge, there are data on electronically-excited states of the RgICl and RgIBr only in the literature (see [22], for example). As to

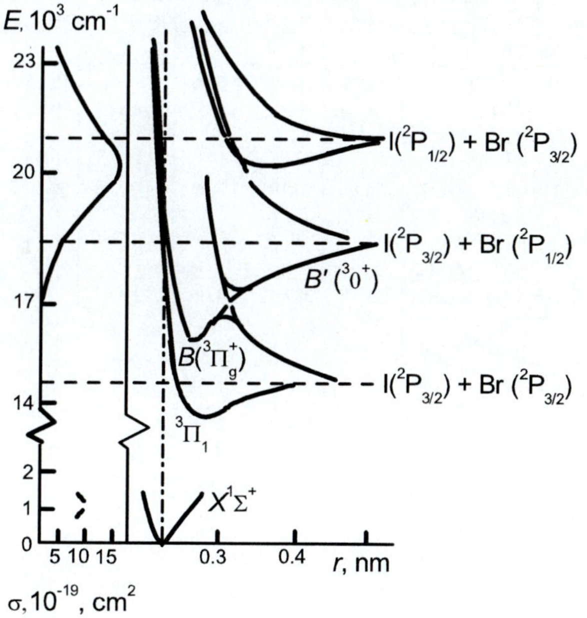

Fig. 6.33 Potential energy curves and the *A*-band of the absorption spectrum of IBr molecules [110]. The broken and full lines represent the diabatic and adiabatic *PEC*s, respectively (see Sect. 4.7 and Fig. 4.23, also)

other RgXY complexes, ground state RgClF, Rg = He, Ar, and ArBrCl complexes have been studied (see [30, 31] and references).

Mutual perturbation of the $B0^+$ and $Y0^+$ states is strong enough to cause avoided crossing of the states in ICl and IBr molecules. This perturbation forces the B state to correlate with ground-state products, and a new $B'(0^+)$ bound-state is formed above the crossing point (Fig. 6.33).

The interactions within RgXY complexes are slightly different than those within the RgX_2 complexes due to the influences of the asymmetric electron distributions and, to some extent, the dipole-induced dipole forces arising from the XY permanent dipole moment, 1.207 D for ICl(X) (see [111] and references). Up to three wells with varying depths are anticipated for the RgXY complexes with the deepest well, being in the $\theta = 0^o$, Rg···X–Y, orientation, where X is the larger, more polarizable halogen atom. The three wells, $\theta = 0^o$, near T-shaped and $\theta = 180^o$, Rg···Y–X, orientations are separated by isomerization barriers.

6.3.3.1 The RgICl vdW Complexes

The HeICl vdW complexes are most well researched, and we start with them.

The HeICl complexes. Data on some spectroscopic characteristics of the $HeI^{35}Cl$ $(X,v_X= 0)$ complexes are given in Table 6.10.

Table 6.10 The dissociation, D_e, and binding, D_0, energies (units, cm^{-1}) as well as equilibrium distances R_e (Å) of the $RgI^{35}Cl(X,\ v_X= 0)$ collinear ($\Theta = 0°$), near T-shaped and antilinear ($\Theta = 180°$) conformers

		Collinear, $\theta = 0$			Near T-shaped			Antilinear, $\theta = 180$		
Rg	Reference	D_e	D_0	R_e	D_e	D_0	R_e/θ_e	D_e	$D_{0\text{-}}$	R_e
He	[112] (ab initio)	58.62	18.29	3.86	38.96	15.15	3.82/111	38.03	12.33	5.12
	[113] (exp)		20.7			16.8	3.8/102			
	[28] (exp)		22.0			16.6				
Ne	[112] (ab initio)	106.8	76.19	3.96	84.79	62.59	3.78/109	76.74	55.73	5.17
	[34] (exp)				~ (72.4 –74.7)	48.2 ±0.5	3.52			
Ar	[114] (ab initio)	328.	282.52	4.06	230.3	199.3	3.94	192.46	163.65	5.36
	[115] (exp)		341.8$^{(1)}$	4.08						

(1)—$\theta \approx 5^o$

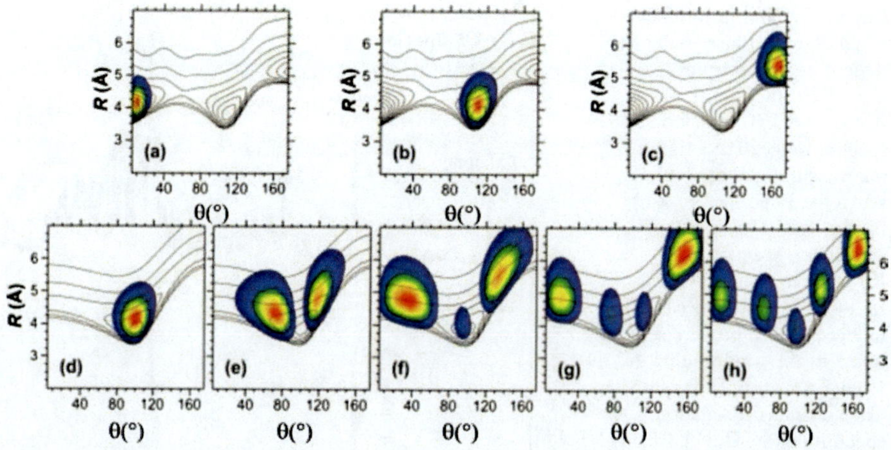

Fig. 6.34 Plots of the J = 0 He⋯I^{35}Cl probability amplitudes for the n_X = 0-2 vdW modes of the the HeI^{35}Cl(X,v_X= 0) state, (a)-(c) panels, respectively, and the n_B = 0-4 vdW modes the HeI^{35}Cl (B,v_B= 3) state, (d)-(h) panels, respectively (see [116]) (Reproduced from A.B. McCoy, J.P. Darr, D.S. Boucher, P.R. Winter, M.D. Bradke, R.A. Loomis, J. Chem. Phys. **120**, 2677-2685 (2004). https://doi.org/10.1063/1.1636693 with the permission of AIP Publishing)

According to [113, 116], the n_X = 0, 1 and 2 vdW modes are located in the collinear (He⋯I–Cl), T-shaped and antilinear (He⋯Cl–I) wells, respectively (Fig. 6.34).

Their energies relative to the HeI^{35}Cl(X, v_X= 0) dissociation limit are equal to −18.29, −15.14 and −11.76, units, cm^{-1}, respectively. The three energy minima are separated by the isomerization barriers: θ = 0 and T-shaped wells, with the energy of −15.3 cm^{-1} (~42.9 cm^{-1} above the collinear well), whereas the T-shaped and antilinear wells by the barrier with energy of −2.8 cm^{-1} (~35.4 cm^{-1} above the collinear well) [113].

The locations of the n_B vdW modes are shown in (d)-(h) panels in Fig. 6.34. The n_B = 0 lowest energy vdW level is bound by 13.30 cm^{-1} and is localized in the near T-shaped well. The next higher n_B = 1 level has a node in the angular coordinate, but is still localized in the T-shaped well. The n_B = 2, 3 and 4 levels, at −6.21, −5.09 and −3.63, units, cm^{-1}, have angular nodes, become delocalized in the angular coordinate and thus are free-rotor levels [116].

The excitation spectrum of the HeICl(B,3) VP product luminescence, experimental and calculated action spectra of the HeI^{35}Cl(E,11) VP products [116] are given in Fig. 6.35.

The assignments of the action spectra bands are correct, though calculated and observed transition energies do not coincide. Besides, the T-shaped HeI^{35}Cl(B,3, n_B= 1 ← X,0,n_X = 1) transition has not been observed in the experiment.

Excitation spectra of the luminescence of the HeI^{35}Cl(E,11, β,1) VP and EP products are given in Fig. 6.36.

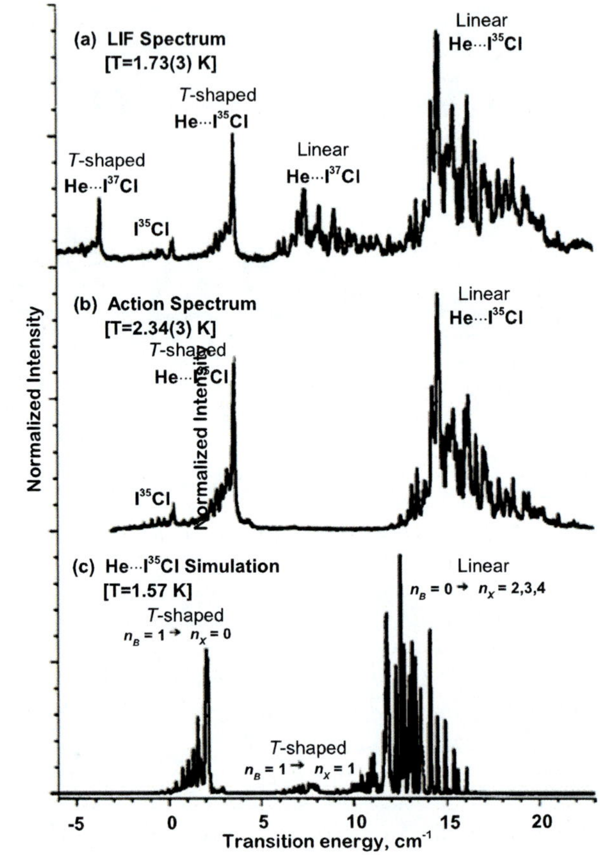

Fig. 6.35 High-resolution ($\sim$ 0.06 cm^{-1}) laser-induced fluorescence spectrum **a** recorded in vicinity of the ICl(B,3 $\leftarrow$ X,0) transition spectral range. The origin of the transition energy scale is set to the $I^{35}Cl$(B,3 $\leftarrow$ X,0) band origin, 17827.49 cm^{-1}. The action spectrum (**b**) collecting the $I^{35}Cl$ ($E \rightarrow X$) luminescence induced with a probe laser tuned to the bandhead of the $I^{35}Cl$(E,11 $\leftarrow$ B,2) transition. The calculated excitation spectrum of the $HeI^{35}Cl$(E,11) VP products (c) (see [116]) (Reproduced from A.B. McCoy, J.P. Darr, D.S. Boucher, P.R. Winter, M.D. Bradke, R.A. Loomis, J. Chem. Phys. **120**, 2677-2685 (2004). https://doi.org/10.1063/1.1636693 with the permission of AIP Publishing)

The spectra at v_E = 1, 2 and v_β = 2 levels have features with nearly the same energetic spacings and intensity profiles as those in the spectra given in Fig. 6.36. The similarities of the features in Fig. 6.36a, b (transitions from $HeI^{35}Cl(X,0,n_X\!=1)$ of the T-shaped configuration) suggest that the $HeI^{35}Cl(A,1,n_A\!=0)$ vdW state is also T-shaped, as the $HeI^{35}Cl(B,2,n_B\!=0)$ one (see Fig. 6.34d). One can assume that the $HeI^{35}Cl(E,11,\ n^b_{E'},\ n^{str}_E$ and $\beta,1,\ n^b_\beta,\ n^{str}_\beta)$ states also have significant probability amplitude in the T-shaped region of the intermolecular PESs (see [28]). Energies of the $HeI^{35}Cl(E,v_E\!=11,12,\ n^b_{E'},\ n^{str}_E$ and $\beta,v_\beta\!=1,2,\ n^b_\beta,\ n^{str}_\beta)$ vdW levels (cm^{-1}) relative to the $HeI^{35}Cl(E,v_E$ and $\beta,v_\beta)$ dissociation limits are given in Table 6.11.

The binding energies of the T-shaped $HeI^{35}Cl$(B,3 and A,15) complexes are 13.4 (4) and 13.4(3) cm^{-1}, respectively. Lifetime of the T-shaped $HeI^{35}Cl$(B,2) complexes is 550 ps [117]. The binding energies of the T-shaped $HeI^{35}Cl$(E,12 and β,1) complexes are equal to 41(1) and 39.2(4) cm^{-1}, respectively. Nonadiabatic coupling between specific vdW modes of the $He^{35}ICl(\beta,v_\beta$ and $D', v_{D'})$ states takes place [28].

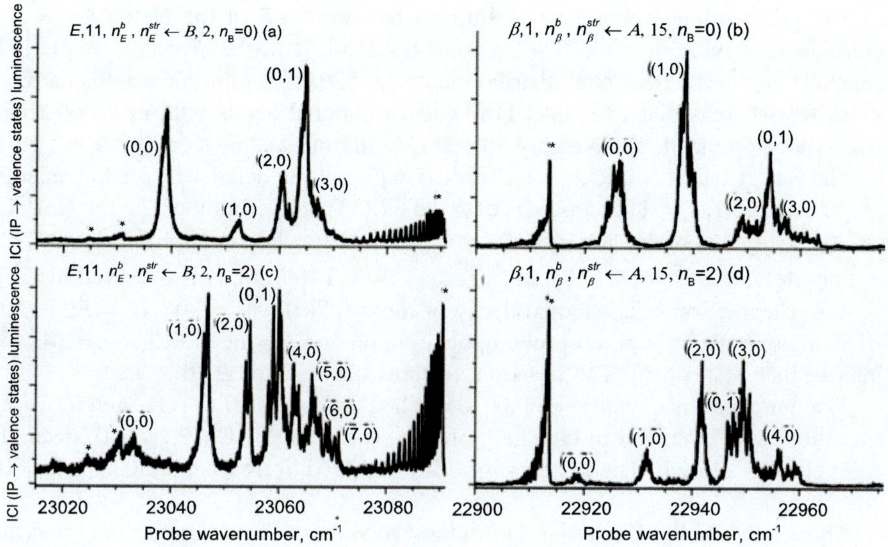

Fig. 6.36 Excitation spectra of the luminescence of the $HeI^{35}Cl(E, \beta)$ VP and EP products, λ_{lum} = 285– 415 nm) recorded in a vicinity of the ICl(E,11 ← B,2 and β,1–A,15) transition spectral ranges. The spectra were acquired with the pump laser fixed on the T-shaped $HeI^{35}Cl$(B,2, n_B= 0 ← X,0,n_X= 1 or A,1,n_A= 0 ← X,0,n_X= 1) transitions (**a**), (**b**) panels. The spectra in (**c**), (**d**) panels were acquired by pumping linear, ground-state complexes to the free-rotor level of the intermediate state, $HeI^{35}Cl$(B,2,n_B= 2 ← X,0,n_X= 0 or A,15,n_A= 2 ← X,0,n_X= 0). The features are labeled with the assigned vdW bending and stretching modes, n_b^i, n_{str}^i, $i = E$, β of the complexes in the IP states. . The asterisks denote either $I^{35,37}Cl$(A/B) VP (T-shaped configuration) or direct dissociation (free-rotor configuration) products that are excited to the $I^{35,37}Cl$ IP states (see [28]) (Reproduced from J.P. Darr, R.A. Loomis, J. Chem. Phys. **129**, 144306 (11 pp) (2008). https://doi.org/10.1063/1.2990661 with the permission of AIP Publishing)

Table 6.11 Energies of the $HeI^{35}Cl$(E, v_E= 11, 12, n_E^b, n_E^{str} and β, v_β= 1, 2, n_β^b, n_β^{str}) vdW levels (cm^{-1}) relative to the $HeI^{35}Cl$(E, v_E and β, v_β) dissociation limits (see [28])

n_b^i, n_{str}^i	v_E		v_β	
	11	12	1	2
0, 0	−40.3(4)	−39.2(4)	−41(1)	−41(1)
1, 0	−26.8(4)	−26.1(2)	−28(1)	−29(1)
2, 0	−18.6(4)	−18.5(2)	−18(1)	−19(1)
0, 1	−14.5(4)	−14.4(2)	−13(1)	−13(1)
3, 0	−13.2(6)	−12.4(2)	−10(1)	−10(1)

The NeICl complexes. Data on some spectroscopic characteristics of the $NeI^{35}Cl$ (X,v_X= 0) complexes are given in Table 6.10. The binding energy of the $NeI^{35}Cl$(A, v_A= 10-23) complexes lies in the (45– 41) cm^{-1}interval [34], and that of $NeI^{35}Cl$(B, v_B= 2) is 42 cm^{-1}; its VP lifetime is $\tau > 50$ ps [118]. The $NeI^{35}Cl$(A,v_A= 14) VP lifetime was found to be large, (3 ± 2) ns [118].

The pump-probe technique was utilized for study VP of the NeICl(A,v_A= 23) vibronic state lying near the ICl(A dissociationlimit. It was shown that Δv_A = 1, 2 channels open. The rotational distribution of the ICl(A,22) vibronic state is narrow, J_A = 1-8, whereas that of ICl(A,21) is wide, rotational levels with up to 69 cm^{-1} energy are populated. The NeICl(A,v_A= 23) VP lifetime is 2.3 $\leq \tau \leq$ 50 ps [119].

The $NeI^{35}Cl$(IP) complexes were studied in [34–36] in detail. The binding energy of the $NeI^{35}Cl$(E,v_E= 1) complex is equal to 87.6(8) cm^{-1}, and those in the $NeI^{35}Cl$ (A, v_A= 10–25) complex are (45–42) cm^{-1}. The $NeI^{35}Cl(E,0,n_E^{str} = 0 - 3 \leftarrow A,11,$ n_A) progression, $\omega_e^E = 52.5(1)\,\text{cm}^{-1}$, $(\omega_e x_e)^E = 6.14(4)\,\text{cm}^{-1}$, is observed in [34].

T.A. Stephenson et al. studied decay of the NeICl(IP,v_{IP} = 0-4), IP = $E0^+$, $D'2$, $\beta 1$ complexes using a pump-probe optical population via the NeICl(A,v_A= 14, 15) intermediate states [36]. The principal features observe are as follows:

The luminescence which can be described as the ICl($D' \rightarrow A'$ and $E \rightarrow X$) transitions was observed in the energy range where the NeICl(IP,v_{IP} = 0) decay is energetically closed. The authors of [36] ascribed it to the NeICl(IP,v_{IP} = 0) complex itself.

There is no NeICl(IP,v_{IP} = 1-4) luminescence due to fast complex VP, and the ICl($D' \rightarrow A'$, $\beta \rightarrow A$ and $E \rightarrow X$) transitions occur. T.A. Stephenson et al. [36] believed that this feature is due to D', β and E mutual perturbation induce by presence of Ne atoms.

In the author's opinion, the NeICl($\beta 1 \leftarrow A$) transition is the most probable, and then ICl(D' and E) state are populated in the NeICl(β1) EP.

The ArICl complexes. Data on some spectroscopic characteristics of the $ArI^{35}Cl$ (X,v_X= 0) complexes are given in Table 6.10. Some spectroscopic characteristics of the ArICl($X0^+$, $A1$, $D'2$, $\beta 1$, $E0^+$) states and the decay of the complex IP states have been studied in [120]. The binding energies of the ArICl(X,0 and A,v_A= 12-15) states are $\leq$ 213.4 cm^{-1}, and (208.3– 205.7) cm^{-1}, respectively. The binding energies of the ArICl(β,0) state is larger, 452.7 cm^{-1}. Luminescence of the ArICl ($\beta 1$, $E0^+$) and, maybe, ArICl($D'2$) complexes themselves have been observed and studied. The radiative lifetime of the ArICl(E,0) state is equal to 23 ns, like those of ICl($\beta 1$, $E0^+$).

The RgIBr complexes. As far as the author knows, there are no data on electronically excited HeIBr and ArIBr state in the literature. The NeIBr(A) vdW complexes were observed and studied, only [121, 122]. The binding energy of the NeIBr(X,v_A= 0) and NeIBr(A,v_A= 16) complexes is 71.8 ± 1.2 and 65.5 ± 1.2, units, cm^{-1}, respectively [122], and NeIBr(A,v_A) VP lifetime decreases from 23 ps (v_A = 12) to 3 ps (v_A = 19) [122].

6.3.4 Molecule-I_2 vdW Complexes

There are several papers in which results of studies of the molecule-X_2 vdW complexes have been reported.

6.3.4.1 H_2I_2 vdW Complexes

J.E. Kenny et al. [123] studied fluorescence excitation as well as fluorescence spectra of ortho- and para-$H_2I_2(B0_u^+, v_B = 16, 20\text{-}23, 26,27)$ and $D_2I_2(B0_u^+, v_B = 16, 20\text{-}23, 26,27)$ complex VP products. They determined Morse potential parameters for the ground- and *B*-state of the H_2I_2 and D_2I_2 complexes (Table 6.12).

The vdW frequencies, ω_e, and the energy gaps between the $n = 0$ and 1 vdW levels, ΔE_{01}, are much larger than those for the NeI_2 valence and *E* state, for example (see Sect. 'The NeI_2 complexes' and Table 6.9). The authors of [124] believed that this feature is due to small H_2, D_2 masses (one should note that the He atoms mass is small, also). The $H_2I_2(B)$ state VP lifetime is short, ~ 18 ps, since, as the author believed, vdW stretching frequency is well matched to that of the $I_2(B)$ stretching frequency (ω_e = 128 cm^{-1}) that leads to good coupling between the initial bound state and the final dissociative state. According to [124], VP lifetime is ~ 29 ps for the $H_2I_2(B, v_B = 17)$ state.

6.3.4.2 N_2I_2 vdW Complexes

According to [125], the binding energies of the $N_2I_2(X, v_X = 0)$, and $N_2I_2(B, v_B = 21)$ states are 218 $cm^{-1} \leq D_0^{X,0} \leq$ 313 $cm^{-1}$and 198 $cm^{-1} \leq D_0^{B,21} \leq$ 284 cm^{-1}, respectively.

Detailed studies of some spectroscopic characteristics as well electronically-excited state decay of the $N_2I_2(X, B$ and $E)$ complexes were carried out in the author's scientific group [126]. An algorithm of the first experiments was the following. Initially, we measured the excitation spectra of luminescence of the $I_2(B,19)$ and products of VP of the $HeI_2(B,19)$ and $N_2I_2(B,19)$ complexes. One sees the uncomplexed $I_2(B,19)$ band, bands corresponding to the *T*-shaped and free-rotor

Table 6.12 Morse potential parameters for the ground- and *B*-state of the H_2I_2 and D_2I_2 complexes. Dissociation (D_e) and binding (D_0) energies of the states, and energy gaps between the $n = 0$ and 1 vdW levels (ΔE_{01}) [123]

	$X, v_X = 0$	$B, v_B = 28$
ω_e (H_2)	130.8	106.3
ω_e (D_2)	92.5	75.2
$\omega_e x_e$ (H_2)	28.4	23.2
$\omega_e x_e$ (D_2)	14.2	11.6
D_e (H_2)	150.6	121.8
D_e (D_2)	150.6	121.8
D_0 (H_2)	92.3	74.4
D_0 (D_2)	107.9	87.1
ΔE_{01} (H_2)	74.0	59.9
ΔE_{01} (D_2)	64.1	52.0

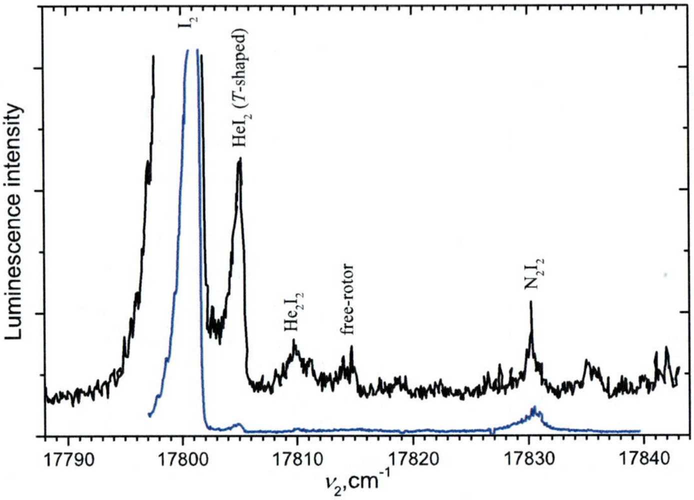

Fig. 6.37 Excitation spectra of luminescence of the $I_2(B,19)$ and products of VP of the $N_2I_2(B,19)$, *T*-shaped, free-rotor $HeI_2(B,19)$ complexes and $He_2I_2(B,19)$ cluster (upper curve) and action spectra measured at the ν_2 = 23469.64 cm^{-1} band, see Fig. 6.38 (lower curve) [126] (Reprinted from Chemical Physics Letters, Vol. 714, V.V. Baturo, R. Kevorkyants, S.S. Lukashov, S.S. Onishcenko, S.A. Poretsky, A.M. Pravilov, The valence and ion-pair states of the N_2I_2 van der Waals complex. p.p. 213-218 (2019) with permission from Elsevier)

$HeI_2(B,19,n_B)$ complexes and $He_2I_2(B,19,n_B)$ clusters in Fig. 6.37. The band displaced ~ 30 cm^{-1} to the short wavelength side (ν_1 = 17830.4 cm^{-1}) can correspond to the $N_2I_2(B,19,n_B)$ complexes [125]. To check this assumption, we measured the excitation spectrum of luminescence in the UV spectral range in the ν_2 = 23300-23500 cm^{-1} spectral range at ν_1 = 17830.4 cm^{-1}. We found some strong bands in the spectrum. Then we measured an action spectrum at one of the bands of the excitation spectrum (Fig. 6.37). The positions of the ν_1 = 17830.4 cm^{-1} band are the same in both spectra given in the figure, and this band has to be assigned to the $N_2I_2(B,19,n_B)$ complex. All the following measurements were carried out at ν_1 = 17830.4 cm^{-1}.

Survey excitation spectrum of luminescence in the UV spectral range, λ_{lum} 2600–3800 Å at ν_1 = 17830.4 cm^{-1} measured at the He + N_2 mixture is presented in Fig. 6.38.

The $N_2I_2(E,v_E,n_E \leftarrow B,19,n_B)$ bands are fitted into several progressions. The most long-wavelength one, $N_2I_2(E,v_E{=}\ 0,n_E \leftarrow B,19,n_B)$, consists of 5, n_E = 0–4, bands. The energy gaps between the first n_E = 0, terms, corresponding to the v_E = 0, 1, has to be close to that of $I_2(E,v_E{=}\ 0,1)$, ~ 101 cm^{-1}, and weak bands of the $N_2I_2(E,v_E{=}\ 1,\ n_E{=}\ 0,1 \leftarrow B,19,n_B)$ transition, $\nu_2 \approx$ 23461 and 23498.8 cm^{-1}, is overlapped with strong $N_2I_2(E,v_E{=}\ 0,n_E{=}\ 3,4 \leftarrow B,19,n_B)$ and $I_2(\beta,1 \leftarrow B,15)$ bands.

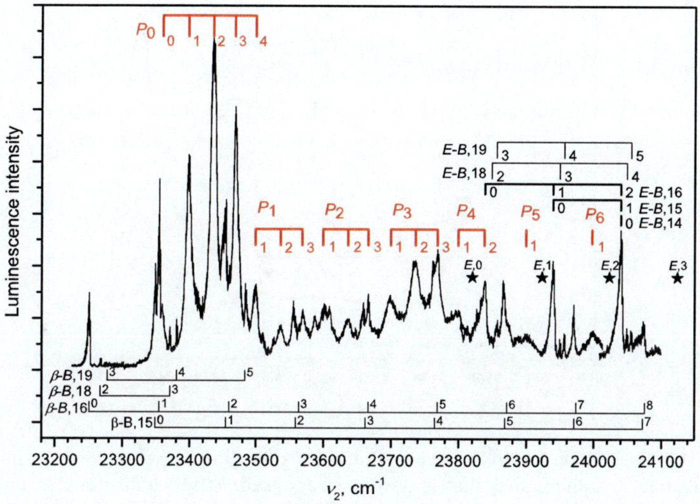

Fig. 6.38 Excitation spectra of luminescence in the UV spectral range, $\lambda_{\text{lum}} \approx$ 2600–3800 Å measured at the He + N_2 mixture at ν_1 = 17830.4 cm^{-1}, ($N_2I_2(B,19,n_B \leftarrow X,0,n_X)$ transition). Assignments of the discovered $N_2I_2(E,\nu_E,n_E \leftarrow B,19,n_B)$ vibrational progressions correlating with the N_2 + $I_2(E,\nu_E$= 0-4) limits) and $I_2(E,\nu_E \leftarrow B,\nu_B,\ \beta,\nu_\beta \leftarrow B,\nu_B)$ transitions are shown. The relative intensities of the $I_2(E,\nu_E \leftarrow B,\nu_B,\ \beta,\nu_\beta \leftarrow B,\nu_B)$ and $N_2I_2(E,\nu_E,n_E \leftarrow B,19,n_B)$ transitions are distorted by the strong saturation in the former (see Sects. 6.2.1.3 and 6.4.1.3 in [18]). Dissociation limits of the $N_2I_2(E,\nu_E,n_E)$ complexes are marked out by filled stars [126] (Reprinted from Chemical Physics Letters, Vol. 714, V.V. Baturo, R. Kevorkyants, S.S. Lukashov, S.S. Onishcenko, S.A. Poretsky, A.M. Pravilov, The valence and ion-pair states of the N_2I_2 van der Waals complex. p.p. 213-218 (2019) with permission from Elsevier)

The $N_2I_2(E,\ \nu_E,\ n_E \leftarrow B,19,n_B)$ progressions are described by equation:

$$E(\nu_E, n_E) \approx T_e(\nu_E) + \omega_e(\nu_E)\left(n_E + \frac{1}{2}\right) - \omega_e x_e(\nu_E) \cdot \left(n_E + \frac{1}{2}\right)^2, \qquad (6.3.20)$$

where T_e, ω_e and $\omega_e x_e$ are spectroscopic constants for the stretching vdW mode. Two ν_E = 0, 3 strongest progressions corresponding to the largest *FCF*s of the $I_2(E, \nu_E \leftarrow B,19)$ transitions were analyzed. The constants obtained are the following: ω_e = 41.9(5) cm^{-1}, $\omega_e x_e$ = 1.38(4) cm^{-1}.

One of the luminescence spectra measured at the strongest $P_0(1)$, $P_0(3)$, $P_3(2)$ and $P_3(3)$ terms to ascertain predissociation channels and determine the population of the predissociation products is shown in Fig. 6.39.

There are two, $\lambda_{lum}^{max} \approx 3180$ Å and 3420 Å bands in Fig. 6.39. The former is simulated as the $I_2(D0_u^+ \rightarrow X0_g^+)$ transition. Two, $I_2(D'2_g \rightarrow A'2_u$ and $\beta 1_g \rightarrow A1_u)$ transitions occur at the $\lambda_{lum}^{max} \approx 3420$ Å band.

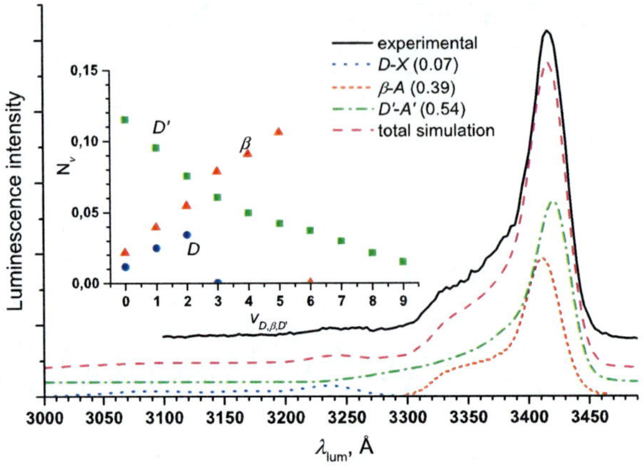

Fig. 6.39 Luminescence spectrum measured at the $P_3(3)$ term (upper curve) and its partial and total simulations assuming that this spectrum is assigned as the transitions in an isolated I_2 molecule (lower curves). Spectral resolution is FWHM = 20 Å. Populations of the $I_2(IP, v_{IP})$ vibronic levels and relative probabilities of different EP channels are given (see [126]) (Reprinted from Chemical Physics Letters, Vol. 714, V. V. Baturo, R. Kevorkyants, S. S. Lukashov, S. S. Onishcenko, S. A. Poretsky, A. M. Pravilov, The valence and ion-pair states of the N_2I_2 van der Waals complex. p.p. 213–218 (2019) with permission from Elsevier)

The best estimations of the binding energies are [126]:

$$284 \le D_0^{X,0} \le 313\,\text{cm}^{-1}$$
$$255 \le D_0^{B,19} \le 284\,\text{cm}^{-1}$$
$$449 \le D_0^{E,v_E} \le 478\,\text{cm}^{-1}$$

The

$$N_2I_2(E, v_E, n_E) \rightarrow N_2(X) + I_2(D', v_D) \tag{6.3.21}$$

and

$$N_2I_2(E, v_E, n_E) \rightarrow N_2(X) + I_2\left(\beta, v_\beta\right) \tag{6.3.22}$$

EP channels have the greatest probability. The former is forbidden in the $C_{\infty v}$ point group (0^+ - 2). Therefore, it proceeds in the C_{2v} point group, ($A_1 - A_1 \oplus B_1$ ($A_1 \oplus B_2$)) and C_s point group, $(A' - A' \oplus A'')$ (bent configuration). The latter is allowed in the C_s point group $(A' - A'' \oplus A')$.

The

$$N_2I_2(E, v_E, n_E) \rightarrow N_2(X) + I_2(D, v_D) \tag{6.3.23}$$

EP has very low probability; it is forbidden in the C_{2v}. point group (A_1 –B_2 (B_1)) [62].

It is interesting to compare the D_0^{E,v_E} values of the $ArI_2(E,v_E,n_E)$ (410–415 cm^{-1}), $KrI_2(E,v_E,n_E)$ (663–683 cm^{-1}) and $N_2I_2(E,v_E,n_E)$ (449– 478 cm^{-1}). The long-range, $\sim R^{-4}$, interaction between the $I_2(E0_g^+ - D0_u^+)$ transition electric dipole moment and the permanent electric quadrupole moment a of N_2 (Θ= 1.5 10^{-26} *esu*) can be strong in these case if the E,v_E and D,v_D vibronic levels are near-resonant (*R* is the distance between centers of mass of the colliding partners). The EP channel (6.3.23) has to prevail in this case (see Sect. 5.3.2.5 in [18]).

Besides, dispersion, $\sim R^{-6}$, interaction has to be strong for all, M = Ar, Kr, N_2, partners due to huge dipole moment of the $I_2(E0_g^+ - D0_u^+)$ transition, $\mu_{E\text{-}D} \approx 25$ 10^{-18} *esu* and, hence, giant polarizability $\alpha_{I_2(E)} \approx 2 \cdot 10^5 \text{Å}^3$ of the $I_2(E)$ state. Binding energies of the complex have to be proportional to the $|\mu_{E-D}|^2 \alpha_M$ products, α_M = 1.64 (Ar), 1.77 (N_2), 2.48 (Kr), units, Å^3, are polarizabilities of the partners (see above and Fig. 6.32).

One sees, that dispersion interaction prevails in the $N_2I_2(E,v_E,n_E)$ complexes since:

the EP channel (6.3.23) has very low probability,
the $N_2I_2(E,v_E,n_E)$ binding energy is larger than that of $ArI_2(E,v_E,n_E)$ and less than $KrI_2(E,v_E,n_E)$.

To examine in what N_2I_2 isomer transitions in the complex occur, calculation of the $N_2I_2(X)$ *PES* taking into account its linear, parallel, *T*-shaped and trapezoidal isomers. It was shown that the parallel $N_2I_2(E)$ isomer is observed in the experiments, probably. The low energy barrier between trapezoidal and parallel isomers results in free rotor vibrational state. Both C_{2v} and C_s symmetry are realized, so EP channels (3.6.21, 3.6.22) are allowed.

6.4 Rare Gas-Halogen Molecule Clusters

6.4.1 $Rg_x^1Rg_y^2I_2$ Clusters

The first studies of the weakly-bound clusters were carried out by D.H. Levy and coworkers [103, 125, 127, 128]. It was shown, in particular, that there are blue shifts in excitation spectra of VP product luminescence

$$Ne_xHe_yI_2(B, v_B \leftarrow X, 0), \tag{6.4.1}$$

$\nu - \nu_0 \approx 6.25x + 3.67y$ (cm^{-1}), $x \leq 4$, $y \leq 4$, v_B= 13–26 [103]

$$Ar_xHe_yI_2(B, v_B \leftarrow X, 0), \tag{6.4.2}$$

$\nu - \nu_0 \approx 13.5x + 3.8y$ (cm^{-1}), $x \leq 3$, $y \leq 2$, v_B= 21, 22 [128] clusters. Here, ν is a wavenumber of the excitation band maximum, and ν_0 is a wavenumber of the excitation band maximum of the uncomplexed I_2 very near to the bandhead). The vdW progressions (n_B= 0,1, and 2) were also identified in $HeNeI_2(B,v_B)$. It was shown that all complexes required at least one $I_2(B)$ stretching quantum per rare gas atom for complete decay. Larger clusters tended to require more vibrational quanta per rare gas atom to decay. The Morse potential parameters, ω_e = 21.71(31), $\omega_e x_e$ = 0.81(0.11), D_e = 88.1(5.4), D_0 = 76.2(5.6) were determined for the $HeNeI_2(B,v_B$= 21) state. The $HeNeI_2(B,v_B$= 12) and $Ne_2I_2(B,v_B$= 12) lifetimes are ~ 100 ps [103]. The probabilities of the

$$Ne_2I_2(B, v_B = 21, 22) \rightarrow \ldots \rightarrow Ne + Ne + I_2(B, v_B - \Delta v_B), \quad (6.4.3)$$

decay are ~ 0.55, 0.41, 0.04, and 0.002 (v_B= 21), and ~ 0.45, 0.47, 0.07, and 0.01 (v_B= 22) for the Δv_B = 2, 3, 4, and 5 channels, respectively [103].

Detailed studies of real-time dynamics of the $Ne_nI_2(B,v_B)$ cluster, n = 2–4, in the picosecond time scale have been carried out by A.H. Zewail and his team [129]. The authors have concluded that the dynamics does not follow a simple sequential mechanism corresponding to direct predissociation:

$$Ne_2I_2(B, v_B) \rightarrow 2Ne + I_2(B, v_B - \Delta v_B); \quad (6.4.4)$$

the presence of IVR has been invoked to account the Δv_B = 3 channel. The Δv_B = 2 channel proceeds via sequential direct VP.

According to [131], the mechanism of the $Ne_2I_2(B,v_B)$ VP can be described, as it is done in Fig. 6.40.

Spectroscopic studies of the $HeNeI_2$ cluster have been carried out using the two-step two-color

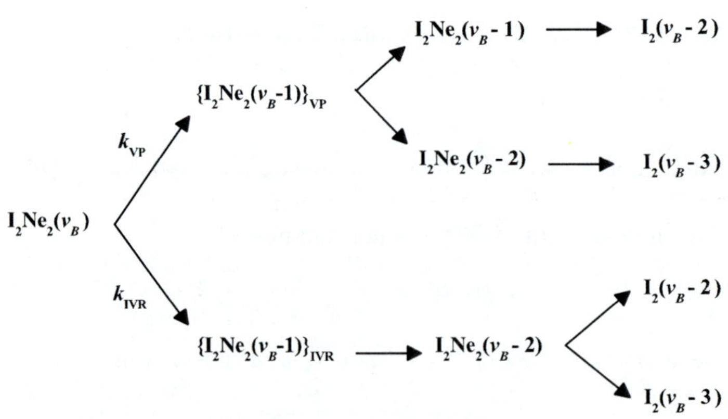

Fig. 6.40 The $Ne_2I_2(B, v_B)$ vibrational predissociation (see [130]) (Reproduced from A. Bastida, J. Zuñiga, A. Requena, N. Halberstadt, J. A. Beswick,. J. Chem. Phys. **109**, 6320–6328 (1998). http://doi.org/10.1063/1.477274 with the permission of AIP Publising)

$$\mathrm{HeNeI_2}\left(E0_g^+, v_E = 0 - 3 \overset{h\nu_2}{\leftarrow} B0_u^+ v_B = 19 \overset{h\nu_1}{\leftarrow} X0_g^+, v_X = 0\right) \qquad (6.4.4)$$

excitation scheme in [133]. The $He_xNe_yI_2(B,19)$, $x + y \leq 5$, complexes and clusters VP products have been observed at the first, $HeNeI_2(B,v_B = 19 \overset{h\nu_1}{\leftarrow} X, v_X = 0)$, step. The $HeNeI_2(B,v_B)$ state decay has been found to be sequential with the formation of the $HeI_2(B,v_B\text{-}1)$ and $NeI_2(B,v_B\text{-}1)$ complexes without intermolecular excitation at the first step:

$$\mathrm{HeNeI_2}(B, v_B) \rightarrow \mathrm{HeI_2}(B, v_B\text{-}1) + \mathrm{Ne}, \qquad (6.4.5a)$$

$$\mathrm{HeI_2}(B, v_B\text{-}1) \rightarrow \mathrm{I_2}(B, v_B\text{-}2) + \mathrm{He} \qquad (6.4.5b)$$

$$\mathrm{HeNeI_2}(B, v_B) \rightarrow \mathrm{NeI_2}(B, v_B\text{-}1) + \mathrm{He}, \qquad (6.4.6a)$$

$$\mathrm{NeI_2}(B, v_B\text{-}1) \rightarrow \mathrm{I_2}(B, v_B\text{-}2) + \mathrm{Ne}. \qquad (6.4.6b)$$

The probability of $HeNeI_2(B,19)$ VP channel (6.4.5a) is ~ 2 times higher than that of (6.4.6a). Besides, the excitation bands can be fully ascribed to the $HeI_2(B, 18, n_B = 0)$ and $NeI_2(B, 18, n_B = 0)$ population, i.e., the loss of one Rg atom does not lead to excitation of the vdW modes. The (6.4.5a and 6.4.6a) decay channels are described in the framework of the direct VP.

The $I_2(D0_u^+, v_D \leq 2, \beta 1_g, v_\beta \leq 3$ and $D, v_D \leq 3, \beta, v_\beta \leq 4)$ have been observed after optical population of the $HeNeI_2(E, v_E = 0$ and $1)$, respectively. No $I_2(E, v_E$ and $D'2_g, v_{D'})$ luminescence have been observed. (Figure 6.41)

These features allow the authors to conclude that $HeNeI_2(E, v_E = 0{,}1)$ decay can be explained under the assumption that the most probable decay channel is EP with loss of He on the first step accompanied by subsequent VP and EP. The $HeNeI_2(X, B, E)$ binding energies have been estimated to be less than 111.8, 101.6, and 117.9 cm^{-1}, respectively.

The $HeNeI_2(X)$ state *PES* based on coupled cluster (CCSD(T)) calculations has been constructed. It has been shown that the potential derived as the sum of HeNe, $HeI_2(X)$, and $NeI_2(X)$ CCSD(T) potentials represents adequately ab initio *PES*. A comparison of the calculated $HeNeI_2(X)$ vibrational energies with the experimental data has shown that observed transitions can be ascribed to the $HeNeI_2(X, B, E)$ cluster in a tetrahedral geometry [132].

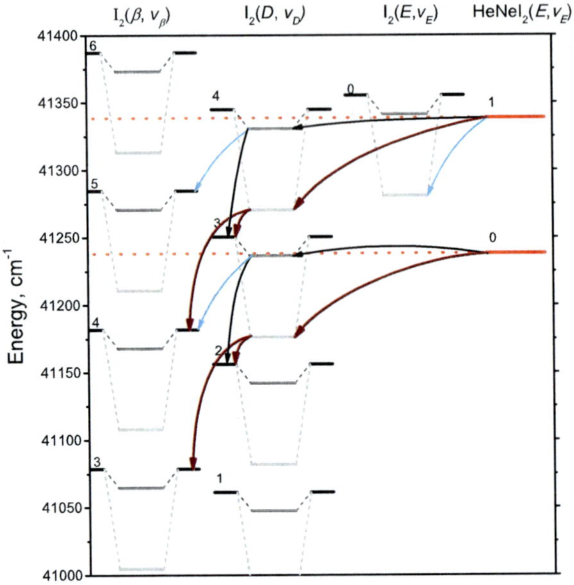

Fig. 6.41 Predissociation scheme for $HeNeI_2(E, v_E= 0, 1)$ decay. On the left side there are $I_2(\beta, D)$ molecule (black) and $RgI_2(\beta, D)$, complex Rg = He, Ne (grey and light grey, respectively) vibrational level energies; the $HeNeI_2(E, v_E=0, 1)$ levels are shown on the right side. All values are calculated relative to the $I_2(X,0,4)$ energy level. The $HeNeI_2(E)$, $HeI_2(IP)$ and $NeI_2(IP)$ binding energies are assumed to be 117.9 cm^{-1}, 14 cm^{-1} [90–92], and 74 cm^{-1} [96, 97] (see Sects. The HeI_2 complexes and NeI_2 complexes, also) (see [132]) (Reproduced from A. S. Andreev, V. V. Baturo, S. S. Lukashov, S. A. Poretsky, A. M. Pravilov, A. I. Zhironkin, J. Chem. Phys. 152, 234307 (10 pp) (2020) https://doi.org/10.1063/5.0008760 with the permission of AIP Publising)

6.4.2 Rg_2Hal_2 Clusters

6.4.2.1 The Ne_2Br_2 Clusters

The authors of [133] have studied the $Ne_2Br_2(B, v_B= 16–23)$ cluster VP utilizing time- and frequency-resolved pump-probe spectroscopy. After optical population of the cluster, the authors have followed the flow of the $Br_2(B, v_B)$vibrational energy to the vdW modes in real time by recording the time-dependent behavior of the $Ne_2Br_2(B, v_B)$, the $NeBr_2(B, v_B\text{-}m)$ intermediates, and the $Br_2(B, v_B\text{-}n)$ VP products. The only $NeBr_2(B, v_B\text{-}1)$ intermediate has been observed for the $Ne_2Br_2(B, v_B= 16\text{-}18)$ clusters, and the majority of the final VP products product is $Br_2(B, v_B\text{-}2)$, i.e., the decay happens via two sequential direct VP steps. The authors have fitted the time-dependent behavior of these species to the sequential $Ne_2Br_2(B, v_B) \xrightarrow{\tau_1} NeBr_2(B, v_B\text{-}m) \xrightarrow{\tau_2} Br_2(B, v_B\text{-}n)$ mechanism and extracted time constants for each step: τ_1 = 32 ps, τ_2 = 82 ps for $Ne_2Br_2(B, v_B= 17$, and decrease with v_B. The decay occurs via multiple pathways for higher v_B levels. The $Br_2(B, v_B\text{-}\Delta v_B)$,

$\Delta v_B > 2$, VP product becomes much more important. For $v_B = 21$, both $NeBr_2(B, v_B\text{-}1$ and $v_B\text{-}2)$ intermediates have been observed. These intermediates have significantly different kinetics, with the $NeBr_2(B, v_B\text{-}1)$ decay rate nearly twice that of the $NeBr_2(B, v_B\text{-}2)$. Both $Br_2(B, v_B\text{-}2)$ and $Br_2(B, v_B\text{-}3)$ VP products are formed in almost equal amounts, but the $Br_2(B, v_B\text{-}2)$ product formation rate is faster than that of $Br_2(B, v_B\text{-}3)$. The broad vibrational product state distributions and multiple dissociation pathways indicate that IVR becomes increasingly important for $v_B > 19$.

6.4.2.2 The Rg_2Cl_2 Clusters

The Ar_2Cl_2 Clusters. $Ar_2Cl_2(X)$ cluster is a distorted tetrahedron with an Ar_2-Cl_2 distance of 3.9(5) Å and an Ar–Ar bond distance of 4.1(6) Å. The binding energy of the $Ar_2Cl_2(X)$ cluster is $447.5 \pm 3.5\ cm^{-1}$, and that of $Ar_2Cl_2(B, 9)$ state is $\sim 426\ cm^{-1}$. Three $Cl_2(B, v_B)$ vibrational quanta are necessary for the $Ar_2Cl_2(B, 9)$ VP [134].

The Ne_2Cl_2 Clusters. The structure of the $Ne_2Cl_2(X)$ state, as well as the structure and dynamics of $Ne_2Cl_2(B)$ state, have been studied using the laser pump-probe spectroscopy in [135]. It has been shown that Ne_2Cl_2 has a distorted tetrahedral structure in both states with a Ne–Ne bond length of 3.23(5) Å and Ne_2 center of mass to Cl_2 center of mass distance of 3.12(3) Å. The $Ne_2Cl_2(B, v_B = 7\text{–}13)$ binding energy is $\sim 147\ cm^{-1}$. For the $v_B < 10$ levels, transfer of one $Cl_2(B, v_B)$ vibrational quantum to the vdW modes is sufficient to dissociate both neon atoms from the complex, but the $Ne_2Cl_2(B, v_B - \Delta v_B)$ VP occurs via $\Delta v_B = 1$, as well as the $\Delta v_B = 2$ channel. For the $\Delta v_B = 2$ channel, the $NeCl_2(B)$ VP product has been observed. This feature indicates that such a sequential mechanism competes with the direct $Ne_2Cl_2(B, v_B)$ VP. The $Ne_2Cl_2(B, v_B)$ lifetime is 123.7, 44.4 and 9.5 (units, ps) for the $v_B = 8$, 11 and 13, respectively.

The He_2Cl_2 Clusters. The action and excitation spectra of the $Cl_2(E \to B)$ luminescence have been measured in [136] to study the VP product state distribution. It has been shown that the $\Delta v_B = 2$ VP channel is the principal one, and it is almost certainly a sequential process:

$$He_2Cl_2(B, v_B) \to \begin{array}{c} He + He_2Cl_2(B, v_B - 1) \\ \downarrow \\ He + Cl_2(B, v_B - 2) \end{array} \tag{6.4.7}$$

The $Cl_2(B, v_B)$ rotational distribution is remarkably similar to that of $HeCl_2(B, v_B)$ observed for the $\Delta v_B = 2$ channel (see Sect. 6.3.2.2). The authors of [137] believed that the reason for the similarity was that vibrational excitation of the $HeCl_2(B, v_B)$ vdW modes had little effect on the product state distribution. He_2Cl_2 provides an example of an extremely floppy, liquidlike cluster. The *ab initio* calculations of the $He_2Cl_2(X, B)$ state structures have been carried out also (see [137] and references).

6.5 RgMe vdW Complexes

As it mentioned in Sect. 6.3.1, the RgMe, Me = Ni, Zn, Cd, Hg, and Al, vdW complexes have weakly bound electronically-excited states. Let us analyze the data concerning these complexes briefly.

The NiAr Complexes. Transitions in the NiAr complex were observed at wavelengths near the Ni($y^1D_2 \leftarrow a^1D_2$) transition, $\tilde{\nu}_2 \approx 34000\,\text{cm}^{-1}$. The $v' \leftarrow v''$, (0–5) $\leftarrow$ 0 bands were assigned in the excitation spectrum, and the binding energy of the upper state was determined, $D_0' = 51.7\,\text{cm}^{-1}$. The upper and lower states were not assigned [44].

The ZnRg Complexes. Some spectroscopic parameters, ω_e, $\omega_e x_e$, D_e, and R_e, of the ZnRg($X0(^1\Sigma^+)$ and $D1(^1\Pi)$), Rg = He, Ne, Ar, vdW complexes have been given and discussed in the monograph [40].

The HgRg Complexes. Some spectroscopic parameters, ω_e, $\omega_e x_e$, D_e, B_e, and R_e, of the HgRg($X0(^1\Sigma^+$, $A0^+(^3\Pi)$, $B1(^3\Sigma^+ + {}^3\Pi)$), and $C1(^1\Pi)$, Rg = He, Ne, Ar, vdW complexes have been given and discussed in the monograph [40].

The CdRg Complexes. Rotational analysis of the CdHe($A0^+$ $(5^3P_1) \leftarrow X0^+$ (5^1S_0) and $B1$ $(5^3P_1) \leftarrow X$) transitions luminescence excitation spectra were carried out in [138, 139]. Some spectroscopic parameters of these states, ω_e, $\omega_e x_e$, D_e, B_e, and R_e, were determined. Data on spectroscopic parameters of CdRg ($X0(^1\Sigma^+$, $A0^+(^3\Pi)$), $B1(^3\Sigma^+ + {}^3\Pi)$), and $D1(^1\Pi)$), Rg = He, Ne, Ar, vdW complexes are given in the monograph [40].

The AlAr Complexes. Rotationally resolved spectra were recorded for six bands of the AlAr($B^2\Sigma^+ \leftarrow X^2\Pi_{1/2}$) transition. The ω_e, $\omega_e x_e$, $\omega_e y_e$, $\omega_e z_e$, B_e, α_e, γ_e, D_e, D_0, and R_e spectroscopic parameters were determined for both states [41].

References

1. Person, W.B.: Theoretical aspects of the study of molecular complexes. In: Spectroscopy and Studies of Molecular Complexes ed. J. Yarwood, 1–104, Springer (1973)
2. Mulliken, R.S., Person, W.B.: Molecular Complexes. John Wiley and Sons, New York (1969)
3. Blaney, B.L., Ewing, G.E.: Van der Waals molecules. Ann. Rev. Phys. Chem. **27**, 553–586 (1976). https://doi.org/10.1146/annurev.pc.27.100176.003005
4. IUPAC Compendium of Chemical Terminology, 2nd ed. (the Gold book). Complied by A.D. McNaught and A. Wilkinson, Blackwell Scientific Publications, Oxford (1997). http://goldbook.iupac.org
5. Arunan, E, Desiraju, G.R., Klein, R.A., J Sadlej, J., Scheiner, S, Alkorta, I., Clary, D.C., Crabtree, R.H., Dannenberg, J.J., Hobza, P., Kjaergaard, H.G., Legon, A.C, Mennucci, B., Nesbitt, D.J.: Definition of the hydrogen bond (IUPAC Recommendations 2011). Pure Appl. Chem. **83**, 1637–1641 (2011). https://doi.org/10.1351/pac-rec-10-01-02
6. Desiraju, G.R., Ho, P.S., Kloo, L., Legon, A.C., Marquardt, R., Metrangolo, P., Politzer, P., Resnati, G., Rissanen, K.: Definition of the halogen bond (IUPAC Recommendations 2013). Pure Appl. Chem. **85**, 1711–1713 (2013). https://doi.org/10.1351/PAC-REC-12-05-10

7. Legon, A.C.: The interaction of dihalogens and hydrogen halides with Lewis bases in gas phase: An experimental comparison of the halogen bond and hydrogen bond: in Structure and Bonding, eds. P. Metrangolo, G. Restani, **126**, 17–64 (2008). https://doi.org/10.1007/430_2007_063
8. Legon, A.C.: A reduced radial potential energy function for the halogen bond and the hydrogen bond in complexes B···XY and B···HX, where X and Y are halogen atoms. Phys. Chem. Chem. Phys. **16**, 12415–12420 (2014). https://doi.org/10.1039/c4cp01444h
9. Springer, S.D., Rivera-Rivera, L.A., McElmurry, B.A., Wang, Z., Leonov, I.I., Lucchese, R. R. Legon, A.C., Bevan, J.W.: CMM-RS potential for characterization of the properties of the Halogen-Bonded OC-Cl_2 Complex, and a Comparison with Hydrogen-Bonded OC-HCl. J. Phys. Chem. A, **116**, 1213–1223 (2012). dx.doi.org/https://doi.org/10.1021/jp209870x
10. Rivera-Rivera, L.A., Scott, K.W., McElmurry, B.A., Lucchese, R.R., Bevan, J.W.: Compound modelmorphed potentials contrasting OC–79Br 35Cl with the halogen bonded OC–35Cl2 and hydrogen-bonded OC–HX (X = 19F, 35Cl, 79Br). Chem. Phys. **425**, 162–169 (2013). https://doi.org/10.1016/j.chemphys.2013.08.016
11. Shaw, R.A., J. Hill, J.G., Legon, A.C.: Halogen bonding with phosphine: Evidence for Mulliken Inner complexes and the importance of relaxation energy. J. Phys. Chem. A. **120**, 8461–8468 (2016). https://doi.org/10.1021/acs.jpca.6b08945
12. Anable, J.P., Hird, D.E., Stephens, S.L., Zaleski, D.P., Walker, N.R., Legon, A.C.: Characterisation of the weak halogen bond in N_2···ICF_3 by pure rotational spectroscopy. Chem. Phys. Letts. **625**, 179–185 (2015). https://doi.org/10.1016/j.cplett.2015.02.023
13. McIntosh, A.L., Wang, Z., Castillo-Chara, J., Lucchese, R.R., Bevan, J.W., Suenram, R.D., Legon, A.C.: The structure and ground state dynamics of Ar—IH. J. Chem. Phys. **111**, 5764–5771 (1999). https://doi.org/10.1021/acs.jpca.6b08945
14. Walton, J.R., Rivera-Rivera, L.A., Lucchese, R.R., Bevan, J.W.: A canonical approach to multi-dimensional van der Waals, hydrogen-bonded, and halogen-bonded potentials. Chem. Phys. **469–470**, 60–64 (2016). https://doi.org/10.1016/j.chemphys.2016.01.011
15. Walton, J.R., Rivera-Rivera, L.A., Lucchese, R.R., Bevan, J.W.: A general transformation to canonical form for potentials in pairwise interatomic interactions. Phys. Chem. Chem. Phys. **17**, 14805–14810 (2015). https://doi.org/10.1039/C5CP01543J
16. Pauling, L.: The Nature of the Chemical Bond. IV. The energy of single bonds and the relative electronegativity of atoms. J. Amer. Chem. Soc. **54**, 3570–3582 (1932)
17. Legon, A.C., Millen, D.J.: Hydrogen bonding as a probe of electron densities: Limiting gas-phase nucleophilicities and electrophilicities of B and HX. J. Amer. Chem. Soc. **109**, 356–358 (1987). https://doi.org/10.1021/ja00236a011
18. Lukashov, S., Petrov, A., Pravilov, A.: The Iodine Molecule: Insights into Intra- and Intermolecular Perturbation in Diatomic Molecules. Springer (2018). https://doi.org/10.1007/978-3-319-70072-4
19. Smalley, R.E., Levy, D.H., Wharton, L.: The fluorescence excitation spectrum of the HeI_2 van der Waals complex. J. Chem. Phys. **64**, 3266–3275 (1976). https://doi.org/10.1063/1.432667
20. Novick, S.E., Davis, P., Harris, S.J., Klemperer, W.: Determination of the structure of ArHCI. J. Chem. Phys. **59**, 2273–2279 (1973). https://doi.org/10.1063/1.1680332
21. Harris, S.J., Janda, K.C., Novick, S.E., Klemperer, W.: Intermolecular potential between an atom and a linear molecule: The structure of ArOCS. J. Chem. Phys. **63**, 881–884 (1975). https://doi.org/10.1063/1.431368
22. Novick, S.: Bibliography of rotational spectra of weakly bound complexes. https://wesfiles.wesleyan.edu/home/snovick/SN_webpage/vdW.pdf
23. Rohrbacher, A., Williams, J., Janda, K.C.: Rare gas-dihalogen potential energy surfaces. Phys. Chem. Chem. Phys. **1**, 5263–5276 (1999). https://doi.org/10.1039/A906664K
24. Chan, K.W., Power, T.D., Jai-nhuknan, J., Cybulski, S.M.: An *ab initio* study of He–F_2, Ne–F_2, and Ar–F_2 van der Waals complexes. J. Chem. Phys. **110**, 860–869 (1999). https://doi.org/10.1063/1.478053

25. Bieler, C.R., Janda, K.C., Hernández-Lamoneda, R., Octavio Roncero, O.: $NeCl_2$ and $ArCl_2$: Transition from direct vibrational predissociation to intramolecular vibrational relaxation and electronic nonadiabatic effects. J. Phys. Chem. A **114**, 3050–3059 (2010). https://doi.org/10.1021/jp906392m
26. Cappelletti, D., Cinti, A., Nicoziani, A., Falcinelli,, S., Pirani, F.: Molecular beam scattering experiments as a sensitive probe of the interaction in bromine–noble gas complexes. Front. Chem., 17 May 2019. https://doi.org/10.3389/fchem.2019.00320
27. Valdés, A., Prosmiti, R.: Preferential stabilization of HeI_2 van der Waals isomers: the effect of energetics and temperature. RSC Adv. **7**, 19273–19279 (2017). https://doi.org/10.1039/c7ra01378g
28. Darr, J.P., Loomis, R.A.: Experimental interrogation of the multidimensional He + ICl(E,$v^\dagger$) and He + IC(β,$v^\dagger$) intermolecular potential energy surfaces. J. Chem. Phys. **129**, 144306 (11 pp) (2008). https://doi.org/10.1063/1.2990661
29. Hussain, A.N., Gareth Roberts, G.: Ultrafast non-adiabatic curve crossing and energy transfer in the IBr BO^+·Ar van der Waals Complex. J. Chin. Chem. Soc. **47**, 821–836 (2000). https://doi.org/10.1002/jccs.200000110
30. Zheng, R., Li, S., Chen, S.-J., Chen, Y., Zheng, L.-M.: Theoretical investigation of potential energy surface and bound states for the van der Waals complex Ar–BrCl dimer. Chem. Phys. **458**, 77–85 (2015). https://doi.org/10.1016/j.chemphys.2015.07.013
31. Higgins, K., Tao, F.-M., Klemperer, W.: The intermolecular potential between an inert gas and a halogen: Prediction and observation of transitions between the linear and T-shaped isomers of HeClF. J. Chem. Phys. **109**, 3048–3061 (1998). https://doi.org/10.1063/1.476896
32. Goldstein, N., Brack, T.L., Atkinson, G.H.: Quantitative absorption spectroscopy and dissociation dynamics of I_2 van der Waals complexes with He, Ar, Kr. Xe. J. Chem. Phys. **85**, 2684–2691 (1986). https://doi.org/10.1063/1.451078
33. Burroughs, A., Kerenskaya, G., Heaven, M.C.: Spectroscopy and dynamics of I_2(B)-Ne. J. Chem. Phys. **115**, 784–791 (2001). https://doi.org/10.1063/1.1378317
34. Drobits, J.C., Lester, M.I.: Optical–optical double resonance of the ICl–Ne complex: Binding energies in the E(0^+), A($^3\Pi_1$), and X($^1\Sigma^+$) states. J. Chem. Phys. **86**, 1662–1669 (1987). https://doi.org/10.1063/1.452164
35. Stephenson, T.A., Hong, Y., Lester, M.I.: Nonadiabatic transitions in the dynamics of the NeICl van der Waals complex. Chem. Phys. Letts. **159**, 549–553 (1989). https://doi.org/10.1016/0009-2614(89)87530-X
36. Stephenson, T.A., Hong, Y., Lester, M.I.: Nonadiabatic electronic interactions in the ion-pair states of NeICl. J. Chem. Phys. **94**, 4171–4181 (1991). https://doi.org/10.1063/1.460650
37. Ray, S.A., McCoy, A.B., Glennon, J.J., Darr, J.P., Fesser, E.J., Lancaster, J.R., Loomis, R. A.: Experimental and theoretical investigation of the He···I_2 rovibronic spectra in the I_2 B-X, 20—0 region. J. Chem. Phys. **125**, 164314 (9 pp) (2006). https://doi.org/10.1063/1.2363378
38. Kalemos, A., Valdés, Á., Prosmiti, R.: Theoretical investigation of the He–I_2($E^3\Pi_g$) ion-pair state: Ab initio intermolecular potential and vibrational levels. J. Chem. Phys. **137**, 034303 (8 pp) (2012). https://doi.org/10.1063/1.4733983
39. Baturo, V.V., Lukashov, S.S., Poretsky, S.A., Pravilov, A.M., Zhironkin, A.I.: The ArI_2(ion-pair states) van der Waals complexes. Chem. Phys. Lett. **647**, 161–164 (2016). https://doi.org/10.1016/j.cplett.2016.01.053
40. Koperski, J.: Van der Waal Complexes in Supersonic Beams. Willey-VCH Verlag GmbH & Co (2003). https://doi.org/10.1002/3527600817
41. McQuaid, M.J., Gole, J.L., Heaven, M.C.: Spectroscopy of the AlAr van der Waals complex: Rotationally resolved $B^2\Sigma^+ \leftarrow X^2\Pi_{1/2}$ electronic transitions. J. Chem. Phys. **92**, 2733–2739 (1990). https://doi.org/10.1063/1.457918
42. Koperski, J., Czajkowski, M.A.: Short-range characterization of the MeAr (Me = Zn, Cd) ground-state potentials from fluorescence spectra. J. Molec. Spectrosc. **212**, 162–170 (2002). https://doi.org/10.1006/jmsp.2002.8550
43. Koperski, J., Atkinson, J.B., Krause, L.: Spectroscopy of the $A0^+$ and B1 states in HgAr and HgNe. Chem. Phys. **186**, 401–407 (1994). https://doi.org/10.1016/0301-0104(94)00170-7

44. Kawamoto, Y., Honma, K.: Fluorescence excitation spectrum of Ni–Ar complex. Chem. Phys. Lett. **298**, 227–232 (1998). https://doi.org/10.1016/S0009-2614(98)01178-6
45. Mulliken, R.S.: Iodine revisited. J. Chem. Phys. **55**, 288–309 (1971). https://doi.org/10.1063/1.1675521
46. Okabe, H.: Photochemistry of small molecules. A Willey Interscience Publication, New York, Chichester, Brisbane, Toronto (1978)
47. Pravilov, A.M.: Fotoprozessy v molekulyarnykh gasakh (Photoprocesses in molecular gases). Energoatomizdat, Moscow (1992). (in Russian)
48. Huber, K.P., Herzberg, G.: Molecular Spectra and Molecular Structure IV. Constants of Diatomic Molecules. Springer (1979). https://doi.org/10.1007/978-1-4757-0961-2
49. Naumkin, FYu., Knowles, P.J.: On the adequacy of pairwise additive potentials for rare gas–halogen systems: The effect of anisotropy of interactions between atoms. J. Chem. Phys. **103**, 3392–3399 (1995). https://doi.org/10.1063/1.470224
50. Coxon, J.A.: The low-lying electronic states of diatomic halogen molecules. Molecular Spectroscopy. **1**, 177–228 (1973). https://doi.org/10.1039/9781847556684-00177
51. Heaven, M.C.: Fluorescence decay dynamics of the halogens and interhalogens. Chem. Soc. Rev. **15**, 405–448 (1986). https://doi.org/10.1039/CS9861500405
52. Tellinghuisen, J., Charkraborty, D.K.: Identification and analysis of the *D*' → *A*' transition in the emission spectrum of Cl_2. Chem. Phys. Lett. **134**, 565–560 (1987). https://doi.org/10.1016/0009-2614(87)87194-4
53. Cline, J.I., Evard, D.D., Thommen, F., Janda, K.C.: The laser Induced fluorescence spectrum of the $HeCI_2$ van der Waals molecule. J. Chem. Phys. **84**, 1165–1170 (1986). https://doi.org/10.1063/1.450507
54. Cline, J.I., Reid, B.P., Evard, D.D., Sivakumar, N., Halberstadt, N., Janda, K.C.: State-to-state vibrational predissociation dynamics and spectroscopy of $HeCI_2$: Experiment and theory. J. Chem. Phys. **89**, 3535–3552 (1988). https://doi.org/10.1063/1.454924
55. Cybulski, S.M., Holt, J.S.: *Ab initio* potential energy surfaces for He–Cl_2, Ne–Cl_2, and Ar–Cl_2. J. Chem. Phys. **110**, 7745–7755 (1999). https://doi.org/10.1063/1.470224
56. Rohrbacher, A., Janda, K.C., Beneventi, L., Casavecchia, P., Volpi, G.G.: Differential scattering cross sections for $HeCl_2$, $NeCl_2$, and $ArCl_2$: Multiproperty fits of the potential energy surfaces. J. Phys. Chem. A **101**, 6528–6537 (1997). https://doi.org/10.1021/jp970738b
57. Williams, J., Rohrbacher, A., Seong, J., Marianayagam, N., Janda, K.C., Burci, R., Szcześniak, M.M., Grzegorz Chałasiński, Cybulski, S.M., Halberstadt, N.: A three-dimensional potential energy surface for He + Cl_2 ($B^3\Pi_{0u+}$): *Ab initio* calculations and a multiproperty fit. J. Chem. Phys. **111**, 997–1007 (1999). https://doi.org/10.1063/1.479190
58. Naumkin, FYu., McCourt, F.R.W.: *Ab initio* based potential energy surfaces, microwave spectrum, and scattering cross section of the ground state Ne–Cl_2 system. J. Chem. Phys. **109**, 1271–1284 (1998). https://doi.org/10.1063/1.476713
59. Cline, J.I., Sivakumar, N., Evard, D.D., Bieler, C.R., Reid, B.P., Halberstadt, N., Hair, S.R., Janda, K.C.: Product state distributions for the vibrational predissociation of $NeCI_2$. J. Chem. Phys. **90**, 2605–2616 (1989). https://doi.org/10.1063/1.456669
60. Evard, D.D., Thommen, F., Cline, J.I., Janda, K.C.: Molecular beam-laser spectroscopy of the $NeCI_2$ van der Waals molecule. J. Phys. Chem. **91**, 2508–2512 (1987). https://doi.org/10.1021/j100294a014
61. Brinza, D.E., Western, C.M., Evard, D.D., Thommen, F., Swartz, B.A., Janda, K.A.: Molecular beam-laser spectroscopy of Ne-CI_2: Observation of a metastable vibrationally excited van der Waals molecule. J. Phys. Chem. **88**, 2004–2009 (1984). https://doi.org/10.1021/j100294a014
62. Buchachenko, A.A., Halberstadt, N., Lepetit, B., Roncero, O.: Ar•••I_2: A model system for complex dynamics. Int. Rev. Phys. Chem. **22**, 153–202 (2003). https://doi.org/10.1080/0144235031000075726

63. Roncero, O., Caloto, D., Yanda, K.C., Halberstadt, N.: From the sparse to the statistical limit of intramolecular vibrational redistribution in vibrational predissociation: $ArCl_2$ as an example. J. Chem. Phys. **107**, 1406–1419 (1997). https://doi.org/10.1063/1.474495
64. Evard, D.D., Cline, J.I., Janda, K.C.: The structure and bond energy of $ArCI_2$. J. Chem. Phys. **88**, 5433–5438 (1988). https://doi.org/10.1063/1.454553
65. Tao, F.-M., Klemperer, W.: The van der Waals potential-energy surfaces and the structures of ArCIF and $ArCI_2$. J. Chem. Phys. **97**, 440–451 (1992). https://doi.org/10.1063/1.463589
66. Xu, Y., Jager, W., Ozier, I., Gerry, M.C.L.: Rotational spectrum, structure, and chlorine nuclear quadrupole coupling constants of the van der Waals complex Ar-CI_2. J. Chem. Phys. **98**, 3726–3731 (1993). https://doi.org/10.1063/1.464050
67. Reid, B.P., Yanda, K.C., Halberstadt, N.: Vibrational and rotational wave functions for the triatomic van der Waals molecules $HeCI_2$, $NeCI_2$, and $ArCI_2$. J. Phys. Chem. **92**, 587–593 (1988). https://doi.org/10.1021/j100314a005
68. Evard, D.D., Bieler, C.R., Cline, J.I., Sivakumar, N., Janda, K.C.: The vibrational predissociation dynamics of $ArC1_2$: Intramolecular vibrational relaxation in a triatomic van derWaals molecule? J. Chem. Phys. **89**, 2829–2838 (1988). https://doi.org/10.1063/1.454986
69. Herzberg, G.: Molecular Spectra and Molecular Structure I. Spectra of Diatomic Molecules. D. van Nostrand Co., Inc, Second Edition (1950)
70. Bieler, C.R., Janda, K.C.: $XeCl_2$: A van der Waals Molecule. J. Amer. Chem. Soc. **112**, 2033–2034 (1990). https://doi.org/10.1021/ja00161a077
71. Bieler, C.R., Spence, K.E., Janda, K.C.: Characterization of the $KrCI_2$ and $XeCI_2$ van der Waals isomers. J. Phys. Chem. **95**, 5058–5064 (1991). https://doi.org/10.1021/j100166a029
72. Prosmiti, R., Cunha, C., Villarreal, P., Delgado-Barrio, G.: Ab initio ground state potential energy surfaces for Rg–Br 2 (Rg = He, Ne, Ar) complexes. J. Chem. Phys. **116**, 9249–9254 (2002). https://doi.org/10.1063/1.1473800
73. Carrillo-Bohórquez, O., Valdés, Á., Prosmiti, R.: Temperature dependence of HeBr 2 isomer's stability through rovibrational multiconfiguration time-dependent Hartree alculations. J. Phys. Chem. A **120**, 9458–9464 (2016). https://doi.org/10.1021/acs.jpca.6b09107
74. Boucher, D.S., Strasfeld, D.B., Loomis, R.A., Herbert, J.M., Ray, S.E., McCoy, A.B.: Stabilization and rovibronic spectra of the *T*-shaped and linear ground-state conformers of a weakly bound rare-gas–homonuclear dihalogen complex: $HeBr_2$. J. Chem. Phys. **123**, 104312 (14 pp) (2005). https://doi.org/10.1063/1.2006675
75. Jahn, D.G., Clement, S.G., Janda, K.C.: Vibrational predissociation dynamics of the He79Br 2 van der Waals molecule near the B state dissociation limit: Binding energies, lifetimes, and implications for the rare gas-halogen potential. J. Chem. Phys. **101**, 283–291 (1994). https://doi.org/10.1063/1.468181
76. van de Burgt, L.J., Nicolai, J.-P., Heaven, M.C.: Laser induced fluorescence study of the HeBr 2 van der Waals complex. J. Chem. Phys. **81**, 5514–5520 (1984). https://doi.org/10.1063/1.447652
77. Buchachenko, A.A., Baisogolov, AYu., Stepanov, N.F.: Interaction potentials and fragmentation dynamics of the Ne…Br 2 complex in the ground and electronically excited states. J. Chem. Soc., Faraday Trans. **90**, 3229–3236 (1994). https://doi.org/10.1039/FT9949003229
78. Thommen, F., Evard, D.D., Janda, K.C.: Rotational resolution of the NeBr 2 B-X 10–0 transition. J. Chem. Phys. **82**, 5295–5296 (1985). https://doi.org/10.1063/1.448610
79. Pio, J.M., van der Veer, W.E., Bieler, C.R., Kenneth C. Janda, K.C.: Product state resolved excitation spectroscopy of He–, Ne–, and Ar–Br_2 linear isomers: Experiment and theory. J. Chem. Phys. **128**, 134311 (10 pp) (2008). http://dx.doi.org/10.1063/1.2885047
80. Nejad-Sattari, M., Stephenson, T.A.: Fragment rotational distributions from the dissociation of NeBr 2: Experimental and classical trajectory studies. J. Chem. Phys. **105**, 5454–5467 (1997). https://doi.org/10.1063/1.473570

81. Naumkin, F.Y., McCourt, F.R.W.: Is the Ar—Br 2(X1Σ + g) van der Waals complex linear rather than T-shaped? A study in terms of ab initio based potential energy surfaces. Mol. Phys. **96**, 1043–1049 (1999). https://doi.org/10.1080/00268979909483047
82. Rohrbacher, A., Ruchti, T., Janda, K.C., Buchachenko, A.A., Hernández, M.I., González-Lezana, T., Villarreal, P., Delgado-Barrio, G.: A combined experimental-theoretical study of the vibrational predissociation and product rotational distributions for high vibrational levels of He79Br 2. J. Chem. Phys. **110**, 256–266 (1999). https://doi.org/10.1063/1.478101
83. Taylor, M.A., Pio, J.M., van der Veer, W.E., Janda, K.C.: Competition between electronic and vibrational predissociation dynamics of the $HeBr_2$ and $NeBr_2$ van der Waals molecules. J. Chem. Phys. **132**, 104309 (8 pp) (2010). http://dx.doi.org/10.1063/1.3353954
84. Cabrera, J.A., Bieler, C.R., Olbricht, B.C., van der Veer, W.E., Janda., K.C.: Time-dependent pump-probe spectra of $NeBr_2$. J. Chem. Phys. **123**, 054311 (8 pp) (2005). http://dx.doi.org/10.1063/1.1990118
85. Hernández-Lamoneda, R., Sanz-Sanz, C., Roncero, O., Pio, J.M., Taylor, M.A., Janda, K.C.: A theoretical study on electronic predissociation in the NeBr 2 van der Waals molecule. Chem. Phys. **399**, 86–93 (2012). https://doi.org/10.1016/j.chemphys.2011.09.015
86. Delgado-Tellez, L., Valdés, Á., Prosmiti, R., Villarreal, P., Delgado-Barrio, G.: HeI_2 interaction potential potential based on an interpolation scheme. Intern. J. Quantum Chem. **112**, 2971–2975 (2012). https://doi.org/10.1002/qua.24160
87. Valdés, Á., Prosmiti, R., Villarreal, P., Delgado-Barrio, G., Werner, H.-J.: Ab initio potential energy surface and spectrum of the $B(^3\Pi)$ state of the HeI_2 complex. J. Chem. Phys. **126**, 204301–204309 (2007). https://doi.org/10.1063/1.2748404
88. Kalemos, A., Valdés, Á., Hea, R.: Theoretical investigation of the $He–I_2(E^3\Pi_g)$ ion-pair state: Ab initio intermolecular potential and vibrational levels. J. Chem. Phys. **137**, 034303 (8 pp) (2012). https://doi.org/10.1063/1.4733983
89. Baturo, V.V., Lukashov, S.S, Poretsky, S.A., Pravilov, A.M., Zhironkin A.I.: The HeI_2 van der Waals complexes in a 'free-rotor' configuration. J. Phys. B: At. Mol. Opt. Phys. **53**, 035101 (8 pp) (2020). https://doi.org/10.1088/1361-6455/ab582b
90. Baturo, V.V., Cherepanov, I.N., Lukashov, S.S., Poretsky, S.A., Pravilov, A.M.: The HeI_2(ion-pair states) van der Waals complexes. Chem. Phys. Lett. **662**, 250–256 (2016). https://doi.org/10.1016/j.cplett.2016.09.021
91. Baturo, V.V., Lukashov, S.S., Poretsky, S.A., Pravilov, A.M.: The RgI_2(ion-pair states) van der Waals complexes. Eur. Phys. J. D **71**, 227(12 pp) (2017). https://doi.org/10.1140/epjd/e2017-80142-6
92. Baturo, V.V., Lukashov, S.S., Poretsky, S.A., Pravilov, A.M.: The population and decay of the $HeI_2(E0_g^+, v_E = 0\text{-}17)$ van der Waals complexes (unpublished data)
93. Delgado-Tellez, L., Valdés, Á., Prosmiti, R., Villarreal, P.: Delgado-Barrio, G: *Ab initio* characterization of the $Ne–I_2$ van der Waals complex: Intermilecular potentials and vibrational bound states. J. Chem. Phys. **134**, 214304–214309 (2011). https://doi.org/10.1063/1.3596604
94. Blazy, J.A., DeKoven, B.M., Russell, T.D., Levy, D.H.: The binding energy of iodine–rare gas van der Waals molecules. J. Chem. Phys. **72**, 2439–2444 (1980). https://doi.org/10.1063/1.439438
95. Baturo, V.V., Lukashov, S.S., Poretsky, S.A., Pravilov, A.M.: Luminescence of the NeI_2 van der Waals complex. Chem. Phys. Lett. **696**, 26–30 (2018). https://doi.org/10.1016/j.cplett.2018.02.031
96. Baturo, V.V., Lukashov, S.S., Poretsky, S.A., Pravilov, A.M.: The *T*-shaped $NeI_2(E0_g^+)$ van der Waals complex. J. Phys. B: At. Mol. Opt. Phys. **52**, 145101 (9 pp) (2019). https://doi.org/10.1088/1361-6455/ab2496
97. Prosmiti, R., Villarreal, P., Delgado-Barrio, G.: An ab initio CCSD(T) potential energy surface for the $Ar–I_2(X)$ complex. Chem. Phys. Letts. **359**, 473–479 (2002). https://doi.org/10.1016/S0009-2614(02)00677-2

98. Wey, J., Makarem, C., Reinitz,A.L., Darr, J.P., Loomis, R.A.: Accurate measurement of the T-shaped and linear Ar . I_2(X,v" = 0) binding energies using vibronic-specific I_2(B, v) fragment velocity-map imaging. Chem. Phys. **399**, 172-179 (2012). https://doi.org/10.1016/j.chemphys.2011.06.039
99. Makarem, C., Loomis, R.A.: Spectroscopic identification of the Ar...I2(B, v') potential. Chem. Phys. Letts. **651**, 119–123 (2016). https://doi.org/10.1016/j.cplett.2016.03.039
100. Baturo, V.V., Kevorkyanz, R., Lukashov, S.S., Poretsky, S.A., Pravilov, A.M., Zhironkin, A.I.: The KrI_2(ion-pair states) van der Waals complexes. Chem. Phys. Letts. **684**, 357–362 (2017). https://doi.org/10.1016/j.cplett.2017.07.007
101. García-Guttierrez, L., Delgado-Tellez, L., Valdés, Á., Prosmiti, R., Villarreal, P., Delgado-Barrio, G.: Intermolecular ab initio potential and spectroscopy of the ground state of HeI_2 complex revisited. J. Phys. Chem. A **113**, 5754–5762 (2009). https://doi.org/10.1021/jp901250u
102. Baturo, V.V., Lukashov, S.S., Poretsky, S.A., Pravilov, A.M. (unpublished data)
103. Kenny, J.E., Johnson, K.E., Shafrin, W., Levy, D.H.: The photodissociation of van der Waals molecules: Complexes of iodine, neon, and helium. J. Chem. Phys. **72** 1109-1119 (1980). https://doi.org/10.1063/1.439252
104. Tscherbul, T.V., Buchachenko, A.A., Akopyan, M.E., Poretsky, S.A., Pravilov, A.M., Stephenson, T.A.: Collision-induced non-adiabatic transitions between the ion-pair states of molecular iodine: A challenge for experiment and theory. Phys. Chem. Chem. Phys. **6**, 3201–3214 (2004). https://doi.org/10.1039/B402655A
105. Baturo, V.V., Lukashov, S.S., Poretsky, S.A., Pravilov, A.M.: The population and decay of the $ArI_2(E0_g^+, v_E{=}\ 0\text{-}16)$ van der Waals complexes (unpublished data)
106. Hirschfelder, J.O., Curtis, C.F., Bird, R.B.: Molecular Theory of Gases and Liquids (Appendix, Table 1c), John Willey & Sons, N.-Y., L (1954). https://doi.org/10.1002/pol.1955.120178311
107. Akopyan, M.E., Bibinov, N.K., Kokh, D.B., Pravilov, A.M., Sharova, O.L., Stepanov, M.B.: The approach-induced $I_2(E0_g^+ - D0_u^+)$ transitions, M = He, Ar, I_2, N_2, CF_4. Chem. Phys. **263**, 459–470 (2001). https://doi.org/10.1016/S0301-0104(02)01028-5
108. Bibinov, N.K., Malinina, O.L., Pravilov, A.M., Stepanov, M.B., Zakharova, A.A.: The "approach-induced" and collision-induced $I_2(E0_g^+ - D0_u^+)$ transitions from low, v_E = 8-23, vibronic levels of the I_2(E) state. Chem. Phys. **277**, 179–189 (2002). https://doi.org/10.1016/S0301-0104(02)01028-5
109. Akopyan, M.E., Pravilov, A.M., Stepanov, M.B., Zakharova, A.A.: The collision-induced I_2(E-D) transitions, M = He, Ar, N_2, CF_4. Chem. Phys. **287**, 399–410 (2003). https://doi.org/10.1016/S0301-0104(02)01028-5
110. Pravilov, A.M.: Spectroscopy and primary process in the photolysis of iodides for iodine photodissociation lasers (review). Soviet J. Quant. Electronics (Eng. transl.) **11**, 847–862 (1981)
111. Bradke, M.D., Loomis, R.A.: Spectroscopic observation of the preferentially stabilized, linear He···ICl(X $^1\Sigma^+$) complex. J. Chem. Phys. **118**, 7233–7244 (2003). https://doi.org/10.1063/1.1562622
112. Prosmiti, R., Cunha, C., Villarreal, P., Delgado-Barrio, G.: The van der Waals potential energy surfaces and structures of He-ICl and Ne-ICl clusters. J. Chem. Phys. **117**, 7017–7023 (2002). https://doi.org/10.1063/1.1506920
113. Boucher, D.S., Darr, J.P., Bradke, M.D., Loomis, R.A., McCoy, A.B.: A combined experimental/theoretical investigation of the He + ICl interaction: Determination of the binding energies of the T-shaped and linear He···I^{35}Cl(X, v'' = 0) conformers. Phys. Chem. Chem. Phys. **6**, 5275–5282 (2004). https://doi.org/10.1039/b411914b
114. Valdés, A., Prosmiti, R., Villarreal, P., Delgado-Barrio, G.: CCSD(T) potential energy surface and bound rovibrational level calculations for the Ar-ICl(X) complex. Chem. Phys. Letts. **375**, 328–336 (2003). https://doi.org/10.1016/S0009-2614(03)00854-6

115. Davey, J.B., Legon, A.C., Waclawik, E.R.: Iodine and chlorine quadrupole coupling in the rotational spectra of Ar···ICl and ICl: intramolecular charge transfer induced in ICl by Ar. Chem. Phys. Letts. **306**, 133–144 (1999). https://doi.org/10.1016/S0009-2614(99)00434-0
116. McCoy, A.B., Darr, J.P., Boucher, D.S., Winter, P.R., Bradke, M.D., Loomis, R.A.: Combined experimental/theoretical investigation of the He + ICl interaction. I. Rovibronic spectrum of He···ICl complexes in the ICl B-X, 3–0 region. J. Chem. Phys. **120**, 2677-2685 (2004). https://doi.org/10.1063/1.1636693
117. Scene, J.M., Drobits, J.C., Lester, M.I.: Dynamical effects in the vibrational predissociation of ICl-rare gas complexes. J. Chem. Phys. **85**, 2329–2330 (1986). https://doi.org/10.1063/1.451080
118. Drobits, J.C., Scene, J.M., Lester, M.I.: Direct lifetime and nascent product distribution for the vibrational predlssociation of lCl-Ne $A(^3\Pi_1)$ state van der Waals complexes. J. Chem. Phys. **84**, 2896–2897 (1986). https://doi.org/10.1063/1.450318
119. Drobits, J.C., Lester, M.I.: Near threshold photofragmentation dynamics of lCl-Ne *A* state van der Waals complexes. J. Chem. Phys. **88**, 120–128 (1988). https://doi.org/10.1063/1.454644
120. Baturo, V.V., Lukashov, S.S., Poretsky, S.A., Pravilov, A.M., Sivokhina, M.M.: Luminescence of ArICl van der Waals complex. Chem. Phys. Letts. 765, 138259 (2021), https://doi.org/10.1016/j.cplett.2020.138259
121. Simpson, W.R., Stephenson, T.A.: The spectroscopy and A state dynamics of the NeIBr van der Waals complex. J. Chem. Phys. **90**, 3171–3180 (1989). https://doi.org/10.1063/1.455867
122. Stephenson, T.A.: Fragment rotational state distributions from the dissociation of NeIBr: Experimental and theoretical results. J. Chem. Phys. **97**, 6262–6275 (1992). https://doi.org/10.1063/1.463688
123. Kenny, J.E., Russell, T.D., Levy, D.H.: Van der Waals complexes of iodine with hydrogen and deuterium: Intermolecular potentials and laser-induced photodissociation studies. J. Chem. Phys. **73**, 3607–3618 (1980). https://doi.org/10.1063/1.440586
124. Gutmann, M., Willberg, D.M., Zewail, A.H.: Real-time dynamics of clusters. II. I_2X_n (n = 1; X = He, Ne, and H_2), picoseconds fragmentation. J. Chem. Phys. **97**, 8037–8047 (1992). http://dx.doi.org/10.1063/1.463426
125. Johnson, K.E., Levy, D.H.: The fluorescence excitation and dispersed fluorescence spectra of the van der Waals molecules I_2N_2 and I_2N_2He. J. Chem. Phys. **74**, 1506–1507 (1981). https://doi.org/10.1063/1.441170
126. Baturo, V.V., Kevorkyants, R., Lukashov, S.S., Onishchenko, S.S., Poretsky, S.A., Pravilov, A.M.: The valence and ion-pair states of the N_2I_2 van der Waals complex. Chem. Phys. Letts. **714**, 213–218 (2019). https://doi.org/10.1016/j.cplett.2018.10.084
127. Sharfin, W., Johnson, K.E., Wharton, L., Levy, D.H.: Energy distribution in the photodissociation products of van der Waals molecules: Iodine–helium complexes. J. Chem. Phys. **71**, 1292–1299 (1979). https://doi.org/10.1063/1.438429
128. Johnson, K.E., Sharfin, W., Levy, D.H.: The photodissociation of van der Waals molecules: Complexes of iodine, argon, and helium. J. Chem. Phys. **74**, 163–170 (1981). https://doi.org/10.1063/1.440868
129. Gutmann, M., Willberg, D.M., Zewail, A.H.: Real-time dynamics of clusters. III. I_2Ne_n, (n = 2–4; X = He, Ne, and H_2), picoseconds fragmentation and evaporation . J. Chem. Phys. **97**, 8048-8059 (1992). http://doi.org/10.1063/1.463427
130. Bastida, A., Zuñiga, J., Requena, A., Halberstadt, N., Beswick, J.A.: Vibrational predissociation of the Ne_2I_2 cluster: A molecular dynamics with quantum transitions study. J. Chem. Phys. **109**, 6320–6328 (1998). https://doi.org/10.1063/1.477274
131. Baturo, V.V., Lukashov, S.S., Poretsky, S.A., Pravilov, A.M., Zhironkin, A.I.: Experimental and theoretical investigations of Ne_2I_2 cluster (unpublished data)
132. Andreev, A.S., Baturo, V.V., Lukashov, S.S., Poretsky, S.A., Pravilov, A.M., Zhironkin, A. I.: Experimental and theoretical investigations of $HeNeI_2$ trimer. J. Chem. Phys. 152, 234307 (10 pp) (2020) https://doi.org/10.1063/5.0008760

133. Pio, J.M., Taylor, M.A., van der Veer, W.E., Bieler, C.R., Cabrera, J.A., Janda, K.C.: Real-time dissociation dynamics of the Ne_2Br_2 van der Waals complex. J. Chem. Phys. **133**, 014305 (15 pp) (2010). https://doi.org/10.1063/1.3456550
134. Bieler, C.R., Edward, D.D., Janda, K.J.: Ar_2Cl_2 and Ar_3Cl_2: Structure, bond energy, and dissociation dynamics. J. Phys. Chem. **94**, 7452–7457 (1990). https://doi.org/10.1021/j100382a027
135. Hair, S.R., Cline, J.I., Bieler, C.R., Janda, K.J.: The structure and dissociation dynamics of the Ne_2Cl_2 van der Waals complex. J. Chem. Phys. **90**, 2935–2943 (1989). https://doi.org/10.1063/1.455893
136. Hernández, M.I., Halberstadt, N., Sands, W.D., Janda, K.J.: Structure and spectroscopy of the He_2Cl_2 van der Waals cluster. J. Chem. Phys. **113**, 7252–7267 (2000). https://doi.org/10.1063/1.1313786
137. Sands, W.D., Bieler, C.R., Janda, K.J.: Spectroscopy and dynamics of He_2C1_2: A quantum liquid cluster? J. Chem. Phys. **95**, 729–734 (1991). https://doi.org/10.1063/1.461078
138. Koperski, J., Czajkowski, M.A.: Excitation spectrum of the $A0^+$ (5^3P_1), $B1$ (5^3P_1) ← $X0^+$ (5^1S_0) transitions in the CdHe van der Waals moleeule: Spectroscopic characterization of the $X0^+$, $A0^+$, and B1 electronic energy states. J. Chem Phys. **109**, 459–465 (1998). https://doi.org/10.1063/1.476584
139. Koperski, J., Czajkowski, M.A.: Interatomic potential parameters of CdHe van der Waals complex in the A^30^+, B^31 and X^10^+ states - revisited. Chem. Phys. Letts. **350**, 367–371 (2001). https://doi.org/10.1016/S0009-2614(01)01298-2

Chapter 7
Chemiluminescence

Abstract Chemiluminescence reactions are discussed in this chapter. In the beginning, definitions of various types of reactions and their rate constants are introduced. The primary attention is paid to considering various types of recombination accompanied by radiation, namely, inverse dissociation, inverse electronic predissociation, and termolecular recombination accompanied by radiation. Some exciting features of these processes are discussed.

7.1 Types of Chemiluminescent Processes. Rate Constants. Definitions

1. *Chemiluminescent reactions* are processes of species collisions, accompanied by a change in their chemical composition and photon emission, and processes of recombination of colliding species into a short-lived complex (quasi-molecule), part or all of the excitation energy of which is disappeared by photon radiation.

Without the second part of the definition, species collision processes lead to the formation of electronically excited complexes, including bound ones, and the radiative transition to the repulsive (therefore, there is no change in chemical composition) state of the complex cannot be considered as chemiluminescence.

Recombination of the $\mathrm{Xe}(^1\mathrm{S}) + \mathrm{O}(^1\mathrm{D})$ (see Fig. 5.21) is an example of such processes.

$$\mathrm{Xe}\left(^1\mathrm{S}\right) + \mathrm{O}\left(^1\mathrm{D}\right) \leftrightarrow \mathrm{XeO}\left(a^1\Sigma^+, b^1\Pi\right) \rightarrow \begin{array}{c} \mathrm{XeO}\left(X^3\Pi\right) + h\nu \\ \downarrow \\ \mathrm{Xe}(^1\mathrm{S}) + \mathrm{O}(^3\mathrm{P}) \end{array} \tag{7.1.1}$$

Under the change in chemical composition in this definition, we should also understand the process of exchange of two identical species, which can be registered in the form of isotopic exchange.

A. Pravilov, *Gas-Phase Photoprocesses*, Springer Series in Chemical Physics 123,
https://doi.org/10.1007/978-3-030-65570-9_7

$$^{n}A + ^{m}AB \rightarrow ^{m}A + ^{n}AB \quad (7.1.2)$$

(here n, m are the masses of isotopes A, $n \neq m$).

Often, chemiluminescent reactions also mean processes of electronic deactivation with energy transfer (see Sect. 5.7), leading to the decay of the excited states of some species, the appearance of vibrational-rotational, electronic or rovibronic excitation in others

$$A^{*} + BC \rightarrow A + BC^{\#} \quad (7.1.3a)$$

$$\rightarrow A + BC^{*} \quad (7.1.3b)$$

and accompanied by the emission of a photon in the IR (reaction 7.1.3a) or visible, UV spectral regions (reaction 7.1.3b). In the author's opinion, this is incorrect methodologically.

There is a thin line between the collision-induced optical transition (Fig. 7.1a, process 1) and a kind of chemiluminescent reaction, namely, inverse dissociation by repulsive state (see below). In the first case, if one of the species is an excited atom, an atomic line broadened due to the collision is observed (the collision weakens the forbiddance of the optical transition). In the second case, a wide molecular band occurs. Obviously, with an increase in the colliding partner kinetic energies, A_2 and B_2 atoms, for example (Fig. 7.1a) the A_2-B_2 internuclear distance can decrease to small values, and along with the atomic line, a molecular band will also be observed. If an excited species is polyatomic one, then it is much more difficult to distinguish these processes.

Chemiluminescent reactions in the gas phase are divided into *exchange chemiluminescence* and *recombination accompanied by radiation* (radiative recombination [1], p. 108).

2. *Bimolecular exchange reactions* are reactions some of the products of which are formed in vibrational-rotational, electronic, or rovibronic excited states:

$$A + BC \rightarrow AB + C^{*} \quad (7.1.4)$$

$$\rightarrow AB^{*(\#)} + C; \quad (7.1.5)$$

$$AB + CD \rightarrow AC + B^{*} + D \quad (7.1.6)$$

$$\rightarrow AC + BD^{*(\#)} \quad (7.1.7)$$

Examples of such reactions are processes [2]

$$Sn(^{3}P) + N_2O \rightarrow SnO(a^{3}\Sigma^{+}, ^{3}\Pi) + N_2\left(X^{1}\Sigma_g^{+}\right) \quad (7.1.8)$$

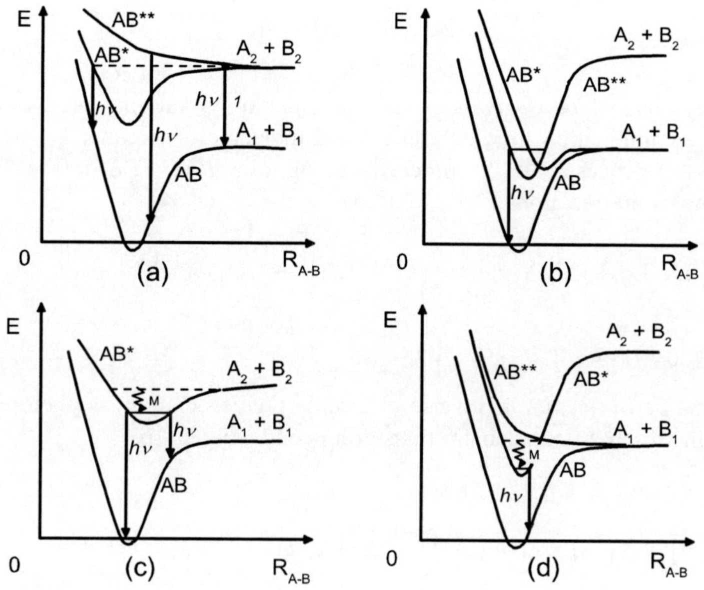

Fig. 7.1 Recombination accompanied by radiation: **a**—inverse dissociation (1—collision-induced optical transition); **b**—inverse predissociation; **c**, **d**—termolecular recombination accompanied by radiation

$$H(^2S) + Cl_2\left(X^1\Sigma_g^+\right) \rightarrow HCl(X^1\Sigma^+, v_X) + Cl(^2P_J) \tag{7.1.9}$$

$$I_2(B0_u^+) + F_2\left(X^1\Sigma_g^+\right) \rightarrow IF(B0^+, v_B) + IF(X0^+, v_X) \tag{7.1.10}$$

$$O_3(\tilde{X}^1A_1) + NO(X^2\Pi) \rightarrow NO_2(^1B_2, ^2B_2) + O_2\left(X^3\Sigma_g^-\right) \tag{7.1.11}$$

3. *Recombination accompanied by radiation* (or radiative recombination) is the recombination of species, accompanied by the emission of the molecule or excimer nascent in the process.

Recombination accompanied by radiation can be divided into two types of processes, namely photorecombination and termolecular recombination accompanied by radiation (three-body radiative recombination [1], p. 114). Photorecombination, in turn, can be divided into inverse dissociation and inverse predissociation.

4. *Photorecombination* is the two species recombination into a third, which rovibronic excitation is disappeared by a photon emission:

$$A + B \rightarrow AB + h\nu \tag{7.1.12}$$

5. *Inverse dissociation*, inversion of rotational or vibrational predissociation is photorecombination in which an excited molecule or excimer is formed in an adiabatic process, i.e., it proceeds along one *PEC* (*PES*) (see Fig. 7.1a). Examples are reactions (7.1.1, 7.1.13)

$$O(^1D) + CO(X^1\Sigma^+) \rightarrow CO_2(^1B_2) \rightarrow CO_2\left(\tilde{X}^1\Sigma_g^+\right) + h\nu, \tag{7.1.13}$$

$\lambda \approx$ 200–300 nm [3, 4], p. 243.

6. *Inverse predissociation* (inverse electronic predissociation) is photorecombination in which a nonadiabatic transition occurs (Fig. 7.1b).

Examples:

$$O(^3P) + N(^4S) \rightarrow NO(a^4\Pi) \rightarrow NO(C^2\Pi) \rightarrow NO(X^2\Pi) + h\nu \tag{7.1.14a}$$

$$\rightarrow NO(A^2\Sigma^+) + h\nu \tag{7.1.14b}$$

λ = 191.5–250 nm, NO δ bands (reaction 7.1.14a) and $\lambda \approx$ 1224 nm (reaction 7.1.1b), p_M < 2 Torr.

7. *Termolecular recombination accompanied by radiation* is the process of recombination of two species, which proceeds with the participation of a species, the third body M. This species carries away a part of the excitation energy 'stored' by a molecule (excimer) formed and stabilizes them. The recombination is accompanied by the radiation of this molecule (excimer). This process may have the kinetics of the third

$$A + B \xrightarrow{M} AB^* \rightarrow AB + h\nu \tag{7.1.15}$$

second and even first orders, as well as an intermediate between them

$$A + B \xrightarrow{M} AB^* \rightarrow AB + h\nu \tag{7.1.16}$$

(the brackets around M means that the kinetics of the process is not termolecular, although M is involved in the reaction) (see Sect. 2.4.1).

Examples:

$$2N(^4S) \xrightarrow{(M)} N_2(B^3\Pi_g, v_B) \rightarrow N_2\left(A^3\Sigma_g^+\right) + h\nu \tag{7.1.17}$$

$\lambda > 500$ nm, the first positive N_2 system. The kinetics of reaction (7.1.17) at M = He, $p_{He} < 20$ Torr is termolecular; at M = He, N_2, $p_{He} = 0.3-1$ Torr, $p_{N_2} \approx 0.1$ Torr, it is intermediate between bi - and termolecular, and at M = N_2, $p_{N_2} > 0.1$ Torr is bimolecular (see [4], p. 96, [5], p. 155 and references);

$$2N(^4S) \overset{(M)}{\rightarrow} N_2\left(A^3\Sigma_g^+, v_A\right) \rightarrow N_2\left(X^1\Sigma_g^+\right) + hv \qquad (7.1.18)$$

$\lambda = 230$–350 nm, the Vegard-Kaplan bands. Reaction (7.1.18) is the first-order one for $[N(^4S)]$ and $[N_2]]$, at $[N^4S] > 10^{12}$ cm^{-3} and $p_{N_2} < 0.13$ Torr. For $[N(^4S)] = (10^{12}-10^{13})$ cm^{-3}, $p_{N_2} > 100$ Torr it is first-order reaction relative to $[N(^4S)]$ and zero-order relative to $[N_2]$
(see [4], p. 96, [5], p. 155, and references).

The complete characteristic of a chemiluminescent reaction is the spectral distribution of the chemiluminescence rate constant k_{chl} (λ).

8. *The spectral distribution of the rate constant of chemiluminescence reactions with*:

unimolecular kinetics:

$$k^{I}_{chl}(\lambda) = I_{chl}(\lambda)/[A]\ \text{photon/species} \cdot \text{s} \cdot \text{nm} \qquad (7.1.19)$$

hereinafter, $I_{chl}(\lambda)$ (photon/cm^3·s·nm) is the intensity of chemiluminescence emitted at 1 cm^3 of the reaction zone and 1 s in a solid angle of 4π sr in the spectral range $\Delta\lambda = 1$ nm; species A is a reagent (molecule, radical), to the concentration of which the rate of the chemiluminescent reaction is proportional;

kinetics intermediate between bi- and uninomolecular:

$$k^{I}_{chl}(\lambda, [M]) = I_{chl}(\lambda)/[A]\ \text{photon/species} \cdot \text{s} \cdot \text{nm}. \qquad (7.1.20)$$

Here the rate constant of chemiluminescence of the first order k^{I}_{chl} (λ, [M]) does not depend linearly on [M];

with bimolecular kinetics:

$$k^{II}_{chl}(\lambda) = I_{ch}(\lambda)/[A] \cdot [B]\ \text{photon} \cdot \text{cm}^3/\text{species}^2 \cdot \text{s} \cdot \text{nm} \qquad (7.1.21)$$

with kinetics intermediate between bi- and termolecular:

$$k^{II}_{chl}(\lambda, [M]) = I_{chl}(\lambda)/[A] \cdot [B]\ \text{photon} \cdot \text{cm}^3/\text{species}^2 \cdot \text{s} \cdot \text{nm} \qquad (7.1.22)$$

with termolecular kinetics:

$$k^{III}_{chl}(\lambda, [M]) = I_{ch}(\lambda)/[A] \cdot [B] \cdot [M]\ \text{photon} \cdot \text{cm}^6/\text{species}^3 \cdot \text{s} \cdot \text{nm} \qquad (7.1.23)$$

9. *Integral chemiluminescence rate constant* of the reaction with unimolecular kinetics is

$$k^{\mathrm{I}}_{\mathrm{chl}} = I_{\mathrm{ch}/}/[\mathrm{A}]\,\mathrm{photon/species}\cdot\mathrm{s} \tag{7.1.24}$$

hereinafter, I_{chl}/[A] photon/species·s is the integral intensity of chemiluminescence emitted in the spectral range where chemiluminescence occurs at 1 cm^3 of the reaction zone and 1 s in a solid angle of 4π sr.

Similarly, as the integral over the spectral range of the spectral distribution of the chemiluminescence rate constant, the chemiluminescence rate constants in reactions with other kinetics are determined. The reaction order can always be understood from the dimension of the constant and its designation (see Sect. 2.1).

7.2 The Basic Patterns of the Recombination Accompanied by Radiation

The recombination accompanied by radiation is very interesting for understanding the properties of free molecule excited states. Having studied the kinetics and spectrum of such a process, one can obtain the spectroscopic characteristics of the emitting state and understand the mechanism of its formation and decay. Consider the basic patterns of these processes.

7.2.1 Inverse Dissociation, Inversion of Rotational or Vibrational Predissociation

This reaction type is the simplest one (for a theoretical description, but not for observation) of the radiative recombination. In the event of a collision of two atoms, it can be calculated in the classical or quantum-mechanical approaches. Two inverse dissociation cases are realized: the reaction proceeds on repulsive or bound *PEC* (*PES*). Inversion of rotational or vibrational predissociation proceeds on a bound *PES* (*PEC*) (Fig. 7.1a).

In terms of formal kinetics, inverse dissociation

$$\mathrm{A} + \mathrm{B} \leftrightarrow (\mathrm{A}\ldots\mathrm{B})^* \rightarrow \mathrm{AB} + h\nu \tag{$-$7.2.1}$$

can be described using the following mechanism (see Fig. 7.1a)

$$\mathrm{A} + \mathrm{B} \leftrightarrow (\mathrm{A}\ldots\mathrm{B})^* \tag{$-$7.2.2}$$

$$(A...B)^* \rightarrow AB + h\nu \quad (7.2.3)$$

$$\xrightarrow{M} \text{products.} \quad (7.2.4)$$

Here:

(A...B)* is a short-lived electronically excited collisional complex, having excitation energy equal to the AB* dissociation energy (without taking into account the kinetic energy of the colliding species);

AB is any other electronic state of this molecule;

Reaction (7.2.4) is collision-induced dissociation, predissociation or electronic deactivation (A...B)* (see Sects. 5.5, 5.7).

The integral intensity of inverse dissociation is proportional to the [A] [B] product

$$I_{2.1} = k_{2.1}[A][B]\,\text{photon/cm}^3 \cdot \text{s}, \quad (7.2.5)$$

where

$$k_{2.1} = \frac{k_{2.2} \cdot k_{2.3}}{k_{-2.2} + k_{2.3} + k_{2.4}[M]}\,\text{photon} \cdot \text{cm}^3/\text{species}^2 \cdot \text{s} \quad (7.2.6)$$

is the integral chemiluminescence rate constant.

Let us evaluate the rate of processes (7.2.2–7.2.4) under ordinary experimental conditions. If we restrict ourselves to a pressure not exceeding one-tenth of atmospheric one, $[M] < 2 \cdot 10^{18}$ cm^{-3}, take the deactivation rate constant equal to the gas kinetic $k_{2.4} = 3 \cdot 10^{-10}$ cm^3/s, then quenching of the photorecombination is noticeable, $k_{2.4}$ [M] > $0.1(k_{-2.2} + k_{2.3})$, only if $(k_{-2.2} + k_{2.3}) < 10^9$ s^{-1}. Obviously, for inverse dissociation on a repulsive *PEC* (*PES*) or on a diatomic molecule bound *PEC*, $1/k_{-2.2}$ is approximately equal to the period of vibration of the molecule, i.e., $k_{-2.2} \approx 10^{13}$ s^{-1}; for dissociation of bound triatomic molecule state, $k_{-2.2} = (10^{10}$–$10^{11})$ s^{-1}. Thus, if a diatomic or triatomic molecule is formed in the process (7.2.2), the process (7.2.4) rate at $[M] < 2 \cdot 10^{18}$ cm^{-3} is much lower than that of dissociation (-7.2.2). Note that in the vast majority of the recombination reactions accompanied by radiation discovered to date, the AB molecule is a di- or triatomic molecule (see [6] as an example). The only exceptions observed are reactions of I $(5p^5\ ^2P_{1/2})$ with perfluoroalkyl radicals, R = CF_3, C_2F_5, C_3F_7, CF_3CFCF_3, C_4F_9, $(CF_3)_3$ [7, 8].

$$I(5p^{5\,2}P_{1/2}) + R \xrightarrow{M} RI + h\nu, \lambda = 700 - 1315\,\text{mm} \quad (7.2.7)$$

However, in the reaction (7.2.7), in the process of vibrational relaxation of the RI* state formed during the recombination of $I(5p^5\ ^2P_{1/2})$ + R, and in the process of its dissociation (a reaction similar to reaction 7.2.4), species M is involved, i.e., reaction (7.2.7) is termolecular recombination accompanied by radiation.

7.2.1.1 Inverse Dissociation on Repulsive *PEC*s

As far as the author knows, no attempt was made to calculate the rate constants of inversed dissociation of polyatomic molecules. For the case of collision of two atoms, such calculations were carried out both in the classical and quantum mechanical approach. The classical approaches for calculating the rate constant of inverse dissociation on repulsive *PEC*s were developed in [9, 10]. They are based on the classical scattering theory [9] and equilibrium statistical mechanics [10] and are valid if optical transitions from a repulsive curve or above the dissociation limit of a bound state. Both approaches show that the chemiluminescence rate constant is equal to:

$$k_{2.1} = \frac{4\pi g^e_{AB^*}}{g^e_A g^e_B \sigma} \int\limits_0^\infty A_{(A...B)^*}(R) R^2 \exp(-[E_{(A...B)^*}(R)/R] dR). \quad (7.2.8)$$

Here:

$g^e_{AB^*}$, g^e_A, g^e_B are the electronic statistical weights of the emitting state and the reagents, σ is the symmetry order (see Sect. 2.2).

$A_{(A...B)^*}(R)$, $E_{(A...B)^*}(R)$ are Einstein spontaneous emission coefficient of the optical transition (7.2.3) and the energy AB* relative to that of AB* dissociation limit, A + B, as functions of internuclear distance $R_{A\text{-}B}$.

The data on $A_{(A...B)^*}(R)$, if available, relate, as a rule, to the zone of Franck-Condon transition from ground state low vibrational levels, i.e., to the zone, which can be achieved only at high kinetic energies of the colliding species, i.e., at very high temperature. For large internuclear distances, one can usually use the results of the $A_{(A...B)^*}(R)$ and $E_{(A...B)^*}(R)$ calculations, only, the accuracy of which is generally low.

At low temperatures in the collision of A and B, an image point can achieve on the AB* repulsive *PEC* points corresponding to large $R_{A\text{-}B}$ values (see Fig. 7.1a). Therefore, the interaction of two atoms takes place along a 'fairly steep' repulsive term, and the chemiluminescence intensity is very low. If one of the atoms is excited, then, as a rule, the intensity of the luminescence induced by collisions is much higher (Fig. 7.1a, process 1)

$$A^* \xrightarrow{B} A + h\nu \quad (7.2.9)$$

(see, for example, data on $O(^1S)$ + He, Ne collisions [11]).

The experimental data on inverse dissociation on a diatomic molecule repulsive *PEC* are available for only one reaction, and even that is not sufficiently accurate and detailed [12]

$$2\mathrm{Cl}(^2\mathrm{P}_{3/2}) \rightarrow \mathrm{Cl}_2(^3\Pi(1_u)) \rightarrow \mathrm{Cl}_2\left(X^1\Sigma_g^+\left(0_g^+\right)\right) + h\nu. \quad (7.2.10)$$

This reaction was considered theoretically. The data obtained in these works agree quite satisfactorily: $k_{2.10} = 8 \cdot 10^{-20}$ (T = 2300 K), 10^{-19} (T = 2800 K) (theory); $1.4 \cdot 10^{-19}$ (T = 2300 K), $1.7 \cdot 10^{-19}$ (T = 2800 K), uncertainty 50% (experiment), all in units photon $\cdot$ cm^3/species^2 $\cdot$ s. It is difficult to evaluate the accuracy of the calculations for the reasons indicated above; experimental data need to be verified.

7.2.1.2 Inverse Dissociation on Bound *PEC*s

It should be noted that observe inverse dissociation on a bound *PEC*, i.e., radiation from the upper vibrational level AB, (A…B)* or inversion of rotational (vibrational) predissociation in its pure form is possible, as a rule, only at low pressures. If one excludes from consideration the inversed dissociation of excimer fragments (He_2, HeH, etc.), then the overwhelming majority of such processes are observed in the near UV and visible spectral regions. The $k_{2.3}$ value for them, as a rule, does not exceed 10^7 s^{-1}. One has to take into account that the (A…B)* excitation energy becomes less than that of corresponding to the dissociation threshold after several collisions with M that makes dissociation (reaction -7.2.2) at T = 300 K is energetically impossible. One should also take into account that the rate constants of vibrational relaxation at high vibrational levels of molecules are large (see Sect. 5.3.4), $k_{2.11} = (10^{-10}–10^{-11})$ cm^3/s,

$$(\mathrm{A}\ldots\mathrm{B})^* \xrightarrow{\mathrm{M}} \mathrm{AB}^*(\mathrm{v}_{max} - \Delta\mathrm{v}) \quad (7.2.11)$$

Therefore, even at $[M] = 10^{17}$ cm^{-3} and $k_{2.8}$ $[M] = k_{2.3} = 10^7$ s^{-1}, collision-induced populations of vibrational levels below the AB* dissociation limit occur at rates comparable to the rate of radiative decay, and inverse dissociation transforms into termolecular recombination accompanied by radiation (see below). The exception is the inverse dissociation, competed with fast predissociation of the emitting state, if the latter is observed from a set of high, up to the dissociation threshold, vibrational levels. In this case, the lifetime of any vibrational level for spontaneous decay is less than the lifetime between collisions with M, and processes (7.2.4, 7.2.11) can be neglected. Therefore, the only reliably assigned inverse dissociation processes on the bound states are reactions (7.1.1, 7.1.13). The mechanism of the first of them is described by the processes:

$$\mathrm{O}(^1\mathrm{D}) + \mathrm{Xe} \leftrightarrow \mathrm{XeO}^*(a^1\Sigma^+, b^1\Pi) \quad (-7.2.12)$$

$$\mathrm{Xe0}^* \rightarrow \mathrm{XeO}(X^3\Pi, A^3\Sigma^-) \rightarrow \mathrm{Xe} + \mathrm{O}(^3\mathrm{P}) \quad (7.2.13)$$

$$\begin{array}{c}\rightarrow XeO\left(X^3\Pi\right)+h\nu\\ \downarrow\\ Xe+O(^3P)\end{array} \quad (7.2.14)$$

(see Fig. 5.21), $k_{1.1}$ = 3.0·10^{-17} photon cm^3/species^2 s, $k^a_{2.12} \approx 5 \cdot 10^{-11}$cm^3/s, $k^b_{2.12} \approx 10^{-10}$cm^3/s, $k^a_{-2.12} \approx 6 \cdot 10^{12}$s^{-1}, $k^b_{-2.12} \approx 1.4 \cdot 10^{12}$s^{-1}, $k^{a,b}_{2.13} = (10^{11}-10^{12})$s^{-1}, $k_{2.14} \approx 5 \cdot 10^5$s^{-1}; the indices a, b correspond to the states $a^1\Sigma^+$, $b^1\Pi$ [4], p. 67, [6].

For the inverse dissociation on a bound *PEC*, the classical approach in the general case does not provide an adequate description, if only because the probability of optical transitions is maximal near the turning points (Frank-Condon principle). Besides, optical transitions to low vibrational levels of the lower state can occur with high probability (see Fig. 7.1a). And finally, optical transitions from quasi-discrete states above the dissociation limit can occur. This process is the reversal of predissociation by rotation, i.e., sub-barrier seepage through the centrifugal potential barrier, which is a consequence of noncollinear collisions of atoms (see Fig. 3.8). As shown in [13], for the reaction

$$Cl(^2P_{3/2})+Cl(^2P_{1/2}) \leftrightarrow Cl_2(B^3\Pi(0_u^+)) \rightarrow Cl_2\left(X^1\Sigma^+\left(0_g^+\right)\right)+h\nu \quad (7.2.15)$$

at low temperatures, the quantum effects mentioned above should make a significant contribution to $k_{2.15}$; at T = 300 K, $k^{class}_{2.15} = 2 \cdot 10^{-21}$, $k^{quant}_{2.15} = 7 \cdot 10^{-21}$ (units, photon · cm^3/species^2 · s).

As mentioned above, up to the date, no 'pure', i.e., uncomplicated by predissociation inverse dissociation, is observed. Maybe, this is the reaction

$$O(^3P)+NO(X^2\Pi) \leftrightarrow NO_2(^2B_1,^2B_2) \rightarrow NO_2(\tilde{X}^2A_1)+h\nu \quad (7.2.16)$$

at p < 0.01 Torr if it proceeds via the $^2B_1$ [14] ($^2B_2$) [15] state, which correlates with O(^{3}P) + NO($X^2\Pi$).

Inverse vibrational predissociation kinetically differs from inverse dissociation only in that the values of $k_{-2.1}$ (predissociation) and $k_{2.1}$ can be less. To the best of the author's knowledge, such a process has not yet described in the literature.

7.2.2 Inverse Electronic Predissociation

The inverse electronic predissociation, as well as the inverse dissociation, can be described by the processes (7.2.1–7.2.4), and the dependence of the intensity and rate constant of this process by the (7.2.5, 7.2.6). However, this description is not accurate and does not describe exactly the kinetics of this process, since inverse electronic predissociation involves two electronic states of AB (see Fig. 7.1b)

$$A + B \leftrightarrow (A...B)^* \leftrightarrow AB^{**} \rightarrow AB + h\nu. \quad (7.2.17)$$

In more detail:

$$A + B \leftrightarrow (A...B)^* \quad (-7.2.18)$$

$$(A...B)^* \leftrightarrow AB^{**} \quad (-7.2.19)$$

$$AB^{**} \rightarrow AB + h\nu \quad (7.2.20)$$

$$\rightarrow A_i + B_i \quad (7.2.21)$$

$$\xrightarrow{M} \text{products} \quad (7.2.22)$$

(reaction (7.2.21) is AB** decay into A_i and B_i states that are different from those involved in reaction (7.2.17). Here we take into account the circumstance that for the overwhelming number of processes, the lifetime of the (A…B)* quasi-molecule is much less than the time between its collisions with M under ordinary experimental conditions. Therefore, the (A…B)* decay in collisions with M can be neglected. From the steady-state conditions (see Sect. 2.4.1) for (A…B)* and AB**, it is easy to obtain:

$$I_{2.17} = k_{2.20}[AB^{**}] = \left[\frac{k_{2.18}k_{2.19}k_{2.20}[A][B]}{(k_{-2.18} + k_{2.19})(k_{-2.19} + + k_{2.20} + k_{2.21} + k_{2.22}[M])}\right] \quad (7.2.23)$$

Let us estimate the rates of the processes (7.2.18–7.2.20, 7.2.22) using the information given in Sect. 7.2.1. In processes of inverse predissociation known up to the date, reactions (7.2.19, –7.2.19) are forbidden by the *spin-conservation rule* (*Wigner rule*), according to which the total spin momentum of the species participating in the process must remain unchanged [16]. Therefore, the rates of these spontaneous processes are low ($\sim 10^9$ s^{-1} in the case of NO, reaction 7.1.14). It is difficult to expect the large value of $k_{2.19}$, $k_{-2.19}$ for unstudied processes for the following reasons: let A and B are atoms, and the symmetry types of AB* and AB** are the same. In this case, the states are 'repelled', and the probability of nonadiabatic transitions (7.2.19, –7.2.19) can be low. Let the types of symmetry of the states are different, and the predissociation is allowed due to the electron-rotational interaction (for $\Delta\Lambda = \pm 1$), and depends on colliding species velocity and impact parameter (see Sect. 3.5). In this case, the nonadiabatic transition occur not for the image point each passage of through the intersection region of the *PEC*s of the AB* and AB** states. The same applies to the case when the spin momenta of the AB* and AB* states are different. If one of the species A or B is a molecule, the probability of nonadiabatic processes is much higher, but the frequency of the passage of an image point favorable for such a process is small due to stochastization of the vibrational modes (see Sect. 4.7.2). All these factors lead to the fact that, as a rule, the probabilities of nonadiabatic transitions per molecule vibration in inverse predissociation are of the order of 10^{-6}–10^{-3}, i.e., $k_{2.19}$, $k_{-2.19}$ = 10^7–10^{10} s^{-1} << $k_{-2.18}$. According to the detailed balance principle (see Sect. 3.1 and (3.1.17)) the $\frac{k_{2.19}}{k_{-2.19}}$ ratio is:

$$\frac{k_{2.19}}{k_{-2.19}} = \frac{g^e_{AB^{**}}}{g^e_{(A...B)^*}} exp - \frac{E_{(A...B)^*} - E_{AB^{**}}}{kT}, \tag{7.2.24}$$

and due to the lack of relaxation, $E_{(A...B)^*} - E_{AB^{**}} = 0$.

The small value of $k_{-2.19}$ leads to the fact that quenching of an inverse predissociation should be a widespread phenomenon. It was found in both currently known similar processes (7.1.14a, 7.1.14b) (see [4], p. 217 and references), and

$$O(^3P) + CO(X^1\Sigma^+) \leftrightarrow CO_2(^1B_2) \rightarrow CO_2\left(\tilde{X}^1\Sigma_g^+\right) + hv$$

(see [4], p. 242, [17] and references). The rate of AB** radiation decay, as a rule, is much lower than the rate of nonradiative decay (–7.2.19).

Given the above facts, the rate constant of inversed predissociation in the absence of quenching is:

$$k_{2.17} = \left[\frac{k_{2.18}}{k_{-2.18}} \frac{g^e_{AB^{**}}}{g^e_{(A...B)^*}} \frac{k_{2.20}}{k_{2.20} + k_{2.21}}\right] \tag{7.2.25}$$

This equation differs from that of for inverse dissociation only in the appearance of the ratio of electronic statistical weight. Therefore, to estimate $k_{2.17}$, one can use considerations about the values of $k_{2.18}$, $k_{-2.18}$, $k_{2.20}$, described in Sect. 7.2.1.

If there is data on the spectroscopic characteristics of the diatomic molecule AB**, then for calculating the $k_{2.17}$ value, one can use the concepts of statistical thermodynamics (see Sect. 2.2, [18] and references) or the resonance scattering theory developed by Carrington [19]. Without taking into account the reaction (7.2.21), the final result of [19] is:

$$k_{2.17} = K_{2.27} k_{2.20}, \tag{7.2.26}$$

where $K_{2.27} = \frac{k_{2.18}}{k_{-2.18}} \frac{g^e_{AB^{**}}}{g^e_{(A...B)^*}}$ is the equilibrium constant of the gross process

$$A + B \leftrightarrow \ldots \leftrightarrow AB^{**} \tag{7.2.27}$$

$$K_{2.27} = \frac{k_{2.27}}{k_{-2.27}} = \frac{g^t_{AB^{**}}}{g^t_{(A...B)^*}} \frac{g^e_{AB^{**}}}{g^e_A g^e_B} \sum_J (2J+1) \exp\left(-\frac{E_{v,J}}{kT}\right). \tag{7.2.28}$$

Here g^t_i is the statistical sum of translational excitation, g^e_i is the electronic statistical sum,

$$(2J+1)\exp\left[-E_{v,J}/kT\right] \tag{7.2.29}$$

is rotational partition sum of the level J with energy $E_{v,J}$ relative to that of A + B. The vibrational partition sum can be taken equal to 1, i.e., one can assume that, in reaction (7.2.19), rotational levels of only one vibrational level are populated. In this case, may remain uncertainty in choosing the values of v and J for a given energy $E_{v,J}$.

A similar approach was used for the reaction (7.1.14a) and gave quite a decent agreement with the experiment [20]. There is no uncertainty mentioned above in this case since the $NO(C^2\Pi, v_C > 0)$ vibronic levels cannot be populated reaction (7.1.14a).

7.2.3 Termolecular Recombination Accompanied by Radiation

Termolecular recombination accompanied by radiation can be roughly described by processes (-7.2.30–7.2.34) if the complex (A…B)*, not (A…M)*, (B…M)* forms during the reaction, i.e., the reaction is described in the framework of the energy transfer mechanism (see Sect. 2.4.1):

$$\mathrm{A}+\mathrm{B} \leftrightarrow (\mathrm{A}\ldots\mathrm{B})^* \tag{$-7.2.30$}$$

$$(\mathrm{A}\ldots\mathrm{B})^* \overset{(\mathrm{M})}{\rightarrow} \mathrm{AB}^* \tag{7.2.31}$$

$$\mathrm{AB}^* \rightarrow \mathrm{AB}+h\nu \tag{7.2.32}$$

$$\mathrm{AB}^* \overset{\mathrm{M}}{\rightarrow} \text{products} \tag{7.2.33}$$

$$\mathrm{AB}^* \rightarrow \mathrm{AB}^{**} \rightarrow \mathrm{A}+\mathrm{B} \tag{7.2.34}$$

(see Fig. 7.1c, d), and the intensity of this process

$$\mathrm{A}+\mathrm{B} \overset{(\mathrm{M})}{\rightarrow} \mathrm{AB}+h\nu \tag{7.2.35}$$

is:

$$I_{2.35} = \frac{k_{2.30}k_{2.31}k_{2.32}[\mathrm{A}][\mathrm{B}][\mathrm{M}]}{k_{-2.30}(k_{2.31}+k_{2.32}+k_{2.33}[\mathrm{M}])}. \tag{7.2.36}$$

At low concentrations of M, at which $k_{2.3}+k_{2.34} > > k_{2.33}$ [M], (7.2.36) transforms into

$$I_{2.35} = \frac{k_{2.30}}{k_{-2.30}} \cdot \frac{k_{2.32}}{k_{2.30} + k_{2.32}}[A][B][M] = k_{2.38}\frac{k_{2.32}}{k_{2.32} + k_{2.34}}[A][B][M] \qquad (7.2.37)$$

where $k_{2.38}$ is the rate constant of the formation of the AB* state from which the radiation transition occurs

$$A + B \overset{M}{\rightarrow} AB^{*} \qquad (7.2.38)$$

Chemiluminescence, in this case, is described in the framework of the termolecular kinetics.

With an increase in [M], the value of $k_{2.33}$ [M] can become comparable with $k_{2.32}$ + $k_{2.34}$, the $k_{2.35}$ value depend on [M], and, if $k_{2.32} + k_{2.34} << k_{2.33}$ [M], $I_{2.35}$ is:

$$I_{2.35} = k_{2.35}[A][B] = \frac{k_{2.30}}{k_{-2.30}} \cdot \frac{k_{2.31}}{k_{2.33}}[A][B] \qquad (7.2.39)$$

Chemiluminescence obeys bimolecular kinetics, but $k_{2.35}$ can depend on the nature of M:

$$k_{2.35} = f(M) \qquad (7.2.40)$$

since the ratio $\frac{k_{2.31}}{k_{2.33}}$ may depend on M. The dependence of a chemiluminescence intensity on [M] is similar to 'canonical' dependence of termolecular recombination rate described in Sect. 2.4 (see Fig. 2.5).

This description are a rather crude approximation, which describes, to some extent, the dependence of the integral rate constant of process (7.2.35) on the M concentration. Equations (7.2.37, 7.2.39) can be used to estimate the rate constants of the termolecular recombination of atoms accompanied by radiation in the presence of inert gases, taking the value $k_{2.31}$ equal to k_{gk}, and calculating $k_{2.30}/k_{-2.30}$ using the approach described in Sect. 7.2.2.

The considerations above are in no way suitable for describing the chemiluminescence rate constants from a specific AB* vibronic levels ($k^{v}_{2.35}$) and the spectral distribution of the chemiluminescence rate constant $k_{chl}(\lambda)$. As is shown in the studies of recombination of N(^{4}S) atoms accompanied by radiation, the reaction (7.1.17) kinetics can be termolecular ((7.2.37) is valid), and chemiluminescence kinetics from individual vibrational levels can be bimolecular, termolecular, intermediate between bi- and termolecular and even depend on $[M]^a$, where $a > 1$, Fig. 7.2 [21].

All these features are consequences of the fact that following the formation of the complex (A…B)* in the presence of M, a set of spontaneous and collision-induced processes in electronic states of AB occur. They are the following: dissociation, emission, spontaneous internal conversion to other electronic states, including predissociation and nonadiabatic processes, electronic

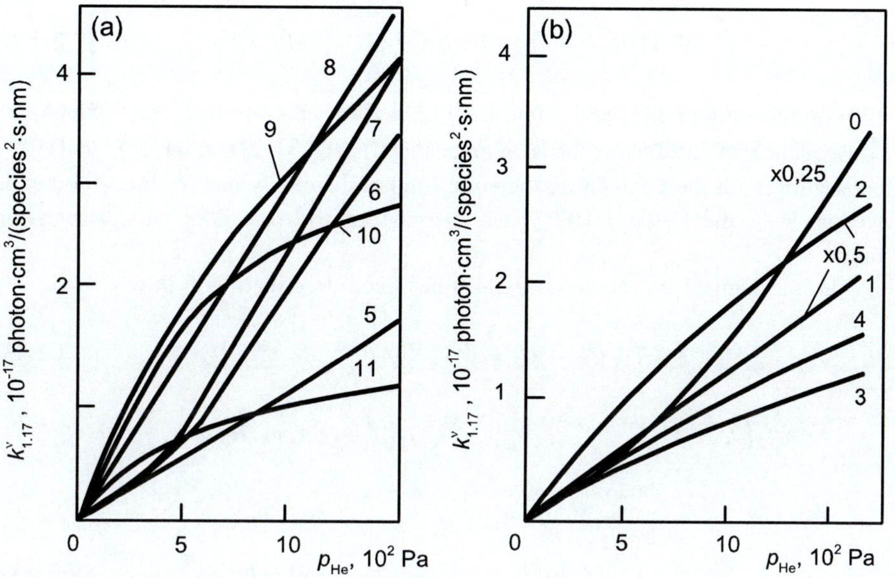

Fig. 7.2 Rate constants of chemiluminescence (7.1.17), $k^v_{1.17}$, for the $N_2(B,v_B)$ vibronic levels, v_B = 5−11 (**a**) and 0−4 (**b**), p_{N_2} <0.01 Torr, T = 293 K [21]

deactivation, collision-induced internal conversion to other electronic states, proceeding with a large or low loss of energy, including collision-induced predissociation, vibrational and rotational relaxation within the same electronic state. The mechanisms of these processes are discussed in Chap. 4.

One must also be aware that $k_{2.32}^{AB^*}$ is not equal to $k_{2.32}^{AB^{**}}$ in the general case. Besides, the collision-induced processes:

$$AB^* \overset{M}{\leftrightarrow} AB^{**} \qquad (-7.2.41)$$

may take place, and several AB*, AB** states can be populated in the process (7.2.30), as is the case in the complex of valence and ion-pair states of halogens (see Figs. 4.4 and 4.10). And, finally, the rate constants of the processes under consideration (both nonradiative and radiation) in the general case depend on the state vibrational quantum numbers and mixing of the states.

Another example: At the 90-s of the last century, the author and his colleague found 'non-canonical behavior' of the dependence of the rate constant of termolecular recombination of $Cl(^2P_{3/2})$ (Cl), atoms accompanied by radiation

$$Cl + Cl \overset{(M)}{\rightarrow} Cl_2\left(A^3\Pi(1_u), v_A\right) \rightarrow Cl_2\left(X^1\Sigma\left(0_g^+\right), v_X\right) + hv \qquad (7.2.42a)$$

$$\overset{(M)}{\rightarrow} Cl_2\left(B^3\Pi\left(0_u^+\right), v_B\right) \rightarrow Cl_2\left(X^1\Sigma\left(0_g^+\right), v_X\right) + hv \tag{7.2.42b}$$

(Figure 7.3) (see [22] and references) on helium pressure (Fig. 7.4), which is a consequence of the factors listed above: the $Cl_2(A')^3\Pi(2_u)$, $A^3\Pi(1_u)$, $B'^3\Pi\left(0_u^-\right)$ correlating with the Cl + Cl dissociation limit collisionally mix with each other as well as with the $Cl_2\left(B^3\Pi\left(0_u^+\right)\right.$ state correlating with the Cl + Cl* dissociation limit.

The mechanism of reactions (7.2.42) has been described as follows:

$$Cl + Cl \leftrightarrow Cl_2\left(A', v_{A'}^{max}; A, v_{v_A}^{max}; B', v_{B'}^{max}\right) \tag{−7.2.43}$$

$$Cl_2\left(A', v_{A'}^{max}; A, v_{v_A}^{max}; B', v_{B'}^{max}\right) \overset{M}{\rightarrow} Cl_2(A', v_{A'}; A, v_A; B', v_{B'}) \tag{7.2.44}$$

$$Cl_2(A', v_{A'}; A, v_A; B', v_{B'}) \overset{M}{\leftrightarrow} Cl_2(B, v_B) \tag{7.2.45}$$

$$Cl_2(A, v_A; B, v_B) \quad \rightarrow Cl_2(X, v_X) + hv \tag{7.2.46a}$$

$$Cl_2(A', v_{A'}; A, v_A; B', v_{B'}; B, v_B) \rightarrow Cl_2(X, v_X) + hv \tag{7.2.46b}$$

$$Cl_2(A', v_{A'}; A, v_A; B', v_{B'}; B, v_B) \overset{Cl}{\rightarrow} \quad \text{products} \tag{7.2.47}$$

$$Cl_2(A', v_{A'}; A, v_A; B', v_{B'}; B, v_B) \overset{Cl_2}{\rightarrow} \quad \text{products} \rightarrow \tag{7.2.48}$$

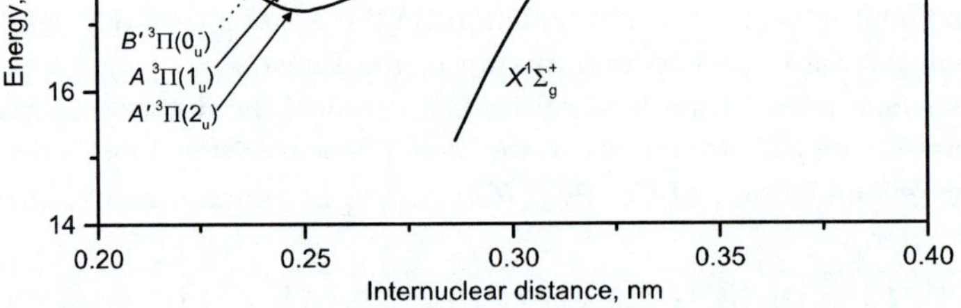

Fig. 7.3 The Cl_2 potential energy RKR (—) and Morse (—) curves for bound states correlating with the $Cl(^2P_{3/2}) + Cl(^2P_{3/2})$ and $Cl(^2P_{3/2}) + Cl(^2P_{1/2})$ dissociation limits. $Cl = Cl(^2P_{3/2})$, $Cl^* = Cl(^2P_{1/2})$. The position of the $Cl_2(B,v_B)$ vibrational levels is marked as lines

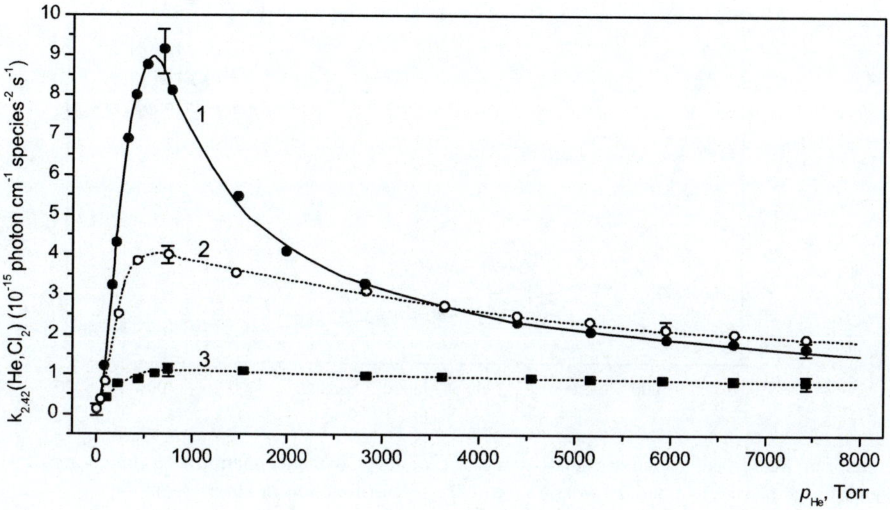

Fig. 7.4 The chemiluminescence rate constant of reactions (7.2.42) $k_{2.42}(Cl_2,He)$ plotted as a function of helium pressure at $p_{Cl2} = 1$ (1), 3 (2) and 10 Torr (3)

$$Cl_2(A', v_{A'}; A, v_A; B', v_{B'}; B, v_B) \overset{He}{\rightarrow} Cl_2(X, v_X) + \Delta E, \quad (7.2.49)$$

the $Cl_2(B,v_B)$ and $Cl_2(A,v_A)$ radiative lifetimes are $3.05 \cdot 10^{-4}$ s [23] and (0.02–0.08) s [24], $k_{2.48} = 8.4 \cdot 10^{-11}$ cm^3/s, $k_{2.49} = 1.6 \cdot 10^{-14}$ cm^3/s. Process (7.2.49) is collisional deactivation (high ΔE) and coupling of the triplet state manifold $Cl_2(A', v_{A'}; A, v_A; B', v_{B'}; B, v_B)$ with the singlet $I_2(X,v_X)$ state (low ΔE) following by fast collisional vibrational relaxation within the $I_2(X,v_X)$ level manifold. This process is strongly forbidden for He atoms due to $\Delta S = 0$ propensity rule (see Sect. 5.7), and its rate constant is low, $k_{2.49} = 2.5 \cdot 10^{-17}$ cm^3/s. The coupling of the Cl_2 triplet states, including the $Cl_2(B, v_B)$ and 'dark' $Cl_2(A', v_{A'}; B', v_{B'})$ states, radiative transitions from which are forbidden, induced by collisions with He atoms is fast. Therefore, the effective reversed radiative lifetime decreases with [He] from ~ 180 s^{-1} (the $Cl_2(A, v_A; B, v_B)$ manifold, process (7.2.46a)) at low He pressure to ~ 20 s^{-1} (the $Cl_2(A', v_{A'}; A, v_A; B', v_{B'}; B, v_B)$ manifold, process (7.2.46b)) at $p_{He} > 5000$ Torr (Fig. 7.5).

Therefore, the chemiluminescence rate constant $k_{2.42}$ decreases due to competition between radiative (7.2.46) and collision-induced (7.2.47–7.2.49) processes.

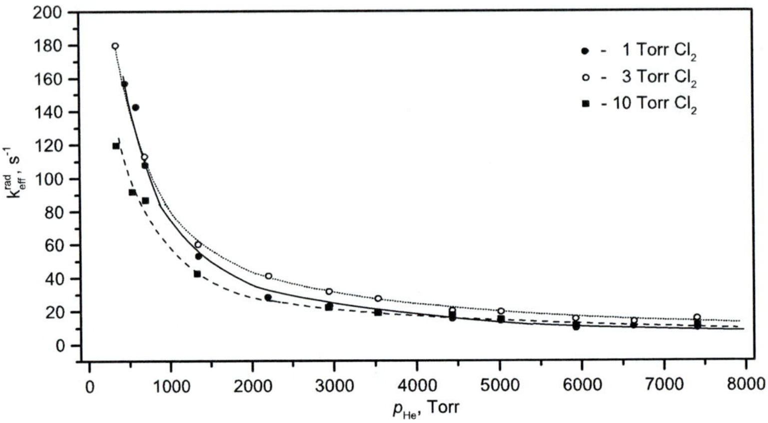

Fig. 7.5 Reciprocal effective radiative lifetime of the Cl_2 triplet manifold as functions of He pressure at $p_{Cl_2} = 1$, 3 and 10 Torr Y-axes k_{eff}^{rad}, s^{-1}, as function of He pressure

References

1. Kondratiev V.N., Nikitin E.E.: Gas-phase Reactions: Kinetics and Mechanism. Springer-Verlag, Berlin Heidelberg New York (1981). https://doi.org/10.1007/978-3-642-67608-6
2. Menzinger, O.: Electronic chemiluminescence in gases. Adv. Chem. Phys. **42**, 1–62 (1980). https://doi.org/10.1002/9780470142615.ch1
3. Pravilov, A.M., Protopopov, C.V., Sidorov, I.I., Skorokhodov, V.A.: Photorecombianation of O(^{1}D) atoms with CO molecules. Sov. J. Chem. Phys. **3**, 2010–2033 (1985)
4. Pravilov, A.M.: Fotoprozessy v molekulyarnykh gasakh (Photoprocesses in molecular gases). Energoatomizdat, Moscow (1992). (in Russian)
5. Pravilov, A.M.: Radiometry in modern scietific experiments. SpringerWien New York (2011). https://doi.org/10.1007/978-3-7091-0104-9
6. Pravilov, A.M., Sidorov, I.I., Skorokhodov, V.A.: Mechanism and kinetics of photorecombination of O(^{1}D) with Xe. Sov. J. Chem. Phys. **3**, 793–807 (1985)
7. Pravilov, A.M., Sidorov, I.I., Skoroknodov, V.A.: Photorecombination of I($^2P_{1/2}$) atoms with C_3F_7 and CF_3CFCF_3 radicals. Kinetics and Catalysis (USSR, Eng.transl.) **26**, 900–905 (1985)
8. Egorov, S.M., Pravilov, A.M., Skorokhodov, V.A.: Recombination of perfluoroalkyl radicals with the I(5p^5 $^2P_{1/2}$) and I(5p^5 $^2P_{3/2}$) atoms accompanied by radiation. Chem. Phys. **173**, 467–477 (1993). https://doi.org/10.1016/0301-0104(93)80161-2
9. Bates, D.R.: Rate of formation of molecules by radiative association. Monthly Notices Roy. Astron. Soc. **111**, 303–314 (1951). https://doi.org/10.1111/j.1365-246X.1951.tb03008.x
10. Palmer, H.B.: Kinetics of two-body radiative association of atoms. J. Chem. Phys. **47**, 2116–2120 (1967). https://doi.org/10.1063/1.1712244
11. Welge, K.H., Atkinson, R.: Collisional quenching of O(^{1}S) by rare gas atoms and collision-induced emission by O(^{1}S)-rare gas eximers. J. Chem. Phys. **64**, 531–538 (1976). https://doi.org/10.1063/1.432242
12. Belokrinitskij, N.S., Kernazitskij, L.A, Shpak, M.T.: The kinetics of chlorine atom photorecombination. Chem. Phys. Letts. **72**, 199–201 (1980). https://doi.org/10.1016/0009-2614(80)80273-9

13. Kurmanalieva, G.P., Shlyapnikov, G.V., Shmatov, I.P.: On some mechanisms of photorecombination in collisions of atoms. Optics Spectrosc. (USSR) **46**, 850–857 (1979). (in Russian)
14. Gangi, R.A., D L. Burnelle, L.: Electronic structure and electronic spectrum of nitrogen dioxide. III Spectral interpretation. J. Chem. Phys. **55**, 851–856 (1976). https://doi.org/10.1063/1.1676154
15. Pitts Jr., J.N., Sharp, J.H., Chan, S.I.: Effects of wavelength and temperature on primary processes in the photolysis of nitrogen dioxide and a spectroscopic-photochemical determination of the dissociation energy. J. Chem. Phys. **40**, 3655–3662 (1964). https://doi.org/10.1063/1.1725067
16. Braslavsky, S.E.: Glossary of terms used in photochemistry, 3rd edition (IUPAC Recommendations 2006). Pure Appl. Chem., **79**, 293–465 (2007). https://doi.org/10.1351/pac200779030293
17. Pravilov, A.M., Smirnova, L.G.: Temperature dependence of the spectral distribution rate constants of the chemiluminescence in the $O(^3P) + CO \rightarrow CO_2 + h\nu$. Kinetics and Catalysis (USSR, Eng.transl.) **22**, 641–648 (1981)
18. Trush, B.A.: Electronic and vibrational energy transfer. In: Spectroscopy of the Excited States, ed. B di Bartolo, N.Y. *e.a.* Plenum press 175–202 (1978)
19. Carrington, T.: Radiative recombination of atoms as a resonance scattering process. J. Chem. Phys. **57**, 2033–2037 (1972). https://doi.org/10.1063/1.1678527
20. Mandelman, M., Carrington, T., Young, R.A.: Predissociation and its inverse, using resonance absorption $NO(C^2\Pi) \leftrightarrow N + 0$. J. Chem. Phys. **58**, 84–90 (1973). https://doi.org/10.1063/1.1678956
21. Bystrov, D.S., Vilesov, A.F., Pravilov, A.M., Smirnova, L.G.: Chemiluminescence kinetics in first positive and Y systems of N_2 during recombination of $N(^4S)$ nitrogen atoms. Sov. J. Chem. Phys. **6**, 2306–2325 (1990)
22. Akopyan, M.E., Grebenev, V.V., Ivanov, V.S., Lugovoj, E.V., Pravilov, A.M.: Dipole moments of the $Cl_2(A^3\Pi(1_u),v'$— $X^1\Sigma_g^+,v''$, $B^3\Pi(0_u^+),v'$—$X^1\Sigma_g^+,v'')$ transitions measured by laser-induced chemiluminescence spectroscopy. J. Phys. B, Atom. Molec.and Optical Phys. **29**, 1369–1379 (1996). https://iopscience.iop.org/article/10.1088/0953-4075/29/7/017
23. Heaven, M.C.: Fluorescence decay dynamics of the halogens and interhalogens. Chem. Soc. Rev. **15**, 405–448 (1986). https://doi.org/10.1039/CS9861500405
24. Barrow, R.F., Long, D.A., Millen, D.J., Coxon, J.A.: Low-lying electronic states of diatomic halogens. Molec. Spectrosc. **1**, 177–228 (1973). https://doi.org/10.1039/9781847556684-00177

Solutions of the Problems

1. The rate constant of the unimolecular process A → products is the reverse lifetime of the species A.
2. The rate constant of the bimolecular process A_1 + A_2 → products is the process rate at the A_1, A_2 species unit concentrations.
3. The rate constant of the termolecular process A_1 + A_2 + A_3 → products is the process rate at the A_1, A_2, A_2 species unit concentrations.
4. Reaction orders and reagent orders can be zero or fractional. One can use termolecular recombination $A + B \xrightarrow{(M)} AB$ as an example. The reaction order relative to [M] can be fractional or equal to zero, and reaction order decrease from 3 to 2 with [M] increase.
5. One should divide $3 \cdot 10^{14} \cdot \exp(-\ 4.8\ (\mathrm{KJ/mol})/RT)$ $\mathrm{cm^3/mol \cdot 1}$ by Avogadro's number $N_A = 6.022 \cdot 10^{23}\ \mathrm{mol^{-1}}$, and take into account that 1 cal = 4.184 J, R = 1.9858 cal/mol. Rate constant is $7.4 \cdot 10^{-11}\ \mathrm{cm^3/species\ s}$.
6. One should divide $3 \cdot 10^{14} \cdot \exp(-\ 4.8\ (\mathrm{KJ/mol})/RT)$ $\mathrm{cm^3/mol \cdot 1}$ by Avogadro's number squared, and take into account that 1 cal = 4.184 J, R = 1.9858 cal/mol. Rate constant is $3.3 \cdot 10^{-34}\ \mathrm{cm^6/species^2 s}$.
7. A stoichiometric reaction order is equal to the sum of the stoichiometric coefficients in the stoichiometric equation, whereas a kinetic (real) reaction order is equal to the sum of reaction kinetic order regarding species which take part in a reaction. The kinetic (real) reaction order can be less than the stoichiometric one (see solution of problem 4).
8. The equilibrium rate constant of the reaction (1)

$$\mathrm{Cl}(^2\mathrm{P}_{3/2}) + \mathrm{Cl}(^2\mathrm{P}_{3/2}) \leftrightarrow \mathrm{Cl}_2\left(X^1\Sigma_g^+\right), \tag{1}$$

A. Pravilov, *Gas-Phase Photoprocesses*, Springer Series in Chemical Physics 123,
https://doi.org/10.1007/978-3-030-65570-9

is:

$$K_1 = \frac{g^e_{Cl_2(X)} \frac{\left(2\pi m_{Cl_2(X)} kT\right)^{3/2}}{h^3} \frac{kT}{B^X_e \sigma h} \left[1 - \exp(-\omega_e/kT\right]^{-1}}{\left[g^e_{Cl} \frac{(2\pi m_{Cl} kT)^{3/2}}{h^3}\right]^2} \exp\left(-\frac{D^0_0}{RT}\right) \text{species/cm}^3.$$

Here,

$$g^e_{Cl_2(X)} = 1, \frac{(2\pi m_{Cl_2(X)} kT)^{3/2}}{h^3} 5.76 \cdot 1026 \frac{\text{cm}^3}{\text{species}}, = \frac{kT}{B^X_e \sigma h}$$
$$= 177.2, 1 - \exp\left(-\frac{\omega_e}{kT}\right]^{-1} = 1.06, \exp\left(-\frac{D^0_0}{RT}\right) = 9 \cdot 10^{-41},$$
$$\{g^e_{Cl} = 4, [g^e_{Cl} \frac{(2\pi m_{Cl} kT)^{3/2}}{h^3}]^2 = 6.6 \cdot 10^{53} \text{cm}^6$$

Therefore, $K_1 = 4.7 \cdot 10^{-65}$ species/cm^3.

9. The steady-state condition for the XeKr($B0^+$,v_{max}) species is

$$k_2\left[\text{Xe}(^3\text{P}_1)\right]\left[\text{Kr}(^1\text{S})\right] = [\text{XeKr}(B0^+, v_{max})] \cdot \left\{k_{-2} + k_{3a} + k_{3b} + (k_4 + k_5 + k_6) \cdot \left[\text{Kr}(^1\text{S})\right]\right\}.$$

Therefore,

$$[XeKr(B0^+, v_{max})] = k_2[Xe(^3P_1)][Kr(^1\text{S})]$$

and

$$k_1 = k_{3a} k_2 \left[\text{Xe}(^3\text{P}_1)\right]\left[\text{Kr}(^1\text{S})\right]$$

10. The steady-state condition for the O(^{3}P) species is

$$r_1 - k_2\left[\text{O}(^3\text{P})\right]_{ss}[\text{O}_2] + k - 2\left[(\text{O}\ldots\text{O}_2)^{\#}\right]_{ss} = 0.$$

So the O(^{3}P) steady-state concentration is:

$$\left[\text{O}(^3\text{P})\right]_{ss} = \left\{r_1 + k - 2\left[(\text{O}\ldots\text{O}_2)^{\#}\right]_{ss} / k_2[\text{O}_2].\right.$$

The steady-state condition for the $(O\ldots O_2)^{\#}$ species is

$$k_2\left[O(^3P)\right]_{ss}[O_2] = \left[(O\ldots O_2)^{\#}\right]_{ss}\{k_{-2} + k_3[M]\}$$

So the $(O\ldots O_2)^{\#}$ steady-state concentration is:

$$\left[(O\ldots O_2)^{\#}\right]_{ss} = k_2\left[O(^3P)\right]_{ss}[O_2]/\{k_{-2} + k_3[M]\}$$

Index

A. Pravilov, *Gas-Phase Photoprocesses*, Springer Series in Chemical Physics 123,
https://doi.org/10.1007/978-3-030-65570-9